Liberal Education in Twenty-First Century Engineering

*To Patrick —
Welcome to the LED,
ASEE + WPI — all of
which contributed to this
volume!
And all the best for 2004,
Lance*

WPI*Studies*

Lance Schachterle
General Editor

Vol. 23

PETER LANG
New York • Washington, D.C./Baltimore • Bern
Frankfurt am Main • Berlin • Brussels • Vienna • Oxford

Liberal Education in in Twenty-First Century Engineering

Responses to ABET/EC 2000 Criteria

Edited by
**David F. Ollis, Kathryn A. Neeley,
& Heinz C. Luegenbiehl**

PETER LANG
New York • Washington, D.C./Baltimore • Bern
Frankfurt am Main • Berlin • Brussels • Vienna • Oxford

Library of Congress Cataloging-in-Publication Data
Ollis, David F.
Liberal education in Twenty-first century engineering /
David F. Ollis, Kathryn A. Neeley, Heinz C. Luegenbiehl.
p. cm. — (WPI studies; v. 23)
1. Technical education. 2. Humanities—Study and teaching (Higher). I. Neeley, Kathryn A.
(Kathryn Angelyn). II. Luegenbiehl, Heinz. III. Title. IV. Series.
T65.3.O45   607'.1—dc21   2003010728
ISBN 0-8204-4924-5
ISSN 0897-926X

Bibliographic information published by Die Deutsche Bibliothek.
Die Deutsche Bibliothek lists this publication in the "Deutsche
Nationalbibliografie"; detailed bibliographic data is available
on the Internet at http://dnb.ddb.de/.

Cover design by Lisa Barfield

The paper in this book meets the guidelines for permanence and durability
of the Committee on Production Guidelines for Book Longevity
of the Council of Library Resources.

© 2004 Peter Lang Publishing, Inc., New York
275 Seventh Avenue, 28th Floor, New York, NY 10001
www.peterlangusa.com

All rights reserved.
Reprint or reproduction, even partially, in all forms such as microfilm,
xerography, microfiche, microcard, and offset strictly prohibited.

Printed in the United States of America

*To the engineering graduates
of the ABET/EC 2000 curricula*

Contents

Acknowledgments   xi

Editors' Preface: A Sea Change in Engineering Education   xiii

Prologue: The Civilized Engineer
SAMUEL C. FLORMAN   xix

Centerpiece: The Eleven Commandments of Twenty-first Century Engineering Education: ABET Engineering Criteria 2000   xxxi

PART ONE
CHALLENGE AND AN INSTITUTIONAL RESPONSE

Liberal Education in Engineering: Challenge and Response

Chapter 1.   Liberal Education and Engineering Criteria 2000   3
EDWARD ALTON PARRISH

Chapter 2.   Liberal Education Responds: Discussing ABET 2000 within a Humanities Division   12
LANCE SCHACHTERLE

PART TWO
SPECIFIC RESPONSES

Effective Communication

Chapter 3. Reuniting Wisdom and Eloquence within the Engineering Curriculum — 41
CAROLYN R. MILLER

Chapter 4. To Arrive Where We Started and Know the Place for the First Time? Re-visioning Technical Communication — 51
KATHRYN A. NEELEY

Chapter 5. Creating a Communication-intensive Undergraduate Curriculum in Science and Engineering for the Twenty-first Century — 64
LESLIE PERELMAN

Chapter 6. Refashioning the First-year Introductory Course on Communication Skills and Engineering Practice — 82
JOHN BROWN

Liberal Education in Problem Formulation and Solution

Chapter 7. The Role of the Humanities in Distinguishing Science from Engineering Design in the Minds of Engineering Students — 91
CHARLES C. ADAMS

Chapter 8. The Role of Values in Teaching Design — 113
HEINZ C. LUEGENBIEHL AND DONALD L. DECKER

Professional and Ethical Responsibility

Chapter 9. Engineering Ethics Education for the Twenty-first Century: Topics for Exploration — 119
HEINZ C. LUEGENBIEHL

Chapter 10. Integrating Engineering, Ethics, and Public Policy: Three Examples — 129
JOSEPH R. HERKERT

Chapter 11. Using Detailed, Multimedia Cases to Teach
Engineering Ethics   145
MICHAEL E. GORMAN, JULIE M. STOCKER,
AND MATTHEW M. MEHALIK

Liberal Education and Contemporary Issues

Chapter 12. Teaching Engineering as a Social Science   158
EDWARD WENK, JR.

Chapter 13. Orienting Engineering Students to Contemporary Issues
through a Broader Perspective   165
CRAIG GUNN

Engineering in Liberal Education

Chapter 14. Reaching Out across Campus: Engineers as Champions
of Technological Literacy   171
JOHN KRUPCZAK

Technology in Art / Aesthetics in Design

Chapter 15. The Museum in the Classroom: Technology in Art   189
ANN BROWN

Chapter 16. The Aesthetics of Engineering: Toward an Integrated View
of Engineering Design   201
KATHRYN A. NEELEY

Integrating Engineering and Humanities

Chapter 17. Integrating Humanities and Engineering: Two Models
for Achieving ABET Criteria 2000 Goals   217
BARBARA M. OLDS AND RONALD L. MILLER

Chapter 18. Responding to ABET 2000: A Process Model for the
Humanities and Social Sciences   233
HEINZ C. LUEGENBIEHL

x | Liberal Education in Engineering

Chapter 19. STS for Engineers: Integrating Engineering, Humanities, and Social Sciences through STS Courses and Programs   245
JOSEPH R. HERKERT

Chapter 20. Teaching Students, Not Texts: The Utility of the Humanities in Fulfilling ABET 2000 Criteria   255
SCOT DOUGLASS

Chapter 21. Implementing an English and Engineering Collaboration   270
ANN BROWN, STEVE LUYENDYK, AND DAVID F. OLLIS

Multidisciplinary Student Teams

Chapter 22. The Parable of Baseball Engineering   280
MARSHALL M. LIH

Chapter 23. A Multidisciplinary Course on Technological Catastrophes   283
JOSEPH R. HERKERT

Chapter 24. Prolegomena for Evaluation of Multidisciplinary Student Teams   296
CHARLES W. N. THOMPSON

Collaborative Teaching Example

Chapter 25. Professional Development at the University of Virginia: Attributes, Experiences, ABET 2000 and an Implementation   305
JOHN P. O'CONNELL, MARK A. SHIELDS, EUGENE R. SEELOFF, TIMOTHY C. SCOTT, AND BRIAN PFAFFENBERGER

Afterword

Chapter 26. A Century of ASEE and Liberal Education (or How Did We Get Here from There, and Where Does It All Lead?)   320
O. ALLAN GIANNINY, JR.

List of Contributors   347

Acknowledgments

The editors are pleased to thank the McGraw-Hill Companies and the American Society for Engineering Education (ASEE) for permission to reprint the articles indicated.

Prologue, The Civilized Engineer, by Samuel C. Florman, first appeared in his book, *Engineering and the Liberal Arts*, McGraw-Hill, Inc., New York, 1968.

Chapter 8, The Role of Values in Teaching Design, by Heinz Luegenbiehl and Donald L. Decker, was published in *Engineering Education*, January 1987.

Chapter 12, Teaching Engineering as a Social Science, by Edward Wenk, Jr., is from ASEE PRISM, December 1996.

Chapter 22, The Parable of Baseball Engineering, by Marshall Lih, is also from ASEE PRISM, September 1996.

Chapter 26, A Century of ASEE and Liberal Education, by O. Allan Gianniny, Jr., was presented at the 1995 ASEE meeting.

EDITORS' PREFACE

A Sea Change in Engineering Education

The New Gauntlet: Redefine Engineering Education

The important and respected national accreditation commission for engineering education recently freed undergraduate curricula from their disciplinary fetters, and threw down in their place an individual challenge to each school of engineering, to be initiated by the year 2001:

1. (re)Define your education mission,
2. (re)Create your curricula accordingly,
3. Assess your graduating student outcomes to verify mission achievement, and
4. Use ongoing assessment for continuous program improvement.

In place of the former accreditation restrictions which required that students take so many units of engineering design, of humanities and social sciences (HSS), and of basic and engineering sciences, the eleven new criteria, known colloquially as EC 2000, or ABET 2000, present an utterly open-ended set of student outcomes that each engineering curriculum must have begun to meet by the year 2001. In psychological terms, we educators are invited to be freed of our disciplinary blinders. A decade from now, we will know if this freedom brought either invention by and integration of faculty, or simply curricular anarchy and loss of a unique opportunity to bridge the "two cultures" of C. P. Snow. This volume is dedicated to achievement of the first possibility.

Whither Liberal Education in Engineering

We believe that EC 2000 invites the full participation of all faculty involved in education of engineers. Faculty authors contributing to this volume indicate how elements of liberal education may address these new engineering education criteria, individually and collectively.

The volume is structured in an introduction followed by Part One (Challenge and an Institutional Response) and Part Two (Specific Responses). The Introduction begins with Samuel Florman's exhortation for liberal education in engineering. The stage is next set for this volume through stating the new "eleven commandments" of the national engineering accreditation criteria, known in short as ABET/EC 2000, nearly half of which relate to elements found within liberal education in general. Part One presents the evolutionary history of the EC 2000 criteria origins and a sample institutional response.

The second, larger Part Two portion provides a menu of opportunities by example, targeted to specific EC 2000 criteria, and is meant to inspire and instigate further academic reform efforts. An afterword contains a history of the ASEE Liberal Education Division, whose members contribute here twenty-two of the twenty-six chapters.

The variety of curricular possibilities presented in this volume indicate that we are in a formative, perhaps even revolutionary, period in engineering education. While the ideals of liberal education are still relevant to engineering education, the relationship between what have traditionally been termed the "technical" and "nontechnical" elements of engineering education is being reconceptualized, as are the elements themselves and the aims of the enterprise as a whole.

Interpreting the Criteria and Defining Our Mission

Brevity and open-endedness are two of the most striking features of EC 2000 criteria; the latter, therefore, not only permit but demand interpretation. They offer the freedom for engineering educators to define our own mission, consistent with the broad aims outlined, but provide less security because we can be less sure of being right or safe. For many of us accustomed to the old system, this emancipation may not be the kind of freedom we want. For those of us whose primary expertise lies in the humanities and social sciences, the new criteria provide another demand: in addition to establishing the importance of what has traditionally been termed "liberal education" we must also establish our relevance to engineering. Thus, EC 2000 asks more of us all.

Integrating Humanities and Social Sciences (HSS) into Full Membership in Engineering Education

Many of the eleven EC 2000 criteria are directly related to aspects of liberal education. Thus, nothing in these open and inclusive criteria suggests that the ideals of liberal education, which are rearticulated in several of the chapters that follow, will be of lesser importance in the future. If anything, the contrary is possible, but only if faculty trained in liberal disciplines commit to contribute to the new enterprise. The important difference is that HSS elements are now to be seen in relation to, rather than distinctive from, other elements.

The word "integration" is often used in discussion of engineering education design and reform, and seems the best term for characterizing some of the newer roles in which the EC 2000 criteria cast HSS. What does integration mean in this latter context?

The *Oxford Minireference Dictionary and Thesaurus* highlights two useful meanings:

1. Combine (parts) into a whole.
2. Bring or come into full membership of a community.

To reorganize something as integral means to see it as necessary to forming a whole; as fundamental, indispensable, intrinsic, necessary. In engineering education, achieving true integration will require the merging of technical and nontechnical elements within a new conceptual framework. It will require all of us, whether technical or nontechnical, to take seriously our role as engineering educators who are specialists in the particular but also contributors to a larger whole. The very interdisciplinary wording of the criteria is indeed a thinly veiled description of what active faculty, not only graduating students, should reflect.

The Elephant in the Room

A huge and significant entity that decisively shapes our faculty behavior but is treated as invisible and rarely mentioned is a set of issues having to do with power, security, and freedom throughout the community of all those involved in engineering education. The new EC 2000 criteria invite but do not require the shared participation of all in this power, security, and freedom. The old criteria allowed humanists and social scientists to justify their role within engineering education on traditional grounds that sometimes set liberal education in an adversarial relation to engineering education, and presumed that some sort of a hierarchy would be established. Within this framework, the HSS were usually presumed to have the moral "high ground," while the technical subjects claimed greater "hardness" and relevance to engineering. The new criteria posit a different set of justifications and relationships, but they leave room for time and resources to be wasted in traditional power struggles. The process of redefining the role of HSS within engineering education is often figured in one way or another as a power struggle in the chapters that follow, evidently reflecting the realities of institutional life.

These pose obstacles to the kinds of creative thinking and cooperation required to make the transition to EC 2000. Much will depend upon deans and other administrators in engineering, HSS, and other colleges, who are uniquely positioned to provide all their faculty with the suport and freedom that catalyze innovation. At the same time, faculty will need to see themselves as designing new jobs for themselves, jobs that draw on hard-won expertise and long-standing interests, but that also expand individual horizons and their sense of relation and relevance of their expertise to the larger challenge at hand.

Redesigning Our Curricula and Our Professional Identities

The projects, courses, and testimonies in the following chapters attest that there is rarely an exact fit between traditional career patterns (areas of training and expertise and programs of research) and much of what EC 2000 calls for, and this circumstance seems equally true for faculty of engineering or HSS specialties.

New activities are needed. Fulfilling the vision of EC 2000 calls for combining expertise of several individuals at all levels: curriculum design, course design, and classroom (team teaching), and seeing where individual expertise fits into the larger picture. Some old pieces and elements (i.e., courses and major units within courses) will remain useful in the new curricula, but these new structures, as several of the following chapters illustrate, will not simply rearrange or reallocate existing units. We need to rethink the nature of the components, their variety, and their relation to the whole. All of this, it seems, is to be undertaken in the context of rethinking what the whole should be.

Another key to success in integration will be to realize that there is no ideal form, no clearly defined pattern or single course of design that can be held up as a model. While orientational first-year courses and multidisciplinary design capstone courses offer two of the most common integration opportunities, all curriculum components should receive examination for reformulation and delivery in a manner supportive of the new criteria.

Pessimists may suggest that the EC 2000 criteria are but one of an endless series of opportunities or demands to "reinvent ourselves," and as such deserve no special attention or intellectual respect. We counter by observing that the EC 2000 criteria may be viewed as nothing less than a demand to create anew engineers as the Renaissance graduates for the twenty-first century. What champion of liberal education could resist joining such a cause?

New and revised courses will need to serve multiple needs. With courses focused primarily on HSS, the offerings should address both the traditional liberal education objectives and contribute in important ways to achievement of the eleven outcomes specified by EC 2000. The capacity to serve multiple purposes is one of the enduring strengths of liberal education, and the transition to EC 2000 should provide renewed demonstrations of this quality.

The use of the phrase "Engineering as Liberal Education" by Lance Schachterle, series editor, reminds us that, for the phrase to become both meaningful and operative, stereotypical definitions of both "engineering" and "liberal education" will have to be not only revised by small groups of innovators, but eventually be widely disseminated and accepted throughout the community of engineering educators. Such dissemination and acceptance are the crowning steps in successful implementation of education reform.

A final problem we face is the very specific and concrete vocabulary in use that reinforces an unintegrated view of HSS within engineering education. Our phrases such as "technical" and "liberal arts" course are, as the following chapters reflect, a convenient way for those of us involved in engineering education to make ourselves understood and establish a conversation. Unfortunately, this vocabulary conceptually undermines the concept of integration. We will be on our way to achieving such integration when it seems more natural for all involved to see ourselves not as technical or nontechnical faculty, but as engineering educators.

We hope the present volume contributes to such a transition.

PROLOGUE

SAMUEL C. FLORMAN

The Civilized Engineer

According to Webster's Dictionary, "civilization" consists of "progress in education, refinement of taste and feeling, and the arts that constitute culture." If we take this definition literally, the average engineer today is simply not civilized. It is paradoxical that without us civilization could not exist, yet we are somehow isolated from the civilizing influence of the culture of our time. This is a misfortune for us as individuals, for our profession, and for the world.

I am not talking about superficial refinement and the ability to sparkle at cocktail parties. I am talking about something more fundamental. It is not an exaggeration to say that liberal education for engineers could improve the quality of life for the average engineer, contribute to the sound development of the engineering profession, and help to preserve and enrich society as a whole. Let us consider some of the ways in which this is so.

Intellectual Competence and Imagination

First of all, a liberal education enlarges intellectual capacity, develops mental agility, and improves our ability to think. As a noted educator has said, liberal education helps "to cultivate those skills and habits of reasoning which constitute intellectual competence, the capacity to think logically and clearly, the ability to organize one's thoughts on any subject on which essential facts are possessed or obtainable."[1]

Intellectual competence and imagination. If we need the liberal arts to maintain and improve our intellectual competence, we need them even more to develop imagination, for without imagination reason is not equal to even the minimum demands of our exploding technology. Lewis Mumford has warned us that a concentration on pure technical training

> ... might defeat even its immediate purposes by depriving original minds of the stimulus and enrichment of wider interests and activities.[2]

Has not Niels Bohr told us that he arrived at the doctrine of complementarity in physics by speculating on an ancient theological dilemma—the impossibility of reconciling perfect love with perfect justice? Humans flew in dreams and communicated instantaneously in myths and fairy stories long before they achieved the technical apparatus for doing so. But would any chain of discoveries and inventions have produced a balloon or a telegraph if the dream had not first suggested these goals? Many significant inventions, from the helicopter to the motion picture, began as toys for amusing the young. Plainly the self-sufficiency of the specialist's world is a prisoner's illusion. It is time to open the gates.[3]

Liberal Education and Leadership

As liberal education improves our intellectual competence and expands our imagination, it also develops those qualities of intellectual curiosity and general understanding, those traits of grace and wit and poise, that characterize leaders. Too often engineers are found lacking in these attributes. Scientifically made personality studies have revealed engineers to be "socially conforming, impersonal, introverted individuals."[3] In industry, the effectiveness of engineers has been found to be limited by their lack of "people-wisdom," their reliance on "coldly rational judgment," and the recurrent appearance of a "noncommunicative syndrome."[4]

The president of a large corporation has said succinctly what many leaders of American industry have come to recognize as fact:

> The specialist cannot function effectively at the top level of management if all he brings to it is his specialty. At that level, the daily problems call for broad general knowledge, open-mindedness, an understanding of human nature, an insight into human frailties, a fairness of mind, a clarity of thought. ... The qualifications needed for leadership in industry are developed largely through a liberal arts education.[5]

The proposition has scarcely changed in the nineteen centuries since Vitruvius, the great Roman engineer, wrote that those who have acquired

> ... skill without scholarship have never been able to reach a position of authority to correspond to their pains. ... But those who have a thorough knowledge of

both, like men armed at all points, have the sooner attained their object and carried authority with them.⁶

The Good Life

In addition to helping each of us to do our jobs more effectively—and of even greater importance in the last analysis—liberal learning yields great riches to the individual in pursuit of the good life. Knowledge and understanding provide pleasure that needs no practical justification. Beauty evokes joyousness that is its own reward. The most precious treasure awaiting the engineer in the world of the liberal arts is enrichment of his or her personal life—enrichment and the tranquility of spirit that accompanies new insight.

We engineers pride ourselves on being members of a profession that engages our energies and challenges our capacities. We are usually too much absorbed in our interesting work to be overly bothered by the doubts and anxieties that plague many of our less fortunate brethren. This concentration on work is a blessing, but it contains a hidden flaw. Our questioning and doubting are liable to be postponed, only to emerge in later years, sometimes with disturbing effect. It is better surely to expose oneself early and often to the eternal problems of philosophy and art than to be awakened with a start in one's waning years by the sudden asking of the questions, "What is life all about? What have I been living for?" Socrates' admonition still rings true: "The unconsidered life is not worth living."

Mark Van Doren has spoken of the happiness won by the individual who has sought inspiration and enlightenment in the "great tradition" of the liberal arts. "That happiness consists in the possession of his own powers, and in the sense that he has done all he could to avoid the bewilderment of one who suspects he has missed the main thing. There is no happiness like this."⁷

Status for the Profession

As the individual engineer profits from acquaintance with the liberal arts, so will the entire profession. Our lack of "status," our unsatisfactory "image"—these are concerns that gnaw away at our collective professional contentment. Only a vastly increased number of liberally educated engineers can remedy this situation. Self-praising pronouncements emanating from our professional organizations surely will not suffice.

The public relations problems of the engineering profession are nothing new. They already existed in the days of ancient Greece. Xenophon spoke for most of his fellow citizens when he said that "the mechanical arts carry a social stigma and are rightly dishonored in our cities." Technologists, he maintained, "simply have not got the time to perform the offices of friendship or citizenship. Consequently they are looked upon as bad friends and bad patriots."[8]

A hundred years ago Ralph Waldo Emerson looked at the technologists of his day and spoke sadly of "great arts and little men." "Look up the inventors," he wrote. "Each has his own knack; his genius is in veins and spots. But the great, equal symmetrical brain, fed from a great heart, you shall not find."[9] Even Thomas Henry Huxley, nineteenth-century advocate of science and technology, expressed concern about technologists becoming "lopsided men." "The value of a cargo," he pointed out, "does not compensate for a ship's being out of trim."[10]

In the early twentieth century, engineering achieved a certain level of prestige, although the profession was still regarded warily even by its greatest admirers. Thus Thorstein Veblen in 1917:

> Popular sentiment in this country will not tolerate the assumption of responsibility by the technicians, who are in the popular apprehension conceived to be a somewhat fantastic brotherhood of over-specialized cranks, not to be trusted out of sight except under the restraining hand of safe and sane businessmen. Nor are the technicians themselves in the habit of taking a greatly different view of their own case.[11]

Today, in the midst of spectacular engineering achievements, this galling tradition persists. An American science editor informs us that "the image that has been projected of the engineering profession—and images are very hard to change—is of a prejudiced, conservative, non-involved group."[12] An English science editor comments that engineers, "the men who daily make history, are still not accepted as important citizens! . . . Even the word "engineer" has connotations of a man in a boiler suit who is a kind of modern blacksmith."[13] A public relations expert captures the essence of a prevalent attitude in a few deft sentences:

> I grew up in the tradition of the engineer being a man with no verbal skills whatsoever. He was one of these guys who if you gave him an applied kind of problem would go off in a corner and work it out for you but, God forbid, don't ask him to explain it. He's no good with the English language; don't expect him to articulate himself. He is this kind of faceless, anonymous character.[14]

Engineers are certainly not insensitive to public opinion, nor do they engage in self-deception. A study has shown that "engineers themselves are convinced that the general public does not hold them in as great esteem as other professions."[15]

Only liberally educated engineers can bring the profession the esteem it craves and, in so many important ways, deserves. For one thing, only liberally educated engineers will possess the eloquence with which to impress upon their fellow citizens the inherent worth of engineering and its importance to society. It has been charged, and rightly so, that the engineering profession "has not been in touch with the people and by default has permitted a working partner (science) to capture the imagination of the nation."[16]

But there is a more important goal than "telling the story" of engineering to the public. If engineers themselves, as individuals, become truly cultured—that is, become educated in the liberal arts—then the word will spread without a "good press." If we engineers do not want to be known as "bad friends and bad patriots," "little men," "lopsided men," "overspecialized cranks," if we resent being characterized as a "prejudiced, conservative, noninvolved group," if we cringe at being looked at as "a kind of modern blacksmith," as "this kind of faceless, anonymous character," then it is up to us to make sure we resemble these things as little as possible. If engineers become increasingly wise, sensitive, humane, and responsible, we will not need public-relations techniques to sell us to the public.

The Public Good

As the engineering profession gains prestige and authority, society as a whole will benefit. For the world is desperately in need of the leadership that only engineers can give. "The politicians, and even the statesmen," as James Reston of the *New York Times* has put it,

> ... are merely scrambling to deal with the revolutions in weapons, agriculture and industry created by the scientists and the engineers. The latter have transformed man's capacity to give life, to sustain and prolong life, and to take life; and the politicians no longer find that they can deal with all the new complexities and ambiguities.[17]

The world must listen to the engineer or it is doomed. Buckminster Fuller has stated the facts in the simplest terms:

> If humanity understood that the real world problem is that of upping the performances per pound of the world's metals and other resources, we might attempt to solve that problem deliberately, directly and efficiently. . . . But I find that approximately no one realizes what is going on. That is why we have been leaving it to the politician to make the world work. There is nothing political that the politician can do to make fewer resources do sixty percent more.[18]

We engineers already possess most of the technical knowledge required to provide food and shelter in abundance, restore purity to our air and water, heal the blight of our cities, untangle the snarl of traffic, harness our rivers, reap harvests from the oceans, husband our resources, and develop power from the sun and atom. We will—if the world will let us—subdue floods, minimize the danger from storms and earthquakes, and eventually control the weather. We will perform new miracles in the fields of medicine, communication, and transportation, and develop a continuous stream of marvelous fabrics and household appliances. We can—if called upon—contribute to the preservation of peace by assisting the underdeveloped nations and by devising improved means of arms control.

But unless we achieve a position of leadership, our talents will continue to be largely wasted and misdirected. The world will persist in demanding our blast furnaces but not our smoke-control devices, our highways but not our parks, our bombers but not our hospital ships. Running wild and out of control, technological progress will become a disease—a plague of asphalt and armaments, pollution and blight.

Admittedly, there is a school of thought that holds that the world's ills are not attributable to lack of leadership by the technologists, but rather to leadership heading in the wrong direction. "In every country in the world," according to George Orwell,

> ... the large army of scientists and technicians, with the rest of us panting at their heels, are marching along the road of "progress" with the blind persistance of a column of ants. Comparatively few people want it to happen, plenty of people actively want it not to happen and yet it is happening. The process of mechanization has itself become a machine, a huge glittering vehicle whirling us we are not certain where, but probably toward the padded Wells-world and the brain in the bottle.[19]

"It is apparently our fate," echoes a French scholar, "to be facing a 'golden age' in the power of sorcerers who are totally blind to the meaning of the human adventure."[20]

Nor is this sentiment restricted to apprehensive artists and intellectuals. In his farewell address, President Eisenhower warned the nation that its public policy might "become the captive of a scientific-technological elite." Senator Bartlett of Alaska has complained that "faceless technocrats in long, white coats are making decisions today which rightfully and by law should be made by the Congress."[21]

It is true that in government and industry engineers are numerically abundant and potentially powerful.[22] But the extent of our actual influence in directing the course of our society is a moot question indeed. In either case—whether the world is racing toward disaster in spite of us or because of us—clearly what is needed is enlightened engineering leadership.

Someone must step forward to say, "We can afford to make that automobile a little safer," "Let us build a factory that is more attractive," "Let us consider the possible harmful effects of that insecticide before we market it," "Let us develop a plant process that will not pollute the water we use," "Let us make that machine a little quieter," "Let us not demolish that historically precious old building," "Let us locate that dam, not only where it will generate the most power, but also where it will serve the interests of the community—aesthetically, politically, and socially," "Let us build a rapid transit system for this city rather than a freeway that will bring more cars into an area already choked with traffic."

This "someone" cannot be an ordinary citizen of good will. He must be able to bolster his arguments with facts—technical, scientific, and economic. Hunches and sentiments will not be sufficient. His recommendations, in order to be persuasive, must be founded in a knowledge of resources, materials, and energy conversion; statistics, probabilities, and decision theory; computers, controls, and systems engineering. Moreover, this "someone" must be concerned. He must be articulate. He must be esteemed. And he must have a highly developed moral and aesthetic sensibility. In short, he must be a liberally educated engineer. (Note: Were this article written today, Florman's "he" would be our current "he" and "she.")

Liberal education will not make prophets and saints out of engineers, and even if it did, this would not necessarily bring about instant Utopia. But this prudent qualification should not discourage us from moving in what we know to be the right direction.

The Public Philosophy

One thing further. A generation of liberally educated engineers would inevitably play an important role in the debates that are instrumental in shaping the public philosophy. Engineers are already developing techniques of decision making which will enable the leaders of society to choose rationally between alternative courses of action. When we are as wise as we are smart, we will be qualified to talk about new goals as well as methods, ultimate ends as well as means. We will have much of worth to contribute to the public philosophy, both in word and in deed.

Although worldwide conquest of poverty and need is far from being achieved, the more developed nations are already encountering an entirely new spectrum of problems born of affluence. Future generations, having won the battle for survival, will be confronted with the task of maintaining vitality and pride in our communities, conquering boredom, and renewing faith in the inherent value of creative labor.

The response of truly civilized engineers could lead to a renaissance of engineering that would dazzle and inspire the world. In every nation, in every town, there would be a flowering of towers, arches, and domes; concert halls, stadiums, theaters, museums, and libraries; universities and art centers; memorials, shrines, and churches; parks, gardens, fountains, pools, arcades, and promenades. Treasures of antiquity would be recovered, preserved, and reconstructed. Research centers and sanctuaries would flourish in the deserts, the jungles, and on the polar ice, under the sea and in outer space. In all of this, every line, every proportion of structure, vehicle, and machine would spring, not from greed or grim necessity, as so much of it does today, but from an inner human striving for beauty and excellence. Lewis Mumford has spoken hopefully of the eventual coming of "a new technology, one so finely adjusted, so delicately responsive, that it will meet all the needs and occasions of life at a minimum cost of human values."[23] Such a technology will not be beyond the reach of engineers who have been steeped in the liberal arts.

With the coming of a civilized technology, the world will discover that engineering is not merely a means to an end, but rather an inherently worthy way of life. A revitalized and enlightened engineering will someday achieve its rightful place as one of the sublime enterprises of mankind. As historian Lynn White, Jr. has said,

> Technology is a profoundly spiritual form of thought. It has flourished best in the context of the Judaeo-Christian presupposition that the physical universe was created for a good purpose, that it is not to be disregarded or transcended, but rather that, like the body itself, it is to be treasured and controlled as the necessary ground of psychic life.[24]

A Recapitulation of Reasons

These, then, are some of the reasons that we engineers should study the liberal arts:

1. To improve our intellectual competence and expand our imaginative powers.
2. To develop those qualities of character and personality that make for leadership and successful careers.
3. To enrich our personal lives with new knowledge and insight, with a keener appreciation of beauty.
4. To elevate the standards of our profession and to gain it increased esteem in our society.
5. To contribute to the public good—first, by using our status to see to it that our sensible technological advice is taken; second, by applying our enlarged wisdom and sensitivity, not only to the solution of engineering problems but also to the selection of worthy goals for our society; and third, by demonstrating to the world that technology can be more than a tool, can in itself be a revitalizing and profoundly beneficial force in human affairs.

What Are the Liberal Arts?

In what has been said so far it has been assumed that we are all familiar with what is meant by a liberal education. And indeed we are, in a broad sense. Liberal learning, we know, concerns itself with the eternal quest for truth, goodness, and beauty. It grapples with such ultimate concerns as the meaning of justice, liberty, virtue, honor, love, and happiness. Its method is to study the works of scholars and artists, both past and present.

There are some who consider the ultimate purpose of liberal education to be the transmission of our cultural heritage and the creation of good citizens loyal to the ideals of our society. There are others who stress the development of the independent, sensitive, questioning mind, who see the liberal arts as having a critical, almost revolutionary, function. However, all agree that the ultimate goal of liberal learning is wisdom and a reverence for beauty.

Literally, arts that are "liberal" are those studies deemed fit for liberated, or free, individuals. In past ages only a few people were truly free; the vast majority lived as serfs, chained to soil they did not own. So liberal education has an aristocratic tradition. Because it was restricted to the upper classes, it came to serve the purpose of artificially defining the upper classes. This "snobbish" side of liberal learning becomes an anachronism in a democratic society. An engineering educator has put it this way:

> The *Oxford English Dictionary* defines liberal education as education fit for a gentleman. That is still an acceptable definition; it is the idea of a gentleman which has changed.... Modern gentlemen do not belong to the leisured class. Many of them work something like a seventy-hour week.[25]

In the universities of the Middle Ages there were established seven branches of liberal learning: grammar, logic, and rhetoric (in other words, literature and philosophy), and arithmetic, geometry, astronomy, and music (science and the fine arts). Through the years a liberal arts education has come to comprise all those studies that are not technical or vocational by nature, those that are not "useful arts." It is no longer said that the liberal arts are for free men. Instead one hears it said that the liberal arts make men free by liberating their minds from ignorance and dogma.

A truly liberal education includes the study of pure science and mathematics. In these fields the average engineer has received a considerable amount of instruction. Liberal learning also embraces the social sciences: sociology, anthropology, psychology, political science, economics, and the like. These are important and useful subjects, and no person can be considered educated without some knowledge of them.

But when we say that engineers lack an adequate liberal arts education, we do not really mean that they need more training in the social sciences. For one thing, to the extent that engineers take liberal arts courses in college, they often select the social sciences, which appear to be more "useful" than such studies, as, for example, literature. Also the social sciences, whatever their merits, have a certain "illiberal" quality about them. They are, after all, sciences, and their approach is essentially clinical. The social scientist is more a statistician than a philosopher. He or she is liable to make the engineer more "lopsided" rather than less.

We are left, then, with those subjects that constitute the true core of liberal learning: history, literature, philosophy, the fine arts, and music. These are the so-called humanities about which the average engineer has always known little and cared less. These are the subjects with which we must now try to gain some familiarity.

A World to Be Won

There is a world of wisdom and beauty to be won. Obviously, as busy engineers, we cannot explore more than the tiniest portion of it. But to explore even a tiny portion, to discover the coastal outlines and see the vistas, to touch in at port here and there is to change our lives for the good.

Our problem is not unlike that of the world traveler who is constantly torn between the impulse to see as much as he or she can and the desire to linger in one spot in order to know it well. How much shall we survey in breadth and how much can we plumb in depth?

Certainly we want the overall view, superficial as it may be. Perspective and broad understanding are necessary first objectives. But the overall view becomes meaningless unless we stop here and there to investigate in detail, unless we find something to linger over and make our own. Each of us must seek our own balance between the general and the particular.

The problem of selecting particular works of literature and art from the myriads which surround us in this age of mass production is extremely vexing, almost paralyzing. Jacques Barzun has stated the problem well:

> The very bulk of the output kills appetite. Symphonies in bars and cabs, classical drama on television any day of the week, highbrow paperbacks in mountainous profusion (easier to buy than to read), "art seminars in the home," capsule operas, "Chopin by Starlight," "The Sound of Wagner," the Best of World Literature: this cornucopia thrust at the inexperienced and pouring out its contents over us all deadens attention and keeps taste stillborn, like any form of gross feeding. Too much art in too many places means art robbed of its right associations, its exact forms, its concentrated power. We are grateful for the comprehensive repertoire which modern industry for the first time puts within our reach, but we turn sick at the aggressive temptation, like the novice in the sweetshop.[26]

If we live in a sweetshop, we must learn discrimination and self-restraint. We must learn to relish the individual morsel without bewailing the fact that we cannot gorge ourselves. A single work of art, studied with love and understood well, may contain in microcosm and by implication as much as a thousand other works together. At least this must be our hope and our expectation, since we are busy professionals, and cannot delude ourselves into thinking that our experience in the arts can ever be comprehensive.

A further dilemma involves the question of whether we should be guided by what is said to be "worthwhile," or follow our own inclinations. We must do both, of course. Acknowledged masterpieces should not be foolishly ignored; we cannot rely exclusively on our taste, which is, after all, also our ignorance and prejudice. But neither must we follow the herd. Our taste should be challenged, refined, and stretched, but never overwhelmed. Ultimately it must prevail. We are under no obligation to worship unthinkingly at old and musty shrines.

Apologetic or Proud?

And this brings us to the question of how we, as engineers, should best approach the liberal arts. Should we come hat in hand, ashamed, apologetic, and willing to be melted down and molded anew? Or should we come proudly, willing to grow, but only with the understanding that we start from what we are? Eric Ashby, author of *Technology and the Academics*, leaves no doubt about the course to be taken. "The path to culture should be through a man's specialism, not by by-passing it.... The *sine qua non* for a man who desires to be cultured is a deep and enduring enthusiasm to do one thing excellently."[27]

Unless the liberal arts can be approached through engineering they will seem lifeless and frivolous to those of us who are professional engineers. Engineering educators are grappling with this problem now. They are agreed that liberal education for engineers is essential, but they are fearful of having it degenerate into mere "appreciation" or "ornamentation."

References

1. McGrath, Earl J. 1959. *Liberal Education in the Professions*, Teachers College, Columbia University, New York.
2. Mumford, Lewis. 1964. From Erewhon to Nowhere, in Thomas Parke Hughes (ed.), *The Development of Western Technology Since 1500*. New York: The Macmillan Company.
3. New England Consultants, Inc. 1963. *The Engineer Today*, prepared for Esso Research and Engineering Company.

4. Dougherty, J. J. The Engineer Among People, *American Engineer*, July 1966.
5. Chapman, Gilbert W. 1960. Specific Needs for Leadership in Management, in Robert A. Goldwin and Charles A. Nelson (eds.), *Toward the Liberally Educated Executive*, New York: Mentor Books, New American Library of World Literature.
6. Marcus Vitruvius Polloia, 1953. Selection from De Architectura, in Walter J. Miller and Leo E. A. Saidla (eds.). *Engineers as Writers*. Princeton: D. Van Nostrand, Inc., p. 14.
7. Van Doren, Mark. 1959. *Liberal Education*, Boston: Beacon Press.
8. Quoted in Benjamin Farrington, *Greek Science*, 1944, Baltimore: Penguin Books, Inc., pp. 28-29.
9. Emerson, Ralph Waldo. 1963. Works and Days, on Arthur O. Lewis, Jr. (ed.), *Of Men and Machines*, New York: E. P. Dutton & Co, Inc., p. 69.
10. Huxley, Thomas H. 1964. Science and Culture, in John J. Cadden and Patrick R. Brostowin (eds.), *Science and Literature*, Boston: D.C. Heath and Co., p. 11.
11. Veblen, Thorstein. 1963. *The Engineers and the Price System*, New York: Harcourt, Brace & World, Inc., pp. 139-140. Originally appeared as a series of essays in *The Dial* in 1919.
12. Ubell, Earl. 1966. Quoted in Public Relations Seminar for the Engineering Profession, American Engineer.
13. Garratt, Arthur. (ed.), 1966. *Penguin Technology Survey*, 1966, Baltimore: Penguin Books, p. 8.
14. Ruder, William. 1966. A Public Relations Executive Looks at the Engineer and His Role in Today's Society, *American Engineer*. March.
15. New England Consultants. 1963. *The Engineer Today*, prepared for Esso Research and Engineering Company, p. 35.
16. Blanchard, Richard D. 1966. Quoted in Public Relations Seminar for the Engineering Profession, *American Engineer*, March.
17. Reston, James. The *New York Times*, December 13, 1964.
18. Fuller, R. Buckminster. 1966. Vision 65 Summary Lecture, *The American Scholar*, Spring.
19. Orwell, George. 1963. The Road to Wigan Pier, in Arthur O. Lewis, Jr.(ed.), *Of Men and Machines*, New York: E. P. Dutton & Co., Inc., p. 259.
20. Ellul, Jacques. 1964. *The Technological Society*, Trans. John Wilkinson, New York: Alfred A. Knopf, Inc. Passage was reprinted in *The Saturday Review*, February 6, 1965.
21. Quoted in Don K. Price, *The Scientific Estate*, Cambridge: Harvard University Press, 1965, p. 57.
22. See Don. K. Price, *The Scientific Estate*, Cambridge: Harvard University Press, 1965; and Jay M. Gould, *The Technical Elite*, New York: Augustus M. Kelley, 1966.
23. Mumford, Lewis. 1964. From Erewhon to Nowhere, in Thomas Park Hughes (ed.), *The Development of Western Technology Since 1500*. New York: The Macmillan Company, p. 26.
24. White, Jr., Lynn. 1963. Humanism and the Education of Engineers, Studies of Courses and Sequences in Humanities, Fine Arts and Social Sciences for Engineering Students, University of California, Department of Engineering, Los Angeles, p. 53.
25. Ashby, Eric. 1963. *Technology and the Academics*, London: Macmillan & Co. Ltd., p. 81.
26. Barzun, Jacques. 1961. *Classic, Romantic and Modern*, Garden City, N.Y.: Anchor Books, Doubleday & Co., p. 148. Copyright 1943, copyright 1961 by Jacques Barzun, with permission of Atlantic Monthly Press, Boston: Little, Brown and Company.
27. Ashby, Eric. 1963. *Technology and the Academics*, London: Macmillan & Co. Ltd., p. 84.

CENTERPIECE

The Eleven Commandments of Twenty-first Century Engineering Education

ABET Engineering Criteria 2000

(Italicized items indicate education opportunities most pertinent to liberal education)

Engineering programs must demonstrate that their graduates have:

(a) an ability to apply knowledge of mathematics, science, and engineering,
(b) an ability to design and conduct experiments, as well as to analyze and interpret data,
(c) an ability to design a system, component, or process to meet desired needs,
(d) *an ability to function on multidisciplinary teams,*
(e) an ability to identify, formulate, and solve engineering problems,
(f) *an understanding of professional and ethical responsibility,*
(g) *an ability to communicate effectively,*
(h) *the broad education necessary to understand the impact of engineering solutions in a global/societal context,*

(i) *a recognition of the need for, and an ability to, engage in lifelong learning,*
(j) *a knowledge of contemporary issues*, and
(k) an ability to use the techniques, skills, and modern engineering tools necessary for engineering practice.

PART ONE

Challenge and an Institutional Response

CHAPTER 1

EDWIN ALTON PARRISH

Liberal Education and Engineering Criteria 2000

Introduction

The past several years have witnessed many changes to both the practices and the standards used by the Engineering Accreditation Commission (EAC) of the Accreditation Board for Engineering and Technology (ABET). Of particular interest here are the changes to the criteria themselves. To place them in perspective as we consider the implications of Engineering Criteria 2000 for liberal education, it is important to understand from where these changes come.

In the following paragraphs, an overview of the development of the criteria will be presented to provide historical context. Then the newest criteria, EC 2000, will be briefly considered, followed by a discussion of the implications of both the former criteria and EC 2000 for liberal education.

In the Beginning

The forerunner to ABET was the Engineer's Council for Professional Development (ECPD) that was organized in 1932. Within the next year or so, ECPD established "criteria for colleges of engineering which will insure to their graduates a sound, educational foundation for the practice of engineering." These criteria involved both qualitative and quantitative parts, not unlike those in use today.

In addition to establishing minimum standards for accreditation, ECPD professed an earnest desire to assist institutions in maintaining high-quality programs. In these early days, this was carried out through visits to different campuses by the same evaluation team, thereby ensuring a degree of consistency in assessment and guidance that is difficult to obtain in practice today.

Evolution at Work

To afford some measure of how these accreditation standards have evolved into the present time, consider the data of Table 1.1. Therein are rough estimates of the lengths of the criteria (excluding sections dealing with policy and other matters) at different points in time. While hardly a figure of merit, these data do show a rather exponential growth in the number of pages devoted to the criteria over roughly uniform periods. This naturally begs the questions, Why?

Consider the 1957 criteria that claimed about 1 1/4 pages of standard-sized text. The initial section dealt with the faculty of an engineering program. There were brief statements dealing with two sets of issues. First, there were those concerned with faculty qualifications, experience, intellectual interests, attainments, and professional productivity. The second set addressed the standards and qualities of instruction.

Next came the curriculum. Every program was mandated to have:

- One year of mathematics and basic sciences about evenly divided;
- One year of engineering sciences (mechanics of solids; fluid mechanics; thermodynamics; transfer rate mechanisms; electric fields, circuits, and electronics; materials);
- One-half year of engineering design;
- One-half to one year of humanities and social sciences; and
- An integrated sequential study that builds upon mathematics and basic sciences.

Readers familiar with current criteria will recognize considerable parallels. It is also worth noting that forty years ago ECPD recognized the important role the humanities, arts, and social sciences should have in preparing engineering students to contribute in many ways to the world they would join as graduates.

The next section of the criteria dealt with the student body. Institutions undergoing evaluation had to provide data substantiating the scholastic work of students and the records of graduates both in graduate study and in practice. Interestingly,

TABLE 1.1. Length of Criteria

Year	Number of Pages
Pre-1955	1
1957	1 1/4
1967	1 1/2
1977	4
1987	16 1/2
1997	19 1/2

this latter factor is gaining more emphasis today in terms of requirements for more formalized outcomes assessment.

Finally, the 1957 criteria addressed the role of the administration. In particular, evaluators were concerned with the attitude and policy of the administration toward its engineering division and toward teaching, research, and scholarly production. These same matters are examined by modern-day teams.

Given that there are so many common elements between the criteria of 1957 and today, how did the document setting forth the criteria expand so much in the intervening years? A review of the different sets of criteria reveals that gradually, in an effort to be helpful to institutions trying to interpret the criteria, explanatory text was added to provide needed guidance. This guidance took the form of statements that included the word "should," for example, as in

- Laboratory experience should be included, and
- Humanities and social science studies should be planned to reflect a rationale

Later, the instances of "should" were replaced with "must," and the criteria began to grow nearly unabated for many years.

Referring back to the 1957 criteria dealing with faculty as outlined above provides a good example of this phenomenon. Those short phrases that set standards for faculty qualifications and standards of instruction gradually were enlarged, until we achieved six paragraphs covering these factors:

1. heart of the program is the faculty,
2. overall competence may be judged,
3. proper size,
4. stability,
5. teaching loads and professional development, and
6. curricular and career advising

Another major contribution to the increasing length and growing prescriptive nature of the criteria had to do with the inclusion of discipline-specific program guidelines. These were offered by the various professional societies as a way of interpreting the criteria in terms of their particular disciplines. Looking again at Table 1.1, a very large increase occurred sometime in the 1980s. This had to do with the incorporation of these guidelines as bona fide "program criteria," after which the original criteria were referred to as the "general criteria." Today, every program must satisfy both the general criteria and the appropriate program criteria (including the nontraditional program criteria for those programs falling outside the norm). In addition, a program title could imply satisfying more than one set of program criteria.

The result of all this evolutionary change over several decades has been a document that some view as encouraging a "cookie-cutter" approach to engineering education. Despite a section of the criteria permitting innovative programs to be evaluated by other standards, rarely if ever has the clause been invoked. Consequently, the emphasis has been on examining what courses students passed rather than what they learned and could do, as well as a lack of encouragement for experimentation with new pedagogy or curricula.

A New Approach

Around 1990, much concern was focused on the bean-counting approach to accreditation and to the overly prescriptive nature of the criteria. This intensified to the point where in 1993 the ABET Board appointed the Accreditation Process Review Committee with former ABET President John Prados as its chair. The committee included representation from various stakeholders and undertook a close examination of engineering accreditation.

Among the issues identified by the APRC was one directly aimed at the criteria:

> The current accreditation criteria are too long and by their very nature encourage a rigid, bean-counting approach that stifles innovation.

As this statement reflected the views of many members of the EAC itself, it led to a series of workshops funded by the National Science Foundation. Of particular interest here is the *Criteria Workshop* held in New York City, May 21-22, 1994, that was chaired by the author and Dr. Ira Jacobson, now EAC past Chair. Some sixty participants from industry, government, and academia deliberated the efficacy of two proposals presented at the opening session. The first was to revise the existing criteria to reduce their prescriptive nature. (As a starting point, I deleted all statements containing "should," which shortened the document by about

25%.) The second was to start with a clean sheet of paper and completely redesign the criteria. These proposals engaged participants in lively discussions that resulted in several recommendations, the most important of which was that a new set of criteria should be defined. These recommendations were presented in a written report[1] that was distributed to all participants and published on the Internet. In addition, presentations were made to the EAC and the report provided to the EAC General and Program Criteria Committee at the Commission's annual meeting in July 1994.

Over the following year, the Criteria Committee worked diligently to flesh-out the recommendations and, at the July 1995 annual meeting, won the unanimous approval of the EAC for what is now called EC 2000.[2] The author had the pleasure of presenting a motion to the ABET Board at its meeting in November calling for approval of EC 2000 and an associated implementation plan. This too received unanimous support and so the EAC embarked on two years of experimental visits to selected institutions, to be followed by three years of transition during which institutions may choose the criteria (then-current criteria or EC 2000) under which their programs will be evaluated. The pilot visits were conducted under section II.A.7 of the then current EAC criteria, as the program criteria needed to complete EC 2000 were yet to be finalized and approved. Beginning in the 2001–2002 academic year, all engineering programs have been evaluated under EC 2000.

As shown in Table 1.2, EC 2000 consists of eight criteria. These are placed in a context that requires outcomes assessment and continuous improvement for all accredited engineering programs. This approach is nearly orthogonal to previous practice and is why full implementation is being done over five years.

As mentioned above, experimental or pilot visits were conducted during the 1996–1997 and 1997–1998 cycles. By starting out with a very small number of institutions, both public and private and of differing sizes, considerable experience was gained with how to evaluate engineering programs under EC 2000. With funding again from the National Science Foundation,[3] workshops were held each spring preceding the pilot visits. For example, in May 1996, representatives from the University of Arkansas and from WPI (Worcester Polytechnic Institute) were brought together with the two EAC teams scheduled to make campus visits. Presentations were made by university administrators as to how they believed their programs satisfied EC 2000. Subsequent discussions and private meetings between

TABLE 1.2. Engineering Criteria 2000

Criterion 1. Students	Criterion 5. Faculty
Criterion 2. Program Educational Objectives	Criterion 6. Facilities
Criterion 3. Program Outcomes and Support and Assessment	Criterion 7. Institutional Financial Resources
Criterion 4. Professional Component	Criterion 8. Program Criteria

associated university representatives and EAC team members led to clarification of many issues dealing with additional information needed and how the visits would be conducted. The two visits were then carried out in the fall of 1996. A similar workshop was held in May 1997 with representatives from Georgia Tech, Harvey Mudd, and Union College and the EAC team members.

Liberal Education and the Current Criteria

Generally speaking, liberal education implies imparting knowledge in mathematics and science as well as the humanities and social sciences (HSS). Within the present context, the first two are a given, so consider now the HSS aspects of the criteria that provide the breadth normally associated with liberal education.

As indicated previously, forty years ago there was a requirement for somewhere between one-half and a full year of studies in the HSS. Examination of the (1996–1997) criteria reveals considerably more emphasis. In fact, four of the five curricular objectives (criteria section IV.C.2.) imply the value placed upon liberal education:

- Sensitivity to socially related technical problems.
- Understanding of ethical characteristics.
- Need to protect public health and safety.
- Lifelong learning.

Continuing, the curricular content section (criteria section IV.C.3) specifies a minimum of one-half year of study in HSS and goes on to require that:

- HSS studies must meet the objectives of a broad education and those of the engineering profession.
- HSS studies must reflect a rational appropriate to the engineering profession and the institution's educational objectives.
- HSS studies must provide both breadth and depth.

A very strong statement appears in this section that bears repeating here:

> In the interests of making engineers fully aware of their social responsibilities and better able to consider related factors in the decision-making process, institutions must require course work in the humanities and social sciences as an integral part of the engineering program. This philosophy cannot be overemphasized.

It is apparent from these excerpts that the criteria are directed at ensuring engineering students receive a liberal education. The impact of these statements over the years has perhaps not been as great as one would like, primarily because of the overwhelming density of requirements and resulting emphases placed by evaluators on the technical portion of the educational programs.

Liberal Education and EC 2000

One of the stated objectives of accreditation within the context of the new EC 2000 is to assure that graduates are adequately prepared to enter and continue the practice of engineering. This surely will require less in the way of narrow, "stovepipe" curricula that heretofore were concentrated on providing considerable expertise within a given discipline. One can also look to numerous recent studies on how engineering education can be made more relevant to societal needs in the new millennium to find growing support for broader undergraduate educational programs.[4] Finally, the Engineering Education Coalitions funded by the National Science Foundation have worked toward implementing many of the recommendations that resulted from these studies. All of these efforts suggest a recognition of the need and a willingness to replace a system that has become ossified and moribund with one that involves regular assessment and feedback such that programs are continuously improved to assure the liberal education required in the twenty-first century.

Turning now to the specifics of EC 2000, consider the elements of the section on program outcomes and assessment (criterion 3). Here we find that no less than six of the eleven capabilities required of graduates relate directly to liberal education:

- Ability to function on multidisciplinary teams.
- Understanding professional and ethical responsibilities.
- Ability to communicate effectively.
- Understanding impact of engineering solutions in a global and societal context.
- Ability to engage in lifelong learning.
- Knowledge of contemporary issues.

There are numerous opportunities within the HSS studies that are still required as an integral component of the engineering program for students to develop and strengthen their capabilities in these important areas.

The section on professional component (criterion 4) also reflects a concern for liberal education. In particular, this criterion requires "a general education component that complements the technical content of the curriculum and is consistent with the program and institution objectives." Many institutions establish fairly broad requirements for general education which are intended to expand students' intellectual perspectives. While some courses listed therein involve development of personal skills, by and large they are suitable within the framework of EC 2000.

It should be noted that, in a break from the tradition in place for more than forty years, the new criteria do not specify a required minimum proportion of the curriculum that must be devoted to HSS studies. Rather, the engineering faculty is required to assure that curriculum devotes adequate attention and *time* (emphasis added) to each component, consistent with the objectives of the program and institution." Furthermore, the criteria go on to state that students

> must have a major design experience that considers most of the following factors: economic, environmental, sustainability, manufacturability, ethical, health and safety, social, and political. Understanding of these factors certainly will be enhanced by an appropriate experience in the HSS.

It seems clear from these considerations that EC 2000 does not set lower expectations for HSS studies than previously has been the case. Arguably, there is even more emphasis given to the importance of a broad and liberal education.

Closing Observations

In reviewing the historical evolution of the accreditation criteria, it becomes evident that the importance placed upon the HSS component of engineering curricula has been fairly consistent. Even though the criteria in place in the years following 1957 until the present time reduced this component to a minimum of one-half year, the added required rationale for selecting courses that complement the technical studies ensured some faculty attention. Thus, there is evidence to support the assertion that the centrality of liberal education has grown significantly since the days of ECPD.

With the advent of EC 2000, there is continued concern for liberal studies manifested in several sections, as indicated above. In point of fact, EC 2000 stresses liberal education in calling for graduates to be able team members, to possess effective communication skills, to know not only how to solve problems, but when and when not to solve them, to have a global and societal perspective, and to become fully engaged in lifelong learning. In addition, EC 2000 places the burden on the faculty and administration to focus on what students learn rather than which courses they take. In this sense, outcomes assessment becomes an important technique for relating the two and for continuously improving engineering education.

One of the more recognizable names in higher education is that of Ernest L. Boyer who, until his death in 1995, served as president of the Carnegie Foundation. He authored many reports that have signaled the need and proposed mechanisms for significant change. In *College*,[5] for example, he cited the polarizations between studies in the professions and those in the liberal arts as well as conflicts about what is taught and what is learned. In addressing assessment, Boyer asks, "Are there, in short, outcomes greater than the sum of the separate parts?" He also writes, "If there are no transcendent purposes, if there is no vision that goes beyond the individual student and a final accumulation of 128 credits, then progress is best measured incrementally, perhaps, through the separate classes. If, however, there are commonalities appropriate for all, if there are expectations that grow out of, and yet transcend, the separate courses, then assessment can serve a larger purpose."

It is in this spirit that EC 2000 is offered to the liberal education community. Creating professional programs that satisfy the spirit and the intent of these criteria may contribute to reducing the fragmentation and clarifying the confused institutional and program goals that so worried Boyer. As he observed, "In too many cases it appears that the professional schools are *at* the university, but not *in* and *of* the university."[6] Perhaps the momentum gathering around EC 2000 will be sufficient to overcome this plight and contribute to realizing his vision for the New American College.[7]

References

1. Parrish, Edward A., Report of the Criteria Workshop, New York City, May 21–22, 1994 (available from ABET Headquarters).
2. Details of EC 2000 are widely available, including the ABET Web site: http://www.abet.ba.md.us/EAC/eac2000,html.
3. Peterson, George D., Parrish, Edward A., and Rogers, Gloria. Proposal for a pilot study for the implementation of Engineering Criteria 2000, submitted to the NSF, March 20, 1996.
4. For example, the Engineering Education Board of the National Research Council and the American Society for Engineering Education have published reports emphasizing many aspects of a liberal education. In addition, the Aerospace Industry Roundtable organized by Boeing has added its voice to the increasing number of advocates for a new approach to engineering education that emphasizes a similar skill set.
5. Boyer, E. L. 1987. *College: The undergraduate experience in America*. New York: Harper and Row.
6. Pelikan, Jaroslav. 1983. *Scholarship and Its Survival* (Foreword by Boyer, E. L.), The Carnegie Foundation for the Advancement of Teaching.
7. Coye, D. Ernest Boyer and the New American College, *Change*, May/June 1997.

CHAPTER 2

LANCE SCHACHTERLE

Liberal Education Responds

Discussing ABET 2000 within a Humanities Division

Introduction

Within the last decade, the professional organization ABET (the Accreditation Board for Engineering and Technology) has rewritten the rules for accrediting undergraduate engineering programs. As the chapter by Edward Alton Parrish indicates, the ABET requirements evolved in a half century from a mere page of guidelines to almost twenty pages of very detailed requirements of what courses engineering students must pass. This ever-increasing specificity promoted a certain measure of quality throughout U.S. institutions, but increasingly observers grew concerned that this quality gravitated around a low common denominator. Nor were the existing ABET requirements fulfilling their original mission of encouraging responsible experimentation. Quite the opposite: many engineering educators protested being fitted into a Procrustean bed where all that really mattered for accreditation was assuring visitors that every graduate had passed a minimal number of courses distributed among various sorting bins. Among educators and employers, a consensus eventually emerged that vitalizing engineering education required rethinking this "bean-counting" environment. The Engineering Criteria 2000 initiative, designed to nurture responsive innovation, is being phased in, with initial experimental visits in 1996 leading to anticipated nationwide adoption in 2001.

The heart of these new criteria (referred to as EC 2000 henceforth) is the recognition that what matters in education is the performance outcomes that engineering students will use as professionals throughout their lives—both in entry-level positions and throughout careers that increasingly lead to new and unanticipated technologies and directions. Thus EC 2000 is predicated upon measuring the outcomes of academic programs, looking at what graduating students *can actually do* rather than *what courses they have passed*. Not only is such a basic strategy far more in accord with actual professional practice, which demands results, it acknowledges the importance of the historical diversity of American educational institutions. Under EC 2000, every accreditable institution must frame its own mission. In turn, this mission statement creates the standards by which accreditation visitors will judge institutional programs and successes. And for the purposes of this volume, EC 2000 also raises potentially revolutionary challenges to rethink the role within engineering education of liberal studies.

In no other country do citizens have so many opportunities to pursue postsecondary education by means of so many different vehicles as in the U.S. Four-year colleges and universities, a variety of two-year institutions, numerous kinds of continuing education programs, as well as increasing paths for pursuing undergraduate degrees at the workplace—all contribute to a rich and diverse mix of educational choices. Thus it is no surprise that the U.S. has one of the highest percentages in the world of citizens who pursue some form of education after the completion of high school—45.2 percent of the eligible population choosing from among 4096 two- and four-year institutions according to the statistics in the 1999 Almanac issue of *The Chronicle of Higher Education* (27 August 1999, p. 7).

At least two historical and social factors explain this unusual diversity of educational programs. First, in contrast to virtually all other industrial nations, in the U.S. no central authority dictates standards across the land for higher education. Unlike European countries, for example, the U.S. has no national Ministry of Education to direct educational policy, often down to the level of the specific curriculum taught in every classroom. Instead, each of the 50 states competes against other states to strengthen education within state boundaries, to reap the benefits of the ensuing comparative advantages for the regional economic climate. This competition among states helps to foster much experimentation and diversity within the educational fabric.

Second, diversity is further stimulated by the presence in the U.S. of a strong cadre of private institutions, who compete both with the state-sponsored institutions and with each other. Endowed private institutions managed by independent boards are free to invest their resources in any educational experimentation they regard as worthwhile risks—at the cost, of course, of potentially damaging losses.

The remarkably competitive environment for education in the U.S. thus nurtures a strong spirit of experimentation. The result is that, among all of its exports, higher education remains one of the strongest "products" of the American socioeconomic system.

Within this diverse macrocosm, engineering education is a microcosm of many different kinds of programs. In one form or another, engineering education is pursued in virtually all of the institutions enumerated above—two- and four-year universities, proprietarial, private and state colleges, on-the-job, and the like. Further, engineering education responds to, and often benefits from, competition with liberal arts programs. From the founding of Harvard University in 1636, American higher education has been dominated by private, liberally oriented institutions. Much of the history of engineering education in the U.S. is determined by the attempts of engineering educators to chart a course for technologically oriented education, which builds upon, complements, and sometimes rejects this historical preference for a liberal education.

In particular, the widespread adoption in the U.S. of a four-year undergraduate curriculum for engineering emulates the distinctive freshman through senior mindset of liberal arts institutions, in contrast to European developments where the initial engineering degree is normally at a higher level but requires more time. Furthermore, in America engineering did not separate itself from the liberal education model by becoming a graduate program at the professional-entry level, as did the other professions like law, medicine, and theology. Despite over a century of proposals to the contrary, in the U.S. engineering remains the only profession where the entry level is achieved with a four-year undergraduate degree. The different levels of time commitment to engineering and to other professions do not go unnoticed when engineering is compared to other disciplines, such as law and medicine, with which its societal contributions should be equally ranked. And at least in American practice, these other professions build upon the four-year undergraduate experience, often in liberal arts programs. If engineering is not, like law and medicine, to "build upon" a four-year liberal education, must it not in some way offer this liberating education if its graduates are to compete equally with those professionals, starting with the foundations of an undergraduate liberal arts education?

Even the etymologies of the key terms in this discussion sharpen the debate about the relationship of engineering and liberal education. The history of the term "liberal arts" points back to the original medieval sense that a "liberal" education conferred "freedom" (from the Latin "liber") from the servitude of manual labor. A liberally educated person, then and often now, is regarded as one working with the head as opposed to the mechanic, workman, or craftsman, who are largely dependent upon manual skills. This distinction between technically and liberally educated is reinforced, in Anglo-Saxon culture, by the derivation of the word "engineer" from the medieval English word "engin" (sometimes spelled "ingin"). The historical quirk that in English "engineer" became associated with "engine" has perpetuated an association of engineering with manual and mechanical talents. Worse, English culture has largely lost the etymological association most important to "engineer": its linkage with "ingin" and its modern cognate "ingenuity." The perpetuation of the association on the Continent of "ingenuity" with "engineering" has made it far easier to see European engineers ("ingenieurs") for what they are—people who apply ingenuity to solving problems.

Historical Background of Liberal Education within Engineering Education

Historically, the first U.S. organization significantly dedicated to engineering was the West Point Military Academy, a late eighteenth-century institution formally founded in 1802, but which began to take on its current form with the arrival of Superintendent Sylvanus Thayer in 1817. Subsequent early academic foundations in engineering education include Rensselaer Polytechnic Institute in 1824 (the first ongoing civilian institution dedicated to engineering), the coestablishment with existing liberal arts institutions of associated technical colleges at Harvard, Yale, and Dartmouth in the 1840s and 1850s, and the Massachusetts Institute of Technology in 1860.

However, by 1860 only a handful of formal programs existed in fields we would today call engineering. As in England, many professionals who practiced engineering did so with little formal education, having acquired their skills as apprentices engaged in the canal and later railroad booms. More than one historian of American engineering education has observed that the Erie Canal was in reality America's first great school for nonmilitary engineering. Until the successes of applying science for military ends in World War II made clear that all future engineering must firmly be grounded in science, the role of shopwork, practical experience, and apprenticeship in various forms continued to play a significant (if ever diminishing) role in engineering education in the U.S. and England. In contrast, in France and Germany—again in part because of the "ingenieur"/"ingenuity" linkage—engineering education was from the early nineteenth century on firmly based in mathematical and scientific study. This historical difference clearly affected the status of engineering education as a form of general or liberal education, in that engineers are far more widely viewed in Europe (as opposed to the U.S.) as having achieved a rank equal or superior to the liberal arts undergraduate.

The balance between technically and liberally oriented institutions shifted dramatically with the Morrill Act of 1862. The act of Congress creating land-grant institutions was designed specifically to nurture the development in every state of publicly supported universities with practical aims (especially in agriculture and engineering), in specific contrast to the perceived focus on liberal arts at established institutions. The booming economic climate after the Civil War, combined with the private endowing of technological institutions and the largesse of the Morrill Act, led in the decade from 1862 to 1872 to an expansion of engineering schools from six to seventy (history of U.S. engineering education).[1-4]

By the end of the nineteenth century, engineering education was flourishing at those institutions like Rensselaer, MIT, and WPI (the third oldest private technological university), focusing directly on engineering education, at the engineering colleges within land-grant universities, as well as at those few older liberal arts schools with technologically oriented programs (such as Union College). The earliest institution to promote discussion among educators focusing on engineering education was the Society for the Promotion of Engineering Education (SPEE), founded in 1893 as part of the Columbian Exposition in Chicago. From the very beginning of professional attention to engineering education through the predecessor organizations of the American Society for Engineering Education (ASEE), a principal concern was the extent to which engineers should be educated in the liberal arts, and how such nontechnical education could be accommodated within the context of a demanding technical commitment.

Among the papers discussed at the 1893 inaugural conference were two on the role of communications in engineering education, both offered by engineering faculty. Also of concern to early engineering educators was the role of foreign languages, since many papers in areas of professional interest still first appeared either in French or German. Subsequent SPEE meetings, however, challenged the time devoted to nontechnical issues, with the President in 1905 in his inaugural address hoping "that the time would soon come when all course requirements in English and foreign languages could be amended"[5] (Gianniny's paper [Chapter 26] is my source for much of the following discussion of the history of liberal learning in engineering education.) However, the continuing need to prepare engineers to communicate effectively (increasingly focused on English as more professional publications first appeared in that language) guaranteed that English, and later the social sciences, came to be recognized as necessities for engineering education. In 1922, the first separate meeting of English teachers occurred at SPEE, and the English committee was elevated to the status of a division in 1942. Subsequently in 1945, the Humanistic-Social Science Committee also was raised to division status; both units were joined together as the Liberal Studies Division in 1965, renamed as the Liberal Education Division in the 1980s.

The context for this evolution of the separate disciplinary interests in English, humanities, and social sciences within engineering education is provided by the succession of self-reflective reports from leading engineering educators, from the first such report by Charles Riborg Mann in 1918 through Sterling Olmsted's 1968 *Liberal Learning for the Engineer*. Probably the most influential of these studies was the series of "Bulletins" commissioned by SPEE and published as two volumes in 1930 and 1934 as the *Report of the Investigation of Engineering Education, 1923–1929*, prepared by a committee chaired by William E. Wickenden. During and after World War II, a succession of further reports raised concerns about the role of liberal education within engineering education. Among the most important of these reports, all published by SPEE or its successor ASEE and usually referred to by the names of the study committee chairs, are *Aims and Scope of Engineering Curricula*, chaired by Harry Parker Hammond in 1940: the *Report on the Committee on Evaluation of Engineering Education*, chaired by Linton Elias Grinter in 1955; *General Education in Engineering*, chaired by George A. Gullette in 1956; and the Olmsted report of 1968. These reports repeatedly affirmed the present accreditation conventions, by which all students graduating from an accredited program are expected to devote a half-year out of their expected four years of study to a program in the humanities and the social sciences (referred to henceforth by the usual abbreviation used among discussants in these matters, HSS for "Humanities and Social Sciences in Engineering Programs").

The new EC 2000 ABET guidelines applicable to HSS state simply that accreditable programs should have:

> a general education component that complements the technical content of the curriculum and is consistent with the program and institution objectives.

This rationale reflects decades of debate within engineering education between those who value most in HSS the instrumental or vocational abilities presumably conferred through liberal education and those who regard liberal education as an end in itself. The substitution of "general" for "liberal" education reflects the prevalence of general education requirements in most regional accreditation guidelines and the resultant structure at many large universities of both general and departmental/program graduation requirements. In the main, for engineering education, general education comprises science, mathematics, humanities, and social sciences—the current formulation of the liberal arts. However, for the most part, as noted later, what is at issue in general education is specifically HSS—not mathematics and science, which increasingly became part of the mainstream of the "technical content of the curriculum" referred to above.

From the earliest reports in Wickenden's studies from the 1920s, policymakers pondered what the general or liberal component to engineering education should be. Often reports stressed the acquisition of communication tools as a necessity for the professional. Thus, throughout much of the early history of the role of HSS faculty within SPEE, inculcating the skills of technical communications, written as well as oral, was a mainstay in the argument for humanities and social sciences. Exposure in English or humanities classes to literary classics was justified as much or more by the utilitarian ends of acquiring knowledge of how to manage people and to communicate, as for serving cultural or aesthetic ends. Similarly, the social sciences were often presumed to provide "people" or managerial skills, or the tools of economics needed in order to pursue large-scale engineering projects.

At the same time, many statements prepared by Wickenden's numerous committees argue eloquently for conferring upon engineers some of the same enriched and liberating cultural experiences presumably enjoyed by graduates of prestigious liberal arts institutions. (See, for example, "Summary of Fact-gathering Stages," pp. 43–47; "The Curriculum," pp. 71–80; "Structure of Undergraduate Engineering Curricula," pp. 84–93; "Cultural Subjects," pp. 405–409). Wickenden's own 1926 "Preliminary Report to the Board of Investigation and Coordination" reflected upon the packed density of the engineering curriculum, resulting in too few opportunities for students to spend much time reading and reflecting on any topics on their own.

> Mere cutting of requirements will not assure these results [of "a larger zeal and capacity for self-directed learning."] The whole structure of subject matter needs to be closer knit. Some of the teaching methods into which we have slipped quite generally in order to maintain overcrowded schedules may need radical modification. For example, it is the pretty general testimony that engineering students devote comparatively little time to reflective study and thought, do no collateral reading and pursue no independent inquiries out of pure interest, being kept busy in grinding out innumerable, small-dimensioned, set tasks. Much of the mass that emerges at graduation appears to be the true product of this sort of industriousness, proficient in a conventional technique, but stale, unimaginative, lacking in scientific grasp and devoid of intellectual enthusiasms.[5]

Of special concern in the subsequent Grinter (1955), Gullette (1956), and Olmsted (1968) reports was the specific focus of designing liberal education programs around topics with scientific and/or technological emphases. Understandably, both humanistic and engineering educators were optimistic that engineering students would respond best to culturally oriented studies if the topic material was inclined in the direction of their professional studies. The development in the 1970s of Science-Technology-Society programs (the STS movement), along with many other manifestations of interdisciplinary study, provided significant impetus in this direction. The Olmsted report specified in more detail than previous documents the four sets of objectives for HSS studies correlated with the engineering vocation:

1. Utilitarian—"the importance of writing and speaking, along with the skills of managing people."
2. Cultural—"an emphasis on subject matters—concepts, principles, and methodologies to induct the student into some larger concept of society."
3. Developmental—"means for developing in the student certain personal and intellectual qualities presumably not fostered to the same extent by the more technical parts of the curriculum."
4. Contextual—"focus on the student in his [sic] role as an engineer as an agent of social change, who must live his professional life within a human context."[6]

Yet all these reports, precisely because they articulated persuasively for HSS a complementary rather than a lead role in engineering education, invited both friends and foes of the liberal arts within engineering education to refer to HSS studies (but not science or mathematics) as the "second stem" or "soft skills." While such terms doubtless were intended to stake significant claims for the importance of HSS in engineering education, their distinguishing adjectives inevitably made HSS of lesser importance to the primary stem or hard skills at the center of engineering education.

Present Expectations and Requirements

The 1995–1996 ABET requirement recently in force (superseded by EC 2000) with respect to the humanities-social science requirement was as follows:

IV.C.3.d.(2) Humanities and Social Sciences

IV.C.3.d.(2)(a) Studies in the humanities and social sciences serve not only to meet the objectives of a broad education but also to meet the objectives of the engineering profession. Therefore, studies in the humanities and social sciences must be planned to reflect a rationale or fulfill an objective appropriate to the engineering profession and the institution's educational objectives. In the interests of making engineers fully aware of their social responsibilities and better able to consider related factors in the decision-making process, institutions must require course work in the humanities and social sciences as an integral part of the engineering program. This philosophy cannot be overemphasized. To satisfy this requirement, the courses selected must provide both breadth and depth and not be limited to a selection of unrelated introductory courses.

The next section of the requirements specifies acceptable humanities courses of both a traditional kind (philosophy, history, literature, fine arts, and the like) and nontraditional (newer, interdisciplinary fields such as history of technology and professional ethics). Only "routine exercises of personal craft" (presumably activities like the proverbial basket-weaving) are excluded from "courses that instill cultural values." The final section, IV.C.3.d.(2) (c), calls attention to the merits of "applied" social science and management activities, cautioning that despite their acknowledged worth to an engineer, they may not be counted as part of the HSS sequence for ABET purposes:

> Subjects such as accounting, industrial management, finance, personnel administration, engineering economy, and military training may be appropriately included either as required or elective courses in engineering curricula to satisfy desired program objectives of the institution. However, such courses usually do not fulfill the objectives desired of the humanities and social science content.

These requirements reflect political decisions among the competing views discussed above aimed at providing students with the following outcomes:

1. Engineering graduates should be sufficiently immersed in the humanities and social sciences to secure at least some of both the vocational and personal outcomes presumably nurtured by such studies.
2. Programs should be made appropriate to the specific interests of the engineering student, in ways left to the judgment and policies of the institution.
3. Some appropriate balance of depth as well as breadth over several disciplines should be achieved.
4. HSS courses should be historically, esthetically, or culturally grounded—thus the exclusion of the current rules of "routine exercises of personal craft."

At the same time, the recent rules sensibly permitted the study of foreign languages from the introductory level (a relatively new permission), making it possible for those students who wish to pursue a language to start that activity from the introductory course (which in previous ABET guidelines had been banned as insufficiently challenging as a cultural experience).

The Challenge of the New Engineering Criteria 2000

The new ABET criteria proposed as EC 2000 speak in a very different tone about humanities. In keeping with the emphasis throughout the new ABET criteria, the language is far more brief and entirely nonprescriptive. The new requirements simply state the following:

Criterion 4. Professional Component

The professional component requirements specify subject areas appropriate to engineering but do not prescribe specific courses. The engineering faculty must assure that the curriculum devotes adequate attention and time to each component, consistent with the objectives of the program and institution. The curriculum must prepare students for engineering practice, culminating in a major design experience based on the knowledge and skills acquired in earlier coursework and incorporating engineering standards and realistic constraints that include most of the following considerations: economic, environmental, sustainability, manufacturability, ethical, health and safety, social, and political. The professional component must include:

1. one year of college level mathematics and basic sciences (some with experimental experience) appropriate to the discipline;
2. one and one-half years of engineering topics, to include engineering sciences and engineering design appropriate to the student's field of study; and,
3. a general education component that complements the technical content of the curriculum and is consistent with the program and institution objectives.

Absent in the new formulation is any time requirement mandating a fixed time period of HSS study—the half-year floor required for HSS is gone in EC 2000 as it presently stands. (Note, however, that the regional accreditation associations increasingly are adopting time-specific minima for HSS and liberal/general studies as part of their requirements for regional institutional accreditation of specialized professional programs.) The new ABET requirements, viewed optimistically, thus challenge all those in any discipline interested in the role of liberal learning within engineering education to work together to define, implement, and assess an appropriate mission. In keeping with the spirit of EC 2000, specifically the emphasis on what students can demonstrate they have learned rather than simply what courses they have passed, the new criteria speak of appropriate performance outcomes. At the center of EC 2000, "Criterion 3, Program Outcomes and Assessment," specifies eleven outcomes:

1. an ability to apply knowledge of mathematics, science, and engineering;
2. an ability to design and conduct experiments, as well as to analyze and interpret data;
3. an ability to design a system, component, or process to meet desired needs;
4. an ability to function on multi-disciplinary teams;
5. an ability to identify, formulate, and solve engineering problems;
6. an understanding of professional and ethical responsibility;
7. an ability to communicate effectively;
8. the broad education necessary to understand the impact of engineering solutions in a global/societal context;

9. a recognition of the need for and an ability to engage in life-long learning;
10. a knowledge of contemporary issues; and,
11. an ability to use the techniques, skills, and modern engineering tools necessary for engineering practice.

What is especially striking about this enumeration of student outcomes is the merging of professional attributes that come from both technical and nontechnical disciplines. Unlike so much of the previous literature about liberal education within engineering education, Criterion 3 does not separate between "soft skills" and "hard skills," nor does it relegate (as often was the case earlier) humanities and social sciences to a second stem. In so doing, the framers of EC 2000 challenged engineering educators from all disciplines to define ways of assisting students in achieving these outcomes, from all of the different disciplinary perspectives that characterize a modern university.

EC 2000 throws down a tough challenge. The intent of the HSS requirement cannot be met simply by students checking off courses they pass in the current areas described in the 1995-1996 requirements. EC 2000 demands that students demonstrate how they can synthesize into actions what they have learned in courses. They must perform effectively as professionals, and presumably also as citizens (outcomes 6, 8, 10) and as self-reflective humans (outcomes 6-10). They must, as with all areas of their education, demonstrate that they have achieved the capacity to carry out all eleven characterizations of professional activity in Criterion 3.

As with all of the outcomes in the new criteria, at least three steps must be carefully coordinated to respond successfully to evaluating education in terms of student outcomes. Not only the faculty, but all relevant constituencies (including, as is practical, students, graduates, and employers) must work together to:

1. Define the mission of the institution, the department, and the program in terms of outcomes that students and graduates must demonstrate.
2. Determine what performance measures can provide all interested constituencies—students, faculty, employers, and accreditation visitors—with indicators that the desired mission, goals, and outcomes have been achieved.
3. Indicate what steps are taken within the program to provide continuous improvement throughout the curriculum so that when student outcomes clearly indicate areas in the curriculum needing improvement, those outcome measures will be used to make the necessary revisions.

Perhaps an example of how to formulate a program outcome and assess its impact on students would be helpful. An outcome in which both technically oriented and liberal education faculty can participate is outcome 8: "The broad education necessary to understand the impact of engineering solutions in a global/societal context." Under the implementation schedule for EC 2000 this outcome, along with the ten others, will need to be part of the undergraduate education of all engineering students in ABET-accredited programs after the year 2001. Consequently, departments and colleges of engineering will need to begin wrestling with designing curricula that promote such outcomes.

The first step in the process of adopting EC 2000 is to review the mission and/or goal statements for the engineering unit within the institution, which may be the college of engineering or the entire campus of a technological university. As many disciplinary departments as possible should be involved in refining such mission and goals statements, to engage students at multiple points in their careers with opportunities to acquire the skills identified in Criterion 3. Thus, this goal of "understanding the impact of engineering solutions in a global/societal context" very appropriately involves faculty collaboration from many different disciplines. Those disciplines participating in formulating the overall mission should in turn indicate through appropriate mission statements for their own programs the ways in which they will specifically promote the outcomes of greatest interest. For example, an engineering program might indicate its concern to engage students in considering the economic impact of technology across borders, while humanities and social science departments might look at language or cross-cultural issues.

Once the general and specific mission or goals are set out, the next step is crucial. Those engaged in framing the outcomes-based curriculum must define metrics that will indicate how student success in achieving such a general goal is measured. The general goal of outcome 8 is to develop within students the self-starting capacity to assess the relative merits of different solutions to engineering problems within global and/or societal contexts. This admirable broad outcome must be characterized by specific metrics that break down the overall goal into sufficiently specific subsets, so that demonstrable outcomes of student performance can actually be identified and assessed.

In framing the metrics, we must also keep in mind that for this, as for many of the Criterion 3 outcomes, the outcome is what we expect graduates and professionals to do, but the only performance we can measure for students is what they demonstrate as undergraduates. Thus the metrics must be at a level appropriate to undergraduates. In the case of this outcome, observers of student activity could measure their success in achieving the desired results by asking if the students:

- ask questions in classes or projects about societal/global impacts of engineering activities, and use what they learn critically throughout their work.
- participate in international programs, or other activities that involve them in both experiential and academic learning within cultural contexts new to them.

- participate in activities on the campus that in any number of ways provide opportunities for American students to interact with international undergraduate and graduate students, and vice versa.
- engage in activities on campus that involve faculty or visitors from cultures new to many of the students, to talk about ways in which engineering problem formation and solutions take place in their cultures.
- participate in any extracurricular or volunteer activities which sharpen their sense of the relationship of their profession to societal contexts.
- finally, in a capstone design project or similar activity, consider the global and societal dimensions of professional work as part of their overall problem solving.

Clearly, both technological and HSS studies can contribute significantly to helping students achieve these kinds of outcomes.

Once such measurements are agreed upon, it is then necessary to set up a system to record and interpret the measurements. Many institutions in the past have found that student portfolios, reviewed by faculty advisors or teachers, provide a convenient vehicle for measuring and preserving such outcomes. Learning how to conduct self-assessments through portfolios is equally valuable to the student as the beginning of a lifelong professional career record. With such portfolios, students should be encouraged to record reflectively their own assessments, to demonstrate to themselves, to potential employers, and—upon occasion—to accreditation visitors that they have achieved these outcomes. In the case of longer documents, such as project reports or videotapes of presentations, it will be very helpful (especially for accreditation visitors) if the sponsoring institution sets up a checklist of the metrics that have been agreed upon for the goal. Faculty evaluating these results can then make brief statements about the degree to which individual students have achieved each of these outcomes in their work.

The process of defining the outcomes and measuring how they are accomplished through agreed-upon performance metrics will be valuable within the context of EC 2000 only to the degree to which the results are evaluated and used by faculty for improvements. Such assessment should thus lead to continued re-engineering of the curriculum. Generally, institutions using student outcomes and performance metrics recognize that not every student will achieve high success for every measurement. Nor will students all register equal success in different measurements at different times and in different ways in their career. Portfolios and other self-assessment techniques can help them and the faculty assess their progress toward achieving these outcomes. Nonetheless, faculty will be achieving the goals of EC 2000 only if they use evaluations of student performance to assess honestly how well existing curricula are meeting the stated missions and goals statements. Wherever the inevitable slippages occur, it is incumbent upon faculty to record the deliberate steps taken to solve the problem by continuously improving the curriculum.

Different Responses at Different Kinds of Engineering Programs

As of the (1999) ABET annual review, approximately 500 institutions had undergraduate ABET accredited programs, indicating that accredited engineering programs exist at well less than 10 percent of the 3,688 institutions reported in the 1996 *Chronicle Almanac*. These institutions can be divided into four categories, for purposes of considering likely impacts of EC 2000 with respect to liberal education:

1. Large state universities, often founded as land grant colleges, with separate colleges of engineering and of arts and sciences.
2. Major research universities dominated by technological studies but with strong HSS programs.
3. Specialized hybrid programs with dedicated HSS studies existing within some large institutions, usually within the college of engineering.
4. Technologically oriented colleges.

Large State Universities

The first category is large state institutions, most of which originated as land grant institutions, that now have major engineering colleges. Instances of such institutions include Texas A&M, Ohio State, Arizona State, Purdue, Illinois, and other major teaching and research institutions producing large numbers of engineering graduates.

Probably the institutions facing the greatest challenges in implementing EC 2000 are these land grant institutions that have historically provided the largest numbers of engineering graduates to the nation. Because these institutions often enroll undergraduate students by the tens of thousands, and are divided into many separate colleges, all competing for state resources, coordinating the curriculum between engineering and liberal education (usually in a separate college of arts and sciences) proves very difficult. In my own experience of attending ASEE meetings for over two decades, complaints about the ineffectiveness of the liberal education as currently delivered come most often from these institutions.

Nonetheless, these institutions may be able to turn in their favor the interest in defining basic or core curricula. Unlike many smaller or more specialized institutions, these land grant universities often are required by state mandate to establish baseline expectations in general education. Increasingly, the regional accrediting societies are also making definite requirements in general education for all undergraduate degrees, especially for professional degrees. These pressures to require specific allocations within the four-year calendar for humanities and social sciences could lead to establishing well-thought-out missions and outcomes for the HSS requirements under EC 2000.

About a decade ago, the Association of American Colleges (now Association of American Colleges and Universities) undertook a project to assist engineering education, especially at the large land grant institutions, to realize the aims of the post–World War II reform reports by assuring that engineering students were exposed to both breadth and depth in HSS. The project resulted in a national conference and a published report summarizing the debates over liberal learning within engineering. This report, entitled *Unfinished Design*,[7] made the case specifically to engineering deans and department chairs for structuring the advising system for engineering students to encourage them to pursue specific HSS themes of professional interest. The report included examples of exemplary programs, drawn equally from the four categories enumerated above. In addition, the project circulated widely a pamphlet ("An Engineering Student's Guide to the Humanities and Social Sciences")[8] aimed at students and advisors with student-centered discussions of some specific themes. These examples, which are still well worth considering, are of students who articulate their professional goals in relation to pursuing HSS sequences (with specific courses identified) in such areas as management, public policy and the environment, American studies, the arts, great ideas, international studies, and cognitive science.

The emphasis on core curricula and the impetus from the regional accrediting societies may well promote serious discussion of creative ways of adopting EC 2000 in HSS studies at the institutions producing the largest number of engineers. However, the possibility also exists that institutions may opt, given the freedom to frame their own missions and goals, to minimize contact with course material in the College of Arts and Sciences. It is certainly plausible to design a curriculum that imparts to students all of the outcomes under Criterion 3 without drawing significantly upon resources outside the College of Engineering. Such curricula could, for example, employ interdisciplinary courses within the college, stress the potential multidimensionality of many engineering courses themselves, and build upon resources that exist at a number of institutions that already have liberal education faculty embedded within departments of engineering.

Such a path certainly risks the loss of the wealth and diversity of study available in the colleges of arts and sciences. However, given the long history of experience that in many cases engineering students at land grant institutions are not well served by crossing the borders into other colleges without very special guidance, it would not be surprising if some institutions responded to EC 2000 by placing some or all of the support required to achieve all eleven of the outcomes entirely within their colleges of engineering.

Major Research Universities

The second category comprises major technological research universities with several colleges, including in the humanities and/or social sciences. In this category obvious examples include MIT, Georgia Institute of Technology, and Stanford, as well as distinguished medium-size institutions such as Rensselaer and Lehigh.

In responding to EC 2000, these research institutions, which were established primarily to pursue science and engineering studies, will doubtless avail themselves of their abundant resources within their humanities and social science departments. MIT, for example, has strong programs in the humanities, arts, and social sciences, from linguistics to music and from public policy to management. Institutions with these strengths will doubtless build upon their existing conventional interpretation of the engineering criteria. At MIT, students take a humanities or social science course in each of their eight semesters, following a thoughtful pattern that assures both breadth and depth in several areas while also providing the students with opportunities to gain experience within communications.

Other institutions—Lehigh University, Georgia Institute of Technology, and Penn State come to mind—have strong programs in humanities and technology and in STS studies. The resources exist at such institutions to provide interesting solutions, which will be models for others, in responding to EC 2000. Other smaller but technologically focused universities with separate colleges like Rensselaer and Clarkson have the advantage of in-house humanities and social science faculty, whose major reason for affiliation with the institution is to provide HSS study for science and engineering students. At Rensselaer, for example, the HSS faculty offer courses in their fields aimed specifically at science and engineering students, but also have substantial programs for interdisciplinary majors and graduates in areas like technology and policy, STS studies, and technical communication. As private institutions, these colleges have also been very creative in their curricula over a period of decades, and can be anticipated to experiment broadly.

Specialized Hybrid Programs

The third classification of institutions are those with the good fortune to have opportunities provided (often by external sponsors) to craft liberal education programs aimed specifically for their engineering students. These institutions also have been very innovative in terms of the structure of their programs. Examples include specialized programs within engineering colleges, such as the Division of Technology, Culture and Communication at the University of Virginia, and of endowed honors programs, such as the McBride Honors Program in Public Affairs for Engineers at the Colorado School of Mines or the Humanities for Engineers program, funded by Clancy Herbst, at the University of Colorado at Boulder.

The Virginia program supplements HSS courses that engineering students take in the College of Arts and Sciences with specialized, technologically oriented courses taken within the division. The McBride program at CSM is an interdisciplinary, twenty-seven semester hour program available by competitive admission to roughly the top 10 percent of the class. The Humanities for Engineers program at Boulder is probably one of the most innovative in the nation, using a "Great Books" seminar discussion approach to engaging engineering students in fundamental questions such as "what is justice?" or "what is truth?" Readings emphasize the philosophical and literary classics, and challenge students—just as in the Chicago and Columbia programs for liberal arts students—to read the texts in translations of the unadorned originals. The former program director, Athanasios Moulakis, has publicized the kinds of discussions the program promotes in his book *Beyond Utility*.[9] In all these cases, faculty strongly dedicated to very specific objectives for carrying out humanities and social science education for engineering faculty work in highly focused programs, providing specialized curricula that often reinforce the offerings available within the College of Liberal Arts in the larger institution.

Technologically Oriented Colleges

The final category consists of smaller technologically oriented colleges, in which the mission of the entire institution concentrates largely (but not exclusively) on producing engineers. Among such institutions are Cooper Union, Rose-Hulman, Harvey Mudd College, and WPI. A related subset are liberal arts institutions with small, focused engineering programs, as at Swarthmore, Hope, and Calvin colleges. These institutions enjoy the advantage of comparatively small size, which promotes opportunities for interaction among faculty from different disciplines. They are generally not divided into separate colleges, and to some degree all the different faculties participate in the formation and delivery of curricula designed to meet their mission for preparing technologically oriented professionals.

Since I know the program and history at WPI best, I will cite an example of how one of these smaller institutions organized an HSS component in response to the post–World War II reports on HSS, and how such a program now faces the challenges of adapting to EC 2000.

In the late 1960s, the senior faculty at WPI adopted a new project-based curriculum. (See "Liberal Learning in Engineering in Engineering Education," in Hutchings and Wutzdorff[10] for background on the program.) To ensure that students had a full opportunity to develop their potential interests in nontechnical areas, the faculty required that two of the four new degree requirements be in non-technical fields. In addition to a nine-credit hour project and competency examination in a major field, the faculty also require all students to complete a nine-credit hour interactive project and an eighteen-credit hour course sequence in humanities culminating in a capstone project experience. (Subsequently, to address the needs to include social sciences within the curriculum, in 1984 the faculty added an additional six-credit hour requirement in the social sciences.)

The purpose of the two initial requirements, the interactive project (Interactive Qualifying Project or IQP in the local jargon) and the humanities requirement, was to develop within students the capacity to achieve two of the outcomes to which HSS can contribute significantly. First, the IQP was intended to assist students in understanding the societal and cultural embeddedness of technological activity. By engaging students in nine-credit hour projects, where they were encouraged to select topics from off-campus sponsors and also to work on teams, the faculty intended that students would develop an understanding "as citizens and as professionals how their careers will affect the larger society of which they are a part" (2000–2001 WPI Catalog, p. 37).

In contrast, the eighteen-hour humanities requirement was intended to complement the technological emphasis of the rest of the curriculum. The students in the program were expected to pass at least five courses of their own choosing within the humanities or arts, with the proviso that they define a theme relating the five courses to the satisfaction of the faculty member who would then advise the final three-credit hour capstone project. In this final project, the students are expected, in an appropriate way depending on the task and discipline, to demonstrate their ability both to pull together their learning in their previous courses and to do something genuinely new with their self-selected theme. The intent of the faculty through this course sequence with the culminating project was to nurture within students a lifelong interest in their self-selected area of the humanities or arts.

Thus the existing HSS program at WPI already seeks to assist students in achieving many of the goals of EC 2000—especially team work, communications, an understanding of global and cultural contexts and of contemporary issues, and some of the personal values associated with professional practice. But the intent of the WPI program goes beyond just valuing HSS for contributing to instrumental outcomes; the WPI HSS program (like many others, especially at smaller institutions) also seeks to help students cultivate self-reflection on their roles within a technological society in both personal and cultural terms.

To articulate the departmental mission in the humanities and arts in preparation for the new ABET criteria, the humanities faculty defined a mission statement for the department itself. In the 1996–1997 academic year, the department, after debate over a much longer and detailed document, decided upon the following text:

We are committed to helping students develop both a knowledge of, and an ability to think critically about, the humanities and arts. We also seek to foster the skills and habits of inquiry necessary for such learning—analytical thought, clear communication, and creative expression. Such an education, we believe, provides a crucial foundation for responsible and effective participation in a complex world.

The challenge the department faces now is to formulate appropriate performance metrics that will indicate how, when, and to what degree these goals are achieved. The values bespoken in this mission statement of developing critical thinking are difficult to measure, especially in developing students—doubtless more difficult than identifying more specific skills like communications. Fortunately, the requirement that all humanities and interactive degree requirements must culminate with an individual student project should make fairly easy the measurement of these outcomes. Further, the faculty has already undertaken a biennial peer review of completed capstone projects. These results are reported in a written document that has been extensively discussed within the department in terms of perceptions about how the defined aims of the humanities program are being achieved.

Nonetheless, it is incumbent on the faculty now to agree upon some specific measurable outcomes to enable all interested constituencies to assess how successful students are achieving the goals implied by this mission statement. Without such measurements, it will not be possible to discern how well students are performing in the program, and then to design and implement changes to make the curriculum that embodies this mission statement as effective as possible.

Avoiding Being Gored on the Horns of a Dilemma

This overview concerning liberal and engineering education suggests that ever since the formal organization of professional interest in engineering education in 1893, articulating an appropriate role for humanities and social sciences in engineering education has been confined to choosing which horn of a dilemma to cling to, to avoid being impaled by its opposite: *either* viewing liberal education as a means of acquiring communication and other tools appropriate for the engineering vocation *or* regarding the liberal education of engineers as an end in itself, justified by the examples of established liberal education programs.

EC 2000 challenges the engineering education community to overcome this dilemma. I believe it is a false dichotomy, in several senses. Faculty involved in assisting engineering students in achieving the outcomes of EC 2000 should be concerned as appropriate to prepare students for each of the eleven outcomes in Criterion 3. Engineering education will not be well served if the liberal arts faculty or the engineering faculty assume exclusive ownership of any of these goals. The whole fabric of engineering education will be significantly strengthened if everyone engaged in creating this warp and weft works together to assure each of these eleven outcomes. Only if science and engineering faculty encourage students to work in teams, to frame technical issues within global/societal contexts, and to communicate clearly will students take such goals seriously. Similarly, if humanities and social science faculty fail to understand the importance of experimentation, mathematical modeling, and engineering design as fundamental human creative and intellectual activities, it is unlikely that students will take them seriously or reap the full benefits of what the entire institution has to offer.

Some, perhaps many institutions, will respond to EC 2000 as suggested above by continuing to concentrate on the "second stem" approach of selecting those HSS disciplines that fit most comfortably with the perceived professional and vocational ends of engineering. In my view, such a path—while attractive for its obvious capacity to defend the softer skills as part of learning about hard technologies—ultimately will impoverish engineering education by attending only to utilitarian ends and ignoring the capacity of a liberal education to enrich students individually and culturally.

A recent example of such a limiting argument is that of Professor James C. MacKinnon, Dalhousie University, Nova Scotia, who wrote in the "Last Word" column (p. 48) of the *ASEE Prism* magazine in April 1997 concerning "The Engineering Humanities." MacKinnon argues for "a core set of humanities courses" designed to achieve "the basic goal of engineering education [which] is to prepare graduates for effective and responsible careers as practicing professionals." He deems four areas as important: engineering in history; professional engineering ethics; engineering, law, and politics; and engineering economics and management. His stress is entirely on most utilitarian "engineering humanities" as playing a supporting role to the engineer's development of hard technological skills. He expresses no interest in HSS either as a way to develop communication skills, or to nurture self-reflective or cultural awareness.

Under current ABET practice, several of the topics MacKinnon advocates (especially in the last category) would not pass muster in the existing HSS requirements, though of course they could well be made parts of the general requirements. He does not write specifically from the point of view of implementing EC 2000. His restriction of humanities topics to those presumed of most immediate vocational use is advocated in part because of the benefits he presumes would come from being able to "eliminate the current freedom to choose liberal arts electives" (p. 48). His op-ed piece—printed under the auspices of the major U.S. organization dedicated to engineering education—doubtless speaks to the frustration many engineers and engineering educators have felt over the inadequacies of implementing existing ABET policies for HSS studies. Too often engineering educators have traded horror stories about required HSS courses being a hodgepodge of introductory courses indifferently presented to students capable of hard work and disappointed if they are not expected to produce quality results. However, to abandon opportunities to use "the current freedom to choose liberal arts electives" decidedly throws the baby out with the bath water.

Conclusion: Engineering as a Liberal Education

In my view, it is neither necessary nor desirable to be forced to choose between utilitarian-vocational or culturally enriching, self-reflective justifications for HSS. Both horns of this dilemma should be avoided. The best model to grasp for engineering education under EC 2000 is one that views engineering education as a form of liberal education. This is far from saying that engineering is the only form of liberal education. Clearly, the 4,000 plus U.S. institutions of higher learning, most of which offer some form of liberal education, will continue providing the diversity of programs that almost half our adult citizens enjoy. Yet within American higher education as a whole, those engaged in studying and learning the humanities portion of the traditional liberal arts are becoming distinctly fewer. "By all available standards, national performance in the humanities has declined."[11] To preserve most broadly the benefits conferred by study of the humanities—both as means to achieving other goals and as ends in themselves—may well require thoughtful incorporation of such programs into broader curricula. In the several hundred U.S. institutions that offer engineering programs, presenting such programs as a genuinely liberating experience—one that prepares students fully for fulfilling lives as individuals, citizens, and professionals—should be an important goal.

Interestingly, both the Wickenden report and Mark van Doren's *Liberal Education*—the two classic studies in their time of technological and liberal education—look beyond such a painful reduction of HSS (and what it embodies about liberal education) as simply "the engineering humanities." And both studies suggestively acknowledge the value of the complementary school of thought, recognizing the benefits that would flow from drawing upon both strengths.

At several places in *Liberal Education*, van Doren draws comparisons between liberal and scientific/technical education as conventionally defined, to the detriment of the former for its comparative lack of rigor and of underemphasis on learning *any discipline* well. (See, for example, pp. 51–57, 67–69, 136–139. By discipline, he means both "area of learning" and "making a commitment to learn a tough topic thoroughly.") In concluding his chapter on "The Idea of a College," van Doren added a section on "Vocation." His argument (framed at a time when custom permitted the referent to all people always to be masculine) is worth quoting at some length:

Liberal education is sometimes distinguished from useful [e.g., engineering] education, but the foregoing pages should have made it clear that the distinction is unfortunate and false. All education is useful, and none is more so than the kind that makes men free to possess their nature. Knowledge and skill to such an end are ends in themselves, past which there is no place for the person to go. It is both useful and liberal to be human, just as it takes both skill and knowledge to be wise. If education is not practical when it teaches men to do the things which become men, then no education is practical.

The distinction, however, has something else in mind. Any man does particular things, and the question arises whether his education should not be directed to those things alone, first as well as last. There is a thing called vocational education, and the emphasis upon its virtues is great in our time. It is not universally admitted that humanity is a vocation of sufficient importance to justify the expense of colleges devoted solely to its study. So technical studies are preferred.

The distinction is still false. Technique was the Greek word for art, and there is a human art which dominates all other arts, since it is the art that teaches them. It teaches them how men do what they do. To miss this lesson is not to know what human work is. It is not even to be prepared for a profession. But the professions are less intellectual than they once were, and so they make a diminishing demand for prepared persons. They are satisfied with preparatory studies, which are quite a different thing. The result is that they can count on even less competence than used to be the case. *This is because our liberal education is bad; it is not technical enough in its own way; it misses intellectual precision.* [Italics mine.] But one of the reasons it is bad is that the professions do not insist upon its being good. They will do so as soon as they give up trying to be trades. Even the advanced study of literature is a trade. Many who pride themselves upon being liberal because they pursue this trade are as ignorant of most other things as any mechanical engineer could be of metaphysics. They call their periodicals, humorously, trade journals. That is what they are, and they are useless to any world for which literature is made. No antipathy appears between technical and liberal education if we remember that both are concerned with art. It has already been said that liberal education suffers today from its ignorance of the liberal arts. *But technical education suffers also because it does not know the meaning of its name; because it is willing to accept students whose training has not been general enough to make them recognize principles when they appear; because it identifies the useful with the utilitarian, and so blunts that very sense of the concrete which it requires.* [Italics mine.] For not to see the truth in a thing is not to see the thing. If liberal education is concerned with truth, and technical education with things, then the two could teach each other. The first needs to be more conscious of its operations, and the second needs to be more theoretical than it is. Liberal education, that is to say, should discover in what ways it can be technical—how it can teach the arts of reading, writing, measuring, remembering, and imagining—and technical education should acknowledge that it also is intellectual.[12]

Van Doren's argument here is potent in expressing, especially in the last sentences, the benefits of the interplay between liberal and technical education. Today, as I will argue below, the significantly increased intellectual rigor of engineering education (and the significant reduction over the last half century of manual activity) have brought what van Doren's refers to as "liberal and technical education" far closer together.

Before making my final points, let us consider a parallel statement from Wickenden's Report (p. 84), which presents a senior engineer's case that "good engineering education [is] good general education":

> The basic process of engineering education should be an undergraduate curriculum of coherent and integral structure, directed to the grounding of the student in the principles and methods of engineering and to those elements of liberal culture which serve to fit the engineer for a worthy place in human society and to enrich his personal life. To serve its purpose adequately, this undergraduate program of studies should provide a basis for the three types of activity common to all engineering, namely:
>
> 1. The control and utilization of the forces, materials, and energy of nature;
> 2. The organization of human effort for these purposes; and
> 3. The estimation of costs and appraisal of values, both economic and social, involved in these activities.
>
> An educational process appropriate to these purposes cannot be narrowly technical. Given a proper choice of subject matter and activity, reinforced by the incentive of purpose, the inspiration of professional ideals, and the effect of a professional objective in bringing the entire process to a focus, the result of engineering education should be a sound, general development, worthy to be compared with the result of any other collegiate program. In short, good engineering education is good general education; there is no warrant for assuming that the two have different ultimate purposes which necessitate two distinct programs for the engineering student.

Both Wickenden and van Doren come close to the argument with which I would like to conclude: that contemporary engineering education should be recognized as a form of contemporary liberal education. At first this argument might seem to founder on fundamental definitions. Certainly a recurring theme in the study of humanities or cultural history is the idea that the essential human condition does not change much if at all.

Such a static mindset clearly is at odds with much of the intellectual activity of the scientist and/or engineer who by definition is concerned with new discoveries and with the desire to make things new. Here one recalls the "Two Culture" debate initiated by C. P. Snow[13] as to whether scientifically/ technologically and humanistically /artistically oriented creators constitute two different cultures.

This argument failed to appreciate (as Snow himself pointed out) that engineers and scientists—as individuals—experience the personal and societal exigencies and constraints of "the human condition" as fully as anyone else. One may or may not agree with Snow that technically oriented people are more likely than artists to take action to ameliorate specific sufferings. But certainly the case remains (as van Doren and Wickenden both suggest) that engineers and scientists, as much as any other human category, will profit from any study that enriches their capacity to reflect critically upon those conditions that characterize all humans.

Van Doren reiterates forcefully that any education must have a discipline, a set of facts and tools that must be mastered to discipline the student as a more mature thinker and person. Engineering education offers such a set of facts and tools that, *when properly generalized,* forms the basis for at least one kind of liberal education ("liberal" again in the root sense of the word in this context as "liberating the possessor from servitude to irresistible constraints"). The form that engineering education takes as a branch of liberal education is one that directs students to generalize the primary characteristics of existing engineering education. In broad brushstrokes, all engineering education prepares students to become professionals who are able to perform the following tasks because they have learned essential critical thinking skills applicable in various contexts:

1. Identify questions or issues worth pursuing.
2. Examine, experience, and experiment with the issues at hand.
3. Identify ways of gathering data on the problem, and collect such data in order to perform analyses.
4. Use whatever models are best suited (computer, mathematical, social science, or artistic) to discern patterns within the data, and to analyze them.
5. Produce one or more solutions to the problem addressed,
6. Make a case to all of the affected constituencies to adopt an effective solution to addressing the problem, and
7. Implement the solution effectively within anticipated and unanticipated constraints.

In this process, probably the most difficult step is the first. To know what questions are really worth pursuing is the mark of a sophisticated professional. To pursue these questions through the fullest set of implications (technical and societal/ global) requires probably an even higher level of sophistication, one in which liberal education can assist by contributing to nurturing broad and critical thinking. At each of the other stages, engineering students acquire not only tools but *ways of analyzing and problem-solving* that are applicable in contexts close to, or further from, the immediate subject matter at hand. But an education that dwells only upon the subject matter of engineering—not recognizing that the underlying analytical thought processes are eminently and crucially transportable to other spheres of activity—severely impoverishes what engineering education can offer to individuals and society as a whole.

Many engineering students find, in the course of their careers, that they grow into such wider responsibilities. The current emphasis to specialize heavily within the undergraduate degree (which dates as far back as the existing written literature in engineering education) too often fails to enable students to recognize that the studies they pursue so vigorously as undergraduates will soon be outdated. Engineering educators should take well to heart the anonymous saying: "Education is what you have left when you have forgotten everything you have learned." The pith of this aphorism is that what really matters, as professionals advance in their careers, are the ways of understanding how to operate in frequently shifting environments, not simply a commanding knowledge of specific (but always already outdated) techniques. The kind of joint cooperation urged here and throughout this volume in order to respond to EC 2000 will, for many institutions, one hopes, provide an environment in which to realize this kind of education. (One of the best exemplars of such an education is the practicing engineer, author, and essayist Samuel C. Florman, whose major works are cited in the References.)[14–18]

Regarding engineering education as a form of liberal education, rather than as a subset or a contrary, resolves one of the debates running through engineering education from the beginning of its formal history with the founding of SPEE in 1893. This point of view also supports another position long debated within engineering education—that undergraduate engineering education should be as general as possible with specialization being deferred until graduate study. Arguing for deferring specialization is bolstered if more conscious attention is directed, during the undergraduate years, at generalizing the critical thinking skills within engineering programs to broader contexts, some of which are outside the usual provinces of engineering.

Such generalization requires those engaged in the learning and teaching of engineering to recognize that their fundamental activities must be disciplined but need not always remain discipline-specific. Typically engineers learn problem-solving by first addressing specific problems (often first given and later discovered), by conducting experiments (physical or "thought"), by modeling results and framing alternative solutions, and then implementing the solutions. This sequence works well not only for engineering but as a model for any critical problem-solving that requires identifying the problem, gathering evidence, thinking critically about alternative solutions, and then effecting the solution. This model of problem-solving can be applied to many challenges, from those traditionally associated with engineering to those associated with creating and performing works of art.

As engineering education evolves into the next decade, century, and millenium, new directions are opening up for its practitioners. Engineering education journalism has noted that increasingly the best graduates are being pursued by graduate programs and employers not usually associated with entry-level engineering positions—opportunities like finance, law, medicine, and the film, theater, musical, and fine arts. For example, "Hacking for Dollars: Young Engineers on Wall Street," *Technology Review*, Vol. 99, pp. 22–29, January 1996; and "Evolving Paths," *ASEE Prism*, Vol. 6, pp. 23–28, October 1996, offer fascinating examples of how the essential critical thinking skills of engineering, discussed above, prepare undergraduates for unanticipated careers, often in areas traditionally associated with the liberal arts.

Framing and re-framing engineering education revolves around debating the ultimate mission and objectives of such programs. Previous engineering accreditation criteria have been locked into a "bean counting" approach to setting the curriculum, with the dubious assumption that students who pass courses can actually perform outcomes later expected of professionals. A subtle consequence of this paradigm is measuring the projected achievement of future professionals at one point in time—when they pass an undergraduate course. In contrast, EC 2000 is fundamentally not about what engineering students have learned in the past but about the patterns of discipline—the mindsets for seeking out and using information—that are required for future growth.

Unlike previous accreditation schemes, EC 2000 recognizes that engineers move through phases of their career—thus the emphasis on lifelong learning. EC 2000 also recognizes that a "one size fits all" approach fails to acknowledge that different engineers assume different kinds of tasks. Increasingly, many change from one arena to another. Most important, EC 2000 expects professional technologists to demonstrate high standards for applying technology within the global environment as for creating technology in the first place.

By liberating engineering education to devise programs that respond to differing needs and talents of different individuals, and by recognizing that such needs are not static, EC 2000 has issued the challenge to begin rethinking how liberal education relates to engineering. True liberation from manual (or intellectual) servitude increasingly will require both the tools and disciplines requisite for a technological culture. And the self-reflection and judgment nurtured by study of the collective human achievements accessible through both technological and HSS learning must direct all meaningful education in the future.

References

1. *Society for the Promotion of Engineering Education*. [William E. Wickenden, Director of Investigations.] 2 vols. 1930, 1934.
2. McGivern, James Gregory. 1960. *First Hundred Years of Engineering Education in the United States (1807–1907)*. Spokane, WA: Gonzaga University Press.
3. Layton, Edwin T. 1971, 1986. *The Revolt of the Engineers: Social Responsibility and the American Engineering Profession*. Baltimore: Johns Hopkins University Press.
4. Grayson, Lawrence P. 1993. *The Making of an Engineer: An Illustrated History of Engineering Education in the United States and Canada*. New York: John Wiley.
5. Report of the Investigation of Engineering Education 1923–1929. Pittsburgh, University of Pittsburgh.
6. Gianniny, O. Allen. 1993. A Century of ASEE and Liberal Education, the Division and its Predecessors, or How Did We Get Here from There, and Where Does It All Lead? Paper presented at the ASEE Liberal education Division Centennial Meeting, Champaign, IL.
7. Johnston, Joseph S. Jr., Susan Shaman, and Robert Zemsky. 1988. *Unfinished Design: The Humanities and Social Sciences in Undergraduate Engineering Education*. Washington, DC: Association of American Colleges.
8. Association of American Colleges, 1988. An Engineering Student's Guide to the Humanities and Social Sciences Washington, DC.
9. Moulakis, Athanasios. 1994. *Beyond Utility: Liberal Education for a Technological Age*. Columbia: University of Missouri Press. Petroski, Henry. 1985. *To Engineer Is Human: The Role of Failure in Successful Design*. New York: St. Martin's Press.
10. Hutchings, Pat, and Allen Wutzdorff. 1988. Liberal Learning in Engineering Education: The WPI Experience, by William R. Grogan, Lance Schachterle, and Francis C. Lutz, in *Knowing and Doing: Learning through Experience*. Vol. 35 in the New Directions for Teaching and Learning Series. San Francisco. Jossey-Bass.
11. Engell, James, and Anthony Dangerfield. 1998. The Market-Model University: Humanities in the Age of Money. *Harvard Magazine*. May–June 1998. p. 48–55, 111.
12. van Doren, Mark. *Liberal Education*. New York: Holt, 1943. p. 178.
13. Snow, C. P. 1956. *The Two Cultures and the Scientific Revolution*. Cambridge: Cambridge University Press.
14. Florman, Samuel C. 1981. *Blaming Technology: The Irrational Search for Scapegoats*. New York: St. Martin's Press.
15. Florman, Samuel C. 1987. *The Civilized Engineer*. New York: St. Martin's Press.
16. Florman, Samuel C. 1968. *Engineering and the Liberal Arts: A Technologist's Guide to History, Literature, Philosophy, Art and Music*. New York: McGraw-Hill.
17. Florman, Samuel C. 1976. *The Existential Pleasures of Engineering*. New York: St. Martin's Press.
18. Florman, Samuel C. 1996. *The Introspective Engineer*. New York: St. Martin's Press.

PART TWO

Specific Responses

CHAPTER 3

CAROLYN R. MILLER

Reuniting Wisdom and Eloquence within the Engineering Curriculum

In the waning years of the Roman republic, orator and statesman Cicero described a problem with the system of education that prepared young men for public life. He attributed the problem to Socrates, whose teachings of several centuries before had influenced first the Greek and then the Roman educational system:

> Socrates . . . separated the knowledge of wise thinking from that of cogent speaking, though in reality they are closely linked together. . . . This is the source from which has sprung the undoubtedly absurd and unprofitable and reprehensible severance between the tongue and the brain, leading to our having one set of professors to teach us to think and another to teach us to speak.[1]

Cicero did not provide a solution to this problem so much as an ideal to be worked toward; that ideal was the unification of wisdom and eloquence. My claim in this chapter is that ABET's proposed Criteria 2000 can provide a contemporary way of working toward that ideal within the engineering curriculum (Engineering Accreditation Commission).

Wisdom is not just technical knowledge, of course, but also broad learning coupled with perceptive judgment; and eloquence is the ability to make knowledge efficacious and powerful in situations when it needs to be applied or to be used by other people. The ideal that unifies, or integrates, these human faculties is surely as relevant for today's engineer as it was for leaders in ancient Rome.

By focusing on outcomes, rather than on courses and credits, the new ABET criteria allow us to see communication within the context of the full range of abilities that engineers need and encourage us to develop the relationships among these abilities. And by requiring that programs produce in students an ability to communicate "effectively," the criteria emphasize that communication is a strategic, situated enterprise that must be judged in context and with an understanding of the constraints and conventions in play and of the challenges to be met.

In designing a curriculum, then, integrating communication into the student's work in a variety of courses, rather than segregating it into one or two courses, will be the Ciceronian way. In this chapter, I want to pursue this theme of integration down three paths: first, I'll report some survey results that emphasize the need for such integration; second, I'll show how the ABET criteria themselves allow for integration; and third, I'll indicate three ways that the Ciceronian tradition can help us design integrated instruction.

Communication in the Workplace: The Need for Integration

The need for integration of communication with thinking and learning is demonstrated by the results of an interview-survey we performed at North Carolina State University, in which students in our technical and business writing courses interviewed working professionals about their communication responsibilities on the job.[2] The respondents were role models who the students themselves had chosen and therefore, we believe, can represent the futures for which our students are preparing themselves.

The quantifiable information from these 378 interviews is not surprising to anyone familiar with surveys about workplace communication over the past fifteen years:

1. Respondents spend an average of 31 percent of their work time in writing and another significant proportion in high-stakes oral communication.
2. Respondents communicate with a variety of people, including international coworkers and clients; they prepare a wide variety of documents that suggest that their purposes for communication are quite varied, as well.

3. Collaboration, management review, and peer review are common aspects of writing in the workplace.
4. Communication technologies are changing the choices and patterns of communication.

But what was most interesting to us was the qualitative information students reported about the conversations they had with the professionals, which showed that communication in the workplace is understood by those who do it as complex and multidimensional. For example,

1. In addition to undergoing peer or supervisor reviews, writing is often reviewed by clients as well; when this happens, the project design can change as expectations are clarified and problems are solved. Thus, writing interacts with technical work.
2. The credibility of the professional is at stake in both oral and written communication; several respondents pointed out that upper management knows them only through their writing.
3. Engineers, in particular (there were 124 in our sample), noted the importance of being able to convince readers of ideas and findings; much of their writing, although it seems purely informational, is used to set strategies for future action.
4. There is a tension between standardization and creativity. Many respondents felt constrained by company style and formats and by the necessity to meet standards of managers; others noted that there's a place for personality and humanity in their writing. One manager observed that his writing draws on all his managerial skills, another noted that she was able to exert unexpected influence on company policies through strategic communication.

In general, we concluded, problem-solving, critical analysis, strategy, teamwork, creativity, and persuasion are seen by those in the workplace as important dimensions of practical communication tasks, and these tasks are understood as central to the technical work being accomplished. In other words, technical and management professionals implicitly understand communication as an integral part of their work.

While there are dangers in assuming that the needs of the workplace should dictate university curricula, here is one case where those in the workplace may have a more sophisticated and more theoretically defensible understanding of how communication works than many in the academy. I should add, however, that when we directly ask employers what they think should be taught in college technical writing courses, their answers generally do not reflect the rich experiential knowledge I've just summarized but rather the vocabulary of their own educations that focused on grammatical correctness.

The ABET Criteria: Opportunities for Integration

What kinds of curricular integration do the ABET criteria permit, or encourage? The answer depends on how one understands communication. In what follows, I adopt a view of communication that makes it not only a result that can be graded or assessed but also, and more important, a medium of interaction that facilitates other activities, a strategic and situated process, and an evidentiary record of social and intellectual effort. Under this view, although effective communication is only one of eleven required outcomes, it permits integration with at least seven of the remaining ten, to varying degrees (see Table 3.1).

I'll begin with a cluster of the more obvious connections. Item (d), ability to function on multidisciplinary teams, takes place, in large part, through interpersonal, oral communication, and the results of such teamwork often take the form of collaboratively written products. Here, communication is both the medium through which collaboration occurs and the evidence that it has occurred. A similar example is item (f), understanding of professional and ethical responsibility.

Such understandings are not only expressed in communication but are developed through communication with other professionals. Ethical and professional responsibilities must guide actions, and communication is an important type of

TABLE 3.1. List of Program Outcomes from the Proposed ABET Criteria 2000

Engineering programs must demonstrate that their graduates have

  (a) an ability to apply knowledge of mathematics, science, and engineering;
  (b) **an ability to design and conduct experiments, as well as to analyze and interpret data;**
  (c) **an ability to design a system, component, or process to meet desired needs;**
  (d) **an ability to function on multi-disciplinary teams;**
  (e) **an ability to identify, formulate, and solve engineering problems;**
  (f) **an understanding of professional and ethical responsibility;**
  (g) **AN ABILITY TO COMMUNICATE EFFECTIVELY;**
  (h) **the broad education necessary to understand the impact of engineering solutions in a global/societal context;**
  (i) a recognition of the need for and an ability to engage in life-long learning;
  (j) **a knowledge of contemporary issues;**
  (k) an ability to use the techniques, skills, and modern engineering tools necessary for engineering practice.

Those that can be integrated with communication are shown in bold type.
Source: Engineering Accreditation Commission, Accreditation Board for Engineering and Technology, Inc. *Criteria for Accrediting Programs in Engineering in the United States: Effective for Evaluations During the 1996–97 Cycle.*

action for every professional; thus, the ethic that an engineer develops is manifested, both explicitly and implicitly, in linguistic and rhetorical choices.

Items (h), understanding the impact of engineering solutions in a global/societal context, and (j), knowledge of contemporary issues, are similar in requiring an appreciation for consequences and implications and an understanding of the perspectives of other people and groups. These dispositions are fostered by attention to communication practice, which must always take audience and context into account. Since excellence in communication requires attention to other persons and their situations, it cultivates what Thomas Farrell has called "relational goods";[3] these include empathy, a sense of social justice, strategic imagination, and civic responsibility. These are qualities that I believe the ABET criteria should encompass.

The three final criteria I want to make a case for are closer to the practice of engineering as traditionally understood. Item (b), the ability to design and conduct experiments, as well as to analyze and interpret data, requires communication for its fulfillment, because the interpretation of data is useful only through language that is intelligible and persuasive to another. This item also focuses on design, as does item (c), the ability to design a system, component, or process to meet desired needs; item (e), ability to identify, formulate, and solve engineering problems, is related through its emphasis on problem solving. In these respects, engineering is closely related to communication, which is also a process centered on design and problem solving.[4] Both communication and engineering are design arts: the ancient Greeks called communication the *technê*, or productive art, of words and compared it to the arts of medicine, ship-building, pottery, and other ancient technologies. All design arts use the theoretical for practical ends; all are creative, goal-directed activities; all must take account not only of intrinsic principles but also of an exterior environment of use and social effects. Understanding communication and engineering as isomorphic processes can enrich the engineering student's experience of both.

The Ciceronian Tradition: Methods of Integration

In Cicero's time, education in communication was the final course of study, after grammar, literature, and history; this knowledge prepared anyone for participation in public life and decision making. There was no specialization in the sense we know it: in law, military arts, architecture, or engineering, careers were learned primarily through apprenticeship.[5] Nevertheless, as I suggested above, we can share the classical understanding that pervades communication in public and professional life and the Ciceronian goal of integrating eloquent communication with wise thinking. Consequently, I believe that some aspects of the well-elaborated classical tradition of instruction can still be useful.[*,7,8] I want to sketch out three of these in particular.

Effectiveness vs. Clarity

It is significant that ABET targets "effective" rather than "clear" communication. The classical tradition also valued effectiveness over clarity as a way of assessing communication. Twentieth-century teaching practice, however, has promoted the notion that clarity is a universal criterion of excellence that can be judged in isolation from the conditions in which the communication is used. The Shannon-Weaver model of electronic communication, which by misleading analogy has been widely applied to human communication, has contributed to this notion by leading us to value primarily the absence of noise or distortion—that is, to value clarity, or transparency. Our everyday language for talking about communication also embeds this transmission model, which has been called the "conduit metaphor": we say, "I *get* what you mean," or "Your ideas don't *come across*"[6] (emphasis added).

But a great deal of linguistic and rhetorical research suggests that this model, or metaphor, which seems so natural, prevents us from understanding some important aspects of communication, specifically that different groups and communities have different needs and customs; this diversity limits the application of universal criteria. The recent Ebonics debate is just one extreme example. Another example is illustrated by Figure 3.1, which is an internal message sent from one Navy electronic communications specialist to another. It is presented by Suchan and Dulek as an example of "clear" communication because it was understandable by 350 naval officers in less than 45 seconds as a request for help in conducting a technical evaluation of an aircraft.[9] It is "clear" to those in the relevant language community, but this is just to say that it is *effective,* because it provides information that they can use in a form that they find appropriate, in addition to identifying itself *and them* as part of that community.

The classical tradition had a rich understanding of communicative effectiveness that did not rely on clarity as a single governing criterion. Cicero outlined three general types of prose style, for example, ranging from the plain to the ornate, suggesting that each had its uses and powers for different audiences, situations, and communicative purposes. One of Aristotle's students, several centuries earlier, posited that clarity was one of four qualities of effective communication, the others being correctness, appropriateness, and ornamentation; Aristotle himself valued clarity, but he used a Greek term (*saphês*) that can as well be translated as "distinctness," which leads to a much different model of communicative effectiveness than clarity.[10]

FIGURE 3.1. An example of technical communication from the Navy that is clear to those who will need to read it and therefore effective.

COMPACMISTESTCEN PT MUGU CA MAG FOURTEEN INFO COMNAVAIRLANT
NORFOLK VA
CG FMFLANT
HAMS FOURTEEN
CG SECOND MAW
NAVAIRTESTCEN PATUXENT RIVER MD
COMNAVAIRSYSCOM WASHINGTON DC VMAQ TWO
UNCLAS //N13600//
CNAL FOR V. MARSH {CODE 53281}/MAG-14 FOR CAPT WITTENBURG AND SECURITY OFFICER
SUBJ: REQ FOR ASSIS TO CONDUCT OEWTPS TECHEVAL ON EA-6B AN/ALQ-126 A/B CONFIG AIRCRAFT
A. PHONCON CNAL #V. MARSH, CODE 53281}/PMTC
{S. NGUYEN, CODE 4046# OF 3 JUL 86
B. PHONCON MAG-14 {CAPT. WITTENBURG#/PMTC {S. NGUYEN, CODE 4046} 03 JUL 86
                      1.IAW REF A AND B FOL INFO PROVIDED:
                      2.PMTC, CODE 4046, HAS BEEN DIRECTED TO CONDUCT A TECHEVAL
/4046/4040/4000/00/02/0141/60000/6001-2/6500/

Source: Suchan and Dulek 1990.

Persuasion vs. Information

The twentieth-century conviction that language can represent reality clearly is related to the assumption that such language can communicate information without persuading. In transmitting facts, one supposedly informs; when one seeks to persuade, then supposedly one must color or distort those facts. And yet, this assumption makes it hard to teach anything useful about communication. If a writer or speaker fails to provide the information needed by audiences, then the problem must lie in some innate incapacity. If a writer can't see the facts right in front of him, or use the obvious plain words to transmit them, then the teacher has little recourse: remediation becomes the universal response to communication problems—remediation or despair.

In contrast, the classical tradition shows us how to conceive of even the most factual communication as inherently persuasive, as involving strategy, selection, adaptive expression—all of which are teachable. Cicero and others taught their followers to discern the essential issue at the heart of a communication situation; the issue would help determine what evidence, arguments, and strategies would be effective. They found four basic types of issues, or what they dealt with basic factual questions about the geology of the site: What is the pattern of faults and fractures in the rock formations underground? How will they affect the pattern of groundwater flow from the site into nearby watersheds and aquifers? These are *factual* questions, but there are, as yet, no fully persuasive answers to them. These questions have clear implications for policy arguments about the design of the facility or the potential decision to abandon the site or even the siting process.[†]

Looking at communication problems as matters of persuasion rather than as information transfer helps show how communication is connected to technical issues such as the sources and sufficiency of evidence, the design of an investigation, the timing of technical work, and the integrity and credibility of the investigators. I believe this approach can help us design curricula and instruct students better.

Arguing on Both Sides of the Question

Cicero also taught that arguments on both sides of a question should be developed and heard, as a matter not only of public policy but also of pedagogical practice. In the public forum, he held, the way to make the best decision in cases of uncertainty or disagreement where there is no foregone conclusion is to hear the best case that can be made for the competing positions. We still follow this principle in our courts and legislatures, and it underlies the rationale for democracy and freedom of speech and the press.

In education, arguing on both sides of the question is useful for developing several capacities in students. It requires them to explore an issue more fully, to develop more adequate evidence, to anticipate rebuttals, to examine their own and others' positions more critically; it may even cause them to change their minds. It can improve the credibility of the engineer because it helps demonstrate that all relevant information is being accounted for, that all positions are being taken seriously, that conclusions are well considered, and that the engineer is trustworthy and judicious. Further, because arguing on both sides requires the taking of more than one perspective, it can promote attention to the ethical dimensions of issues: the student engineer must put herself in the shoes of the city manager, the assembly-line worker, the NIMBY homeowner, the OSHA inspector, as well as the engineering manager. Such perspective taking can not only improve the responsibility and quality of writing and speaking, it may also enhance the utility and acceptance of technical designs and solutions.

Instruction of this sort can ask students to engage in live debates, to write rebuttals of their own or a classmate's design recommendations, and to include refutations of alternate conclusions in final reports. It should help them understand how claims can be qualified and limited and still be useful. It should insist on adequate evidence and careful reasoning and not permit hasty dismissals. The use of case scenarios can help impress upon students that positions can be well supported and passionately held and still be proven wrong in the long run.

Conclusion

Reuniting wisdom and eloquence in the engineering curriculum means acknowledging that all faculty must be concerned with both what students know and how they express that knowledge as they put it to work in statements and arguments of their own. It means that we can't rely on any firm distinction between form and content—that we can't give a paper A for content, for the "technical part," and C for the form, or the "writing part," for communication cannot be good if it includes faulty technical analysis, insufficient technical evidence, or irrelevant technical assumptions. Reuniting wisdom and eloquence means that faculty in rhetoric and communication must be working closely with faculty in engineering, and vice versa, that students can't be inoculated once for immunity to poor communication, but instead must receive instruction and feedback throughout the curriculum. And it means that assessing communication outcomes in engineering programs can't be done in isolation from assessing the other abilities and knowledge that are essential to engineering education.

Is there still a use for separate communication courses? Curriculum development will vary, of course, but in many curricula a combination of separate courses focused on communication principles and strategies together with engineering courses where those principles and strategies are put to work within specific contexts may prove the best approach. This is only to recognize the advantages to the widespread communication-across-the-curriculum movement in undergraduate education, a movement that aims to incorporate writing and speaking into every program and department across our campuses in order to demonstrate that these are not isolated subjects but rather arts that are relevant to every subject.

There is, after all, much to be taught. In addition to the pedagogical strategies I've noted above, the classical tradition also emphasized the need for ethical integrity in communication as well as broad understanding of culture and history; these goals are also compatible with the ABET criteria. Recently there has been much dissatisfaction with the degraded quality of public discourse—in the courts, the legislatures, the media—and I believe that this situation emphasizes the crucial importance of communication education of the Ciceronian kind, not only for engineers but for all students. Some final words from Cicero will make his point better than I can:

> The stronger this faculty of eloquence is, the more necessary it is for it to be combined with integrity and supreme wisdom, and if we bestow fluency of speech on persons devoid of these virtues, we shall not have made orators of them but shall have put weapons into the hands of madmen. (*de Oratore*, III 55)

Notes

* I should note that my general argument is indebted to earlier efforts by S. Michael Halloran.
† Some of these issues are explored more fully in Katz and Miller.[11]

References

1. Cicero, Marcus Tullius. 1942. *De Oratore*. Trans. H. Rackham. Cambridge: Harvard University Press.
2. Miller, Carolyn R., Jamie Larsen, and Judi Gaitens. 1996. *Communication in the Workplace: What Can NCSU Students Expect?* Center for Communication in Science, Technology, and Management, North Carolina State University. Research Report. 2. http://www.chass.ncsu.edu/ccstm/pubs/.
3. Farrell, Thomas B. 1991. Practicing the Arts of Rhetoric: Tradition and Invention. *Philosophy and Rhetoric* 24: 183–212.
4. Kaufer, David S., and Brian S. Butler. 1996. *Rhetoric and the Arts of Design*. Mahway, NJ: Lawrence Erlbaum.
5. Marrou, H. I. 1982 [1948]. *A History of Education in Antiquity*. Trans. George Lamb. Madison: University of Wisconsin Press.
6. Reddy, Michael J. 1979. The Conduit Metaphor—A Case of Frame Conflict in Our Language about Language. In *Metaphor and Thought*. Ed. A. Ortony. Cambridge: Cambridge University Press. 284–324.
7. Halloran, Stephen M. 1971. "Classical Rhetoric for the Engineering Student." *Journal of Technical Writing and Communication* 1 (1): 17–24.
8. Halloran, S. Michael. 1978. Eloquence in a Technological Society. *Central States Speech Journal* 29: 221–227.
9. Suchan, James, and Ronald Dulek. 1990. A Reassessment of Clarity in Written Managerial Communications. *Management Communication Quarterly* 4 (1): 87–99.
10. Consigny, Scott. 1987. Transparency and Displacement: Aristotle's Concept of Rhetorical Clarity. *Rhetoric Society Quarterly* 17 (4): 413–419.
11. Katz, Steven B., and Carolyn R. Miller. 1996. The Low-Level Radioactive Waste Siting Controversy in North Carolina: Toward a Rhetorical Model of Risk Communication. In *Green Culture: Environmental Rhetoric in Contemporary America*, pp. 111–140. Ed. C. G. Herndl, and S. C. Brown. Madison: University of Wisconsin Press.

CHAPTER 4

KATHRYN A. NEELEY

To Arrive Where We Started and Know the Place for the First Time?

Re-visioning Technical Communication

> *We shall not cease from exploration*
> *And the end of all our exploring*
> *Will be to arrive where we started*
> *And know the place for the first time.*
> T. S. Eliot, "Little Gidding"[1]

Teachers of technical communication are likely to welcome the emphasis that the ABET 2000 accreditation criteria place on effective communication as an integral component of engineering preparation and practice. But we would do well to remember that we are hardly the first to attempt to transform engineering education by giving communication a more prominent and integrated role in the curriculum.

Even a cursory examination of the history of the American Society for Engineering Education (ASEE) makes it clear that one of the most striking features of the "new vision" of engineering education is that many of its elements have been included in previous blueprints. Since the early 1900s, the members of ASEE have shown an enduring concern with the writing and speaking abilities of engineering students.[2,3] An English Committee was formed in 1914.* By 1925, teachers of what was then called "English for engineers" recognized the need for interdepartmental cooperation with their engineering colleagues and the importance of including oral as well as written work. In 1926, over 100 teachers of English attended the annual meeting of the Society for the Promotion of Engineering Education (ASEE's predecessor). In 1932, they held the first of a number of summer schools for teachers of English for engineers, where they exchanged ideas about how to improve their curricula and teaching. By 1936, they recognized that teachers of English for engineers would benefit greatly from some form of specialized training beyond the typical Ph.D. From the beginning, they recognized what seems to many of us a revolutionary insight today: the need to integrate instruction in English into the engineering curriculum.

Thus, we have in a sense arrived "where we started" several times over the last 100 years. Our challenge for ABET 2000 and beyond will be to "know the place for the first time" in a way that will achieve lasting change in engineering education. Our central task in knowing the place for the first time will be to get outside of the conceptual framework and vocabulary that promote an impoverished view of communication and separate communication from other aspects of the engineering curriculum.

One of the most welcome aspects of the new accreditation standards is that they abandon the dichotomy in which courses dealing with language are presumed to be *either* "skills" courses *or* intellectually broadening courses. This dichotomy isolates communication, minimizes its connection to professional development and intellectual activity, and promotes an impoverished view of what successful communication entails. At most institutions, this conceptual isolation of communication has been reinforced by disciplinary and departmental structures and by a vocabulary that presumes sharp distinctions among various kinds of courses and enforces boundaries between faculty trained in different areas. We have spent more than fifty years operating with a vocabulary that separates "technical electives" from "humanities and social science electives," "technical faculty" from "professors of technical communication," and "skills" from "content." This vocabulary reflects a conceptual framework aimed at making distinctions rather than seeing things whole.

By contrast, the discussions related to the new criteria have been dominated by the vocabulary of integration and interdisciplinarity, a vocabulary that reflects a new view of engineers, their expertise, and their relations with others.[4,5] This new vocabulary arises at least in part from cultural changes related to engineering, changes that we need to understand and exploit in order to effect significant change in a curriculum that has proved remarkably resistant to change. To put it another way, we need a new vision of technical communication and a new vocabulary for talking about it in order to revise the curriculum successfully.

The central purpose of this chapter is to articulate a new and richer vision of technical communication and to outline a conceptual framework for thinking about technical communication in the full range of contexts in which it can be taught, that is, (1) in courses devoted primarily to communication; (2) in courses emphasizing various aspects of the humanities and social sciences; and (3) in courses emphasizing what has traditionally been viewed as "technical" content. This framework presumes that communication will be considered in all aspects of the engineering curriculum but emphasized to different degrees and approached in different ways in different courses.

The central proposition underlying this framework is that instruction in communication should be not one thing but many things, and that communication is both a specialized field of research and teaching *and* an area of interest common to many disciplines and activities. Conceptualizing communication this way both recognizes the role of specialists in communication (traditionally conceived as teachers of English) and also helps all engineering and humanities/social science faculty see how they can contribute to helping students learn to communicate effectively.

Because I believe a sense of history can shed light on the challenges we face, the chapter begins with a brief history of ASEE's concern with and approach to communication as an aspect of the engineering curriculum. This history leads into a discussion of what an integrated vision of technical communication would entail. To flesh this out, I discuss risk communication as a model for integrated communication instruction in engineering. I conclude by discussing the prerequisites for achieving lasting change.

History and Its Lessons

The history of communication as a component of engineering education can be traced through two reports produced by members of ASEE: "The English Division of the American Society for Engineering Education: A History," published by Alvin M. Fountain of North Carolina State University in 1961 and "A Century of ASEE and Liberal Education, the Division and Its Predecessors," written by O. Allan Gianniny, Jr. of the University of Virginia in 1993. These two histories reveal that interest in English for engineers began to gather momentum as early as 1907 and seems to have reached its height in the 1930s. That interest has never entirely disappeared, but it has diminished to the point that the idea of creating a "Communication Division" within ASEE seems like a new idea rather than the revival of a tradition. Within this overall pattern of fluctuating but enduring interest in communication, there are several trends and key events that are important for understanding the problems of integrating communication into a revised engineering curriculum.

Writing and Speaking as Skills

Writing and speaking have typically been discussed in terms of "skills," even though most teachers of English for engineers seem to have recognized them as arts, activities that call for experienced judgment and a wide range of knowledge. This customary way of talking about communication has tended to promote a reductionist view of communication practice and pedagogy and to obscure the wide range of knowledge and abilities that constitute the capacity to communicate effectively. Conceptualizing communication in terms of skills has also tended to separate communication from the humanities and social science (HSS) elements of engineering education.

Small Group of Innovators

The calls for reform and other statements of ideals for engineering education that have appeared in various forms over the years have not usually reflected or changed common practices in engineering schools. There has always been a small group of innovators within ASEE who have embraced newly articulated ideals and implemented innovative curricula, but their activities and philosophies have probably influenced the various reports published by the society more than they have affected educational practices in engineering.

Two-stem Approach

An important turning point came in the years following World War II. Since the 1920s and 1930s, engineering educators had been working with what came to be known as a "two-stem" approach to engineering education. The first stem was "scientific-technical," the second stem "humanistic and social." The system of accreditation implemented in the 1950s emphasized the importance of the second stem but at the same time encouraged the separation of the two stems by insisting on a clear distinction between their functions. The new accreditation system created a sanctuary for humanities and social science courses within the engineering curriculum, but sharpened the boundary between writing and speaking and other aspects of the humanities and social sciences.

The result was that writing and speaking "skills" were left in a kind of academic limbo. Although educators still recognized the need for communication instruction, there was no institutional framework to support it. The importance of the second stem was justified in terms to which no self-respecting teacher of English could object. Reformers often spoke of the "engineer as educated man" and focused on efforts to "broaden the engineer's intellectual outlook" and "liberate [him] from provincialism." The result of all this—most likely an unintended one—

was that communication was isolated from the humanities and social sciences. Teachers of technical communication tended to find themselves in what was often an uneasy relationship with their colleagues in the humanities and social sciences.

Efforts to Revolutionize by Integrating

Efforts to integrate the "second stem" go back at least to 1968 and the Olmsted report entitled "Liberal Learning for the Engineer."[6] Although this report contained a number of powerful ideas and inspired both creative thinking and lively debate, it appears to have led to little real change in engineering education. When Gianniny revisited the Olmsted report in 1973 and 1993, he found a number of interesting experiments but no widespread adoption of the integrated approach the Olmsted report called for.[7] These reports reflect a striking resistance to change: although "integration" seems to be a new idea, the integration effort itself is nearly thirty years old.

This brief history reveals the forces that have isolated communication within the engineering curriculum and suggests a richer vision of technical communication that makes it easier to see how integration could be achieved.

First, we must abandon the "skills" vocabulary in discussing communication. Although discussing communication in terms of skills tends to gain ready acceptance from engineering faculty, it also tends to grow out of a rather impoverished view of communication. This is not to say that there are no skills involved in communication; there are, in fact, many aspects of both writing and speaking that can be automated in the sense that the communicator devotes relatively little thought to them once they are acquired.[8] Grammar, punctuation, spelling, typing, formatting, the use of an overhead projector, or the suppression of "ums" in speech are examples of skills, but these kinds of activities play only a relatively small part in effective communication. Using a vocabulary of skills captures only a fraction of what is involved in communication. I suggest, instead, that we should think in terms of "communication ability" rather than "communication skills," and of "professional communication" rather than "technical communication." In doing so, we will help bring to light the strong links between communication and intellectual activity.

Second, we must articulate the whole range of knowledge, skills, and experiences that help people become effective communicators and be more specific about what we mean by integration as it relates to communication in the engineering curriculum. In current discussions of educational reform in engineering, engineering educators, industrial leaders, and others concerned with engineering education have been very clear in their insistence that communication is essential to success in engineering and thus to engineering education. Yet they provide few details about either the ways communication ability is demonstrated or the specific knowledge, skills, and experiences that form the foundation of an ability to communicate effectively. Moreover, although there seems to be widespread agreement that integration is a good idea, there seems to be little explicit understanding of what integration entails or how to achieve it on a large scale. We will need to learn to think critically about both communication and the concept of integration if the current movement for reform is to go beyond a small group of innovators.

Third, in order to overcome the inertia that inhibits widespread reform, we must tap into the larger movements and cultural changes that can support the integrative effort. The current emphasis on communication as a professional engineering skill is often attributed to the perception that lack of communication ability keeps engineers out of corporate boardrooms and executive suites and limits the influence of the engineering profession as a whole.[9, 10, 11] While this perception is undoubtedly an important motivation for putting more emphasis on communication, it offers only a partial explanation. As Martin Marietta CEO Norman Augustine has pointed out, in the post-Three Mile Island era, the American public is less willing than it had been previously to "accept every new innovation that engineers unleash," and regulation of technology-related activity has become more prevalent. In this area as in the others, we will need to go beyond generalizations and ready explanations to inquire into the changes in the organization of engineering work and the broader cultural changes that have led to renewed interest in the communication abilities of engineers.

Fourth, we must recognize that a strong linkage between communication and HSS strengthens the position of both within the engineering curriculum. For engineering educators interested primarily in communication, it might be tempting to stay with the widely accepted vocabulary of communication skills and to emphasize the immediate and obvious utility of their field. But this easy acceptance would come at the cost of obscuring the richness and complexity of their subject. For engineering educators interested primarily in HSS, emphasizing the relevance of their subjects for the development of communication ability provides additional arguments for continuing to include a significant HSS component in engineering curricula.[12] Working with the rich conception of communication presented in this chapter should make it possible to link HSS to communication without diminishing the professional respectability of HSS disciplines.

Integrated Vision of Technical Communication

To create the courses, pedagogical approaches, and assessment tools that will make communication an integral part of engineering education, we must get beyond the idea that communication is somehow the exclusive territory of teachers of English. Like any other traditional academic discipline in the engineering curriculum, communication can be approached as a collection of highly specialized areas of expertise. But it can also be approached from an interdisciplinary perspective that appreciates the wide variety of knowledge, skills, and judgment that goes into successful communication. The most significant strength of this interdisciplinary perspective is that it reflects the world of practice, where communication always brings together expert knowledge, an understanding of context and audience, and the range of abilities necessary for articulating and delivering an effective message.

Thus, an integrative approach to communication must not only establish strong links between writing and thinking, but also locate any act of communication in a network of technical, social, cultural, commercial, and institutional activity. To integrate communication would, in this case, mean treating communication as it arises naturally in the context of engineering practice and research, where it occurs or is planned out of the necessity of getting things done. This does not mean that all communication courses should be the same, but it does mean that communication instruction should be undertaken with the specific contexts of engineering work in mind and treat communication as a normal part of engineering practice and pedagogy, rather than as a special or artificial element to be added or imposed.

This integrated instruction should have several key components:

1. It should be consistent with the notion of career preparation as education rather than training, i.e., it should prepare people to think, lead, and make decisions rather than limit their vision to executing the plans of others. To borrow Marshall Lih's terminology, it should motivate and prepare them to be cathedral builders rather than just brick layers.[11] Accordingly, it should stress a long-term and broad perspective, one that takes the future and a wide range of factors into account. It should locate technical subjects within a big picture view in which technical components are seen in relation to organizational and cultural factors.
2. It should recognize the existence and value of multiple perspectives and help students learn to look at subjects and issues from a variety of viewpoints, as well as recognize the role of nonquantifiable factors in decision making.[13] Specifically, it should abandon the hierarchy implicit in the expert audience/lay audience distinction in favor of a more democratic view in which communicating on equal terms with experts in other fields and with nontechnical audiences is seen as a normal parts of professional practice.[14,15] This kind of instruction would promote both the ability to work in teams and the capacity to communicate with nonexperts.

3. It should recognize that both ethical management of technology and successful marketing of new technology require communicating effectively with users and others—mostly nonexperts—who are affected by the technology in question. For example, obtaining informed consent from those exposed to technology-related risks and winning acceptance for new technology require communicating with people who are not experts on the technology in question and who cannot be assumed or required to accept expert views without question.[16]
4. It should offer students instruction and practice in writing and speaking, reading and listening, as well as formal and informal, visual and verbal communication. It should avoid emphasizing academic forms of writing such as lab reports and expose students to the full range of modes and genres in which they will communicate during their careers.
5. It should exploit the "situatedness" of technical communication. Technical communication is situated in the sense that we know the situations in which, and the subjects about which, our students will be communicating. To exploit situatedness we must take advantage of and respond to what we know about the contexts in which our students will communicate. We must also strike a balance between general principles and specific guidelines to provide students with the perspectives, knowledge, and strategies they need to adapt to a wide range of communication situations.

An integrated approach, then, is one that recognizes the context in which communication occurs and the way activities relate to each other in that context. It may be useful to abstract communication temporarily from context in order to highlight certain features or for pedagogical purposes, but it is important to remember that communication is always undertaken in a particular context and is quite context sensitive. An integrated approach exploits this context sensitivity and helps students see the ways that successful communication weaves together a number of different kinds of expertise.

Risk Communication as a Model

Risk communication, by which I mean simply communication about technology-related risks, provides a concrete example of an integrated approach and exhibits a number of characteristics that make it an appropriate model for integrated communication instruction. It is a topic that might appropriately and profitably be taken up in any class within the engineering curriculum in the sense that the expertise being developed in that course would have relevance in developing a deeper understanding of some aspect of risk communication. It could provide the focus for a course in communication, be a major unit within a thematic humanities/social science course, or serve as a benchmark or ongoing point of reference in a specialized technical course.

One of the most significant characteristics of risk communication is that it clearly links technical, social, and communication elements and is inherently interdisciplinary. Whether we are dealing with the safety of microwave ovens or the siting of nuclear reactors, there are many factors to be considered: scientific and technical principles; mathematical probabilities; social norms and roles; political processes; ethical principles; psychological factors having to do with the ways people assimilate new information; and strategic problems of how to explain and help readers visualize complex and remote phenomena, to name just a few. There is hardly any discipline that does not contribute to a better understanding of risk communication, and not a single discipline that provides a complete view. Thus, risk communication provides a good example of how an interdisciplinary approach can be used to gain a better grasp of complex situations and problems.

Another strength of risk communication is that it is interesting and relevant to engineers and can only be discussed effectively through the use of specific examples. Engineers tend to understand its importance and to feel that they have something to contribute to discussions of it. Risk is a factor in all branches of engineering, provided the technical systems in question are scaled up to the point of implementation. This is a particularly important consideration if we are thinking of integrating instruction in communication into existing engineering courses. Moreover, risk is an important aspect of many contemporary issues, especially those related to the environment.

Risk communication also provides a window into culturally and historically significant changes. The recent emphasis on risk communication grows in part out of increased public concern with risks, which in turn reflects doubts about the ways experts are likely to perceive and manage risk. It also reflects a move away from an expertise-based view of technology management toward more democratic management of technology.

Another advantage of risk communication is that it is intellectually engaging in a way that more abstract forms of technical communication such as process description or oral presentation are not. Risk communication also offers the opportunity to observe the ways that knowledge is transformed as it moves outward from those most expert in the technology in question to increasingly broader audiences. The rich body of recent scholarly work in the area of risk communication provides testimony to its capacity for intellectual engagement.[17, 18, 19] Faculty who pursue this area have a lot to draw on, both as teachers and as researchers.

In addition, the study of risk communication is engaging and appropriate at all levels of engineering education from first-year through graduate study. Students who study risk communication at lower levels get insight into the ways that educated nonexperts must grapple with scientific and technical questions. As they advance in their fields of study, they gain further insight into the problems that experts face as they attempt to communicate with wider audiences on matters of importance.

Another major benefit of the study of risk communication is that it provides a forum for helping students explore and understand ethical and philosophical concepts that are important in engineering.[20–23] Dealing with risk communication requires students to examine the differences between expert and nonexpert perceptions of risk and to begin unpacking the layers of assumptions that make up common perceptions of those differences, as well as to consider how those differences can be dealt with effectively in developing public policy. It also gives students concrete experience in thinking about the importance and limitations of scientific evidence. Ideals of persuasion in science and engineering are based on the principle that we start with evidence and see what position it leads us to, rather than starting with a position and seeing what evidence we can find to support it. This is an important difference from much of the persuasive discourse students are exposed to, and it is a particularly important one for them to understand.

In proposing this model, I am not suggesting that all communications courses, much less all courses in the engineering curriculum, should necessarily deal with risk communication. What I am suggesting is that the study of risk communication as I have defined it here reflects a rich view of professional technical communication. It provides both students and faculty with a clear and convincing example of the many factors that must be brought together in order for effective communication to occur. It demonstrates the ways that social, ethical, and rhetorical issues are intimately related to technical issues, rather than being peripheral to them. It also provides a clear example of an area of engineering practice where the experts in communication have a lot to offer but that is clearly not the province of English teachers alone. There are undoubtedly other aspects of engineering practice that offer similar opportunities, and they can be identified by faculty who approach communication as it arises in context and from an interdisciplinary perspective.

Conclusion: Waiting for the Revolution?

The integrated vision of technical communication sketched out above is far from being realized in current engineering curricula, and perhaps the majority of engineering educators have yet to envision it as a possibility. We are, it seems, still waiting for the revolution. I have argued here that the revolution would entail two forms of integration. First, it would grow out of a rich, interdisciplinary view of communication. Second, it would make all communication instruction sensitive to, and reflective of, the contexts in which engineering students will be working and communicating, with the important proviso that "context" should be very broadly conceived.

Neither of these arguments is particularly new, but there are several reasons why the revolution may really happen this time. The first is that academic institutions seem to be becoming increasingly aware of the limitations of disciplinary thinking and structures. Interdisciplinary organizations and fields of study are much more common than they were thirty years ago, and the value of crossing disciplines is increasingly recognized. This larger movement toward interdisciplinarity should support efforts to integrate the humanities, social sciences, and communication into engineering education.

The second reason for optimism is the change to the ABET 2000 accreditation system. The last major change in the 1950s had profound consequences for communication. The current change in accreditation puts communication on at least as solid ground as that occupied by HSS. The instability created by the implementation of the ABET 2000 criteria brings risks, but it should also create openings and incentives for innovation in engineering curricula that would not have existed otherwise.

Third, although the majority of engineering faculty may not yet be committed to change, the commitment extends beyond the community of teachers of communication and HSS and is broadening. In a number of publications, broad-based advisory groups to the NSF have emphasized the need to change the culture of engineering. The changes they propose strongly support an integrated approach to communication.

The prerequisites for revolutionary change include persuasion and participation. Those of us who specialize in teaching engineering students to communicate effectively must practice what we teach and take on the challenge of persuading our colleagues to make the changes that will be necessary. Perhaps most significantly, we have to help them see what they have to offer to the process of teaching students to communicate effectively, as well as how their views of their own subjects can be enriched by considering the challenges and motivations for communicating those subjects to audiences beyond the intimate circle of experts in that same subject. We also have to be willing to participate in the revolution ourselves—to change in some rather radical ways by broadening our own views of communication and recognizing what experts in other fields have to offer.

It would be possible and perhaps reasonable to view the history of communication instruction in engineering as a cyclical story like that of Sisyphus and his rock, as a record of successive failures punctuated by moments of illumination and hope. I see this history as testimony to the strength of an idea—the idea that communication is essential to success in engineering and that the task of teaching engineers to communicate is an intellectually engaging and rewarding one. This history is also a history of exploration, which started with an instinctive understanding of the importance of communication and has evolved through successive articulations of the myriad factors that contribute to successful communication.

The time the journey has taken bears witness to the complexity of the terrain we are exploring and the richness of our experience of it. The complex set of interactions with which we deal as we attempt to understand professional communication in engineering offers us many opportunities to "know the place for the first time." Perhaps this sense of richness is one of the most valuable capacities we can cultivate in our students. As we move to ABET 2000 and beyond, an interdisciplinary view of communication can take all of us as engineering educators to places we have not been before.

Notes

* This committee became a division of ASEE in 1942 and merged with the Humanistic-Social Sciences Division to become the Liberal Studies Division in 1965. The name of the Liberal Studies Division was later changed to Liberal Education Division.

References

1. Eliot, T. S. 1950. Little Gidding, In *The Complete Poems and Plays*. New York: Harcourt, Brace, and Company.
2. Fountain, Alvin M. 1961. The English Division of the American Society for Engineering Education: A History. Unpublished paper produced at North Carolina State College on June 26, 1961, and distributed to members of the English Division of ASEE.
3. Gianniny, O. Allan. 1993. A Century of ASEE and Liberal Education, the Division, and Its Predecessors: or How Did We Get Here from There, and Where Does It All Lead? Paper presented to the Liberal Education Division of the American Society for Engineering Education, Champaign, Illinois, June 22, 1993.
4. Perrow, Charles. 1984. *Normal Accidents: Living with High Risk Technologies*. New York: Basic Books.
5. National Science Foundation, Directorate for Education and Human Resources. 1996. *Shaping the Future: New Expectations for Undergraduate Education in Science, Mathematics, Engineering, and Technology* (NSF 96-139).
6. Humanistic-Society Research Project, Sterling W. Olmsted, Chairman. 1968. Liberal Learning for the Engineer. *Journal of Engineering Education*, 59 (4):310-337.
7. Gianniny, O. Allan, ed. 1975. Liberal Learning for the Engineer: An Evaluation Five Years Later. *Engineering Education*, 4 (65): 301-324.
8. Hillocks, George, Jr. 1986. *Research on Written Composition: New Directions for Teaching*. Urbana, IL: ERIC Clearinghouse on Reading and Communication.
9. CEO in Focus: The Philosopher Engineering. 1995. Interview with Norman R. Augustine. *Graduating Engineer*, 16 (3): 28-32 and 41.
10. Griffiths, Phillip A. 1995. Breaking the Mold. *Prism*. November 1995.
11. Lih, Marshall M. 1997. Educating Future Executives. *Prism*, January 1995: 30-34.

12. Johnston, Joseph S., Jr., Susan Shaman, and Robert Zemsky. 1988. *Unfinished Design: The Humanities and Social Sciences in Undergraduate Engineering Education*. Washington, DC: Association of American Colleges.
13. Pacey, Arnold. 1983. *The Culture of Technology*. Cambridge, MA: MIT Press.
14. Bruner, Ronald D. et al. 1994. Science, *Technology and Democracy: Research on Issues of Governance and Change*. Prepared for the National Science Foundation.
15. Sclove, Richard E. 1995. *Democracy and Technology*. New York: The Guilford Press.
16. Cilengir, Erika N. 1992. Controlling Technology Through Communication: Redefining the Role of the Technical Communicator. *Technical Communication*, 2nd quarter, 1992: 166–174.
17. Covello, Vincent T., Peter M. Sandman, and Paul Slovic. 1991. Guidelines for Communicating Information About Chemical Risks Effectively and Responsibly. In *Acceptable Evidence*, pp. 66–90, ed. New York: Mayo and Hollander.
18. Herkert, Joseph R. 1994. Ethical Risk Assessment: Value Public Perception, *IEEE Technology and Society Magazine*, 13(1): 2–12.
19. Morgan, Granger et al. 1992. Communicating Risk to the Public. *Environmental Science and Technology*, 26 (11): 2048–2056.
20. Mayo, Deborah G., and Rachelle Hollander, eds. 1991. *Acceptable Evidence: Science and Values in Risk Management*. New York: Oxford.
21. Shrader-Frechette, Kristin. 1991. Reductionist Approaches to Risk. In *Acceptable Evidence*, ed. New York: Mayo and Hollander 218–248.
22. Slovic, Paul. 1991. Beyond Numbers: A Broader Perspective on Risk Perception and Risk Communication. *In Acceptable Evidence*, ed. New York: Mayo and Hollander 48–65.
23. Valenti, John, and Lee Wilkins. 1995. An Ethical Risk Communication Protocol for Science and Mass Communication. *Public Understanding of Science*, 4:177–194.

CHAPTER 5

LESLIE PERELMAN

Creating a Communication-intensive Undergraduate Curriculum in Science and Engineering for the Twenty-first Century

Introduction

At the end of the Second World War, MIT conducted a two-year extensive review of its educational programs, which resulted in a major and highly influential reconception of its undergraduate programs in science and engineering.[1] MIT's Committee on Educational Survey developed a new curriculum that focused on giving students a lifelong framework for learning. Now, a half-century later, MIT is beginning a similar in-depth review and restructuring of its undergraduate programs by reapplying this same general principle.

Fifty years ago, MIT restructured its curriculum to address the rapid pace of technological change. Technical areas were evolving too quickly for a university to give students a stable platform of acquired industrial knowledge and skills. MIT realized that it could not give its students all the knowledge necessary for successful engineering practice, but it could give them a strong grounding in basic science to understand and use effectively these scientific and technological innovations.

Now, MIT, like other science and engineering schools, is assessing what basic knowledge and skills will best provide our graduates with a viable framework for adaptability and success in the twenty-first century. At MIT, the primary impetus for this reassessment is not the changing ABET standards themselves, but the changing world that has motivated the ABET 2000 accreditation criteria. Because technological change is now even more rapid than it was fifty years ago, providing an education based on a stable platform of knowledge is now even less possible.

Furthermore, the end of the Cold War, the comparative scarcity of resources available for research and development, the rapid transformation and restructuring of many traditional large technology corporations, and the globalization of the world economy have made communication skills an essential and integral part of any scientific or technical activity. Most engineers no longer expect to work for one firm for most of their careers, nor do they expect to stay in a single technical area. Our graduates become managers, entrepreneurs, and even financiers. In these varied professional roles, success often depends on the ability to write and speak effectively to a wide range of audience in order to persuade, to listen, to work cooperatively, and to lead. Fifty years ago, science and technical education had to prepare students for rapid changes in what they would be required to know; today, it also must prepare students to adapt to rapid changes in what they will be required to do.

In response to these rapid and continuing changes in professional life, the MIT Faculty unanimously voted in April 1997 to include the following statement of principle:

> The Faculty believes that the ability to communicate clearly is fundamental; that students should receive instruction and feedback in writing and speaking during each undergraduate year; and that responsibility should be distributed across the entire MIT undergraduate curriculum.

This declaration was the culmination of a three-year preliminary assessment and design process, and the beginning of a three-year curricular development phase to inform the specific formulation and implementation of a new undergraduate communication-intensive curriculum.

Overall, MIT will spend over ten years in formulating, designing, evaluating, and refining exactly how it will integrate writing and speaking into its undergraduate programs. Such a long period for development and implementation is necessary because the goal, although stated easily, is difficult to achieve. Indeed, the process itself has been and continues to be primarily a complex design problem possessing imposing boundary conditions and competing criteria. The general characteristics of this problem are not unique to MIT. All undergraduate engineering programs need to address these curricular issues, not just to meet the new ABET standards but to equip students with the abilities necessary for professional growth and success in the next century. Moreover, viewing this type of long-term educational reform as an exercise in design provides a useful and highly flexible framework in which to develop these essential curricular elements within the substantial constraints inherent in undergraduate technical programs.

In addition, these constraints pose more than just a problem in design. They also reveal differing visions of the role of liberal education within a technical university. Are MIT and similar institutions places where the teaching of writing and speaking is seen primarily as the responsibility of humanities faculty, who exist solely to serve the needs of the scientific and engineering curricula? Or are they places that acknowledge different varieties of communicative competence and believe that the responsibilities for developing them are shared among all divisions of the faculty? Our experience has made it clear that before developing any effective communication-intensive curriculum, any technically oriented university must also answer these fundamental questions.

Background

The Writing Requirement

In 1982, MIT instituted an undergraduate writing requirement. Superficially, the writing requirement may appear to have been ahead of its time. It is outcome based, requiring students to show proficiency in expository writing by the beginning of their sophomore year and in the technical discourse of their major field by the middle of their senior year. Although students can display this proficiency by passing certain courses, most fulfill each of the two parts by submitting a paper for evaluation. Almost from its inception, there has been criticism that this kind of proficiency requirement has done little to help our students become better writers and has done nothing to help them become better speakers. Without any instructional space mandated within the current curriculum, students often have viewed the acquisition of minimal proficiency in writing as a curricular afterthought, something to be done after all the problem sets were completed.

Initial Discovery Process

In the fall of 1994, MIT's Committee on the Undergraduate Program (CUP), in response to these concerns, commissioned a special Institute-wide faculty subcommittee to study the effectiveness of the writing requirement. Its charge was to identify what skills in writing and speaking are necessary for undergraduate academic achievement and for professional success and growth in the twenty-first century and to assess how well MIT develops these abilities in its undergraduates. It was also commissioned to survey current efforts under way at MIT to provide students with effective instruction in writing and speaking and to identify successful programs and experiments that could be adapted and expanded. Finally, the CUP charged the subcommittee to identify key issues and recommend, if appropriate, general guidelines for substantially revising the current writing requirement. This subcommittee's discovery work led to a subsequent charge from the CUP to the faculty Committee on the Writing Requirement (CWR) to continue these efforts by first articulating specific design criteria and then by developing a detailed proposal for a new curriculum, in consultation with interested faculty, department heads, and school deans. Originally, the two committees planned to spend about a year designing the new curriculum. The complexity of the overall design problem and the unresolved issues surrounding the respective roles of liberal arts and technical faculty in teaching various types of communication, however, led to a quick realization that a more lengthy and comprehensive design process was necessary.

Findings

Effective Communication for Professional and Personal Success

Fewer MIT engineering majors begin work as bench engineers than did ten years ago. Some engineering graduates now immediately begin careers in consulting or finance. Many of those graduates who do choose engineering careers quickly become managers, devoting most of their time to administrative functions. In all of these roles, the abilities to write and speak clearly and effectively are increasing in importance. Moreover, technology is producing rapid and continuous changes in the basic forms of communication, and the ability to become proficient in these new media is crucial. E-mail, for example, was largely unknown ten years ago. Now it is rapidly becoming ubiquitous in most technical work environments. Laurence Zuckerman quotes Michael Murray, vice president for human resources and administration at Microsoft, as describing his corporation as "an E-mail run company. The primary means of communication is the written word. If you are unable to express yourself concisely in one paragraph or less, you are not going to be effective."[2]

Effectiveness of the Current Curriculum

The subcommittee employed various and complementary methods to assess the effectiveness of the MIT undergraduate program in developing the various abilities necessary for both academic success at MIT and for subsequent professional growth. Assessment procedures included collecting useful anecdotal information; interviewing faculty, writing instructors, and students; analyzing writing requirement statistical data and preexisting information from surveys of MIT seniors and alumni; and conducting a carefully designed study that ranked the writing quality of thirty-two expository essays from a group of randomly selected juniors and then compared the quality of writing exhibited in each essay with each author's overall academic profile.

In addition to investigating how well the current curriculum in general developed effective communication abilities, the subcommittee paid particular attention to assessing how well the current outcome-based writing eequirement accomplished its twin objectives of ensuring that students graduate proficient in both general exposition and in the specialized discourses of their professional fields. The subcommittee also identified the current levels of proficiency in writing and speaking that students need to succeed at MIT, and compared these levels to those it had already identified as necessary for future professional growth and development.

These investigations produced several important and disturbing findings. First, the subcommittee's review of the junior essays led it to conclude that a quarter to a third of MIT students have inadequate writing abilities in their junior year, while another third possess writing abilities that are only marginally adequate. Second, a comparison of these junior writing samples to essays written by the same students when they were freshmen revealed that the writing abilities of most of the students in the study had not improved noticeably during their time at MIT. Third, most students receive no instruction and practice in oral communication. Although one department requires a minimum of six formal oral reports, most undergraduate programs require, at most, one oral presentation and provide little, if any, instruction or feedback. Fourth, the data show no correlation between a student's writing ability and his or her overall grade-point average. This last finding emphasized that the present culture at MIT clearly does not reward student attention to communicating clearly and effectively, even though this ability is essential after graduation.

Indeed, the alumni survey data indicate that while MIT does an excellent job in developing three of the four abilities considered currently essential or very important by 85 percent or more of respondents, only 25 percent of these alumni reported that MIT contributed significantly to their development of the fourth ability highly ranked in importance, the capacity to write clearly and effectively. The following written response from one of the alumni surveyed provides an eloquent description of some of the typical consequences of this common deficiency in undergraduate technical education.

The general problem solving skills that I learned at MIT have been very useful in analyzing a wide range of business problems. Upon graduation from MIT, I went to work in consulting, an industry dominated by "ivy league" type graduates. While my college education probably prepared me better for analyzing problems, my counterparts were better at presenting their ideas and working with others. I was kept available for behind-the-scenes analytically intensive assignments while others had more opportunity to meet with clients and become exposed to higher level issues.

Responses such as this one have been prime motivators of the current efforts at MIT to improve our students' communicative and interpersonal skills. Because such experiences are fairly common, alumni, rather than current students or faculty, have generally been the most persistent and persuasive advocates for change.

Causes

The primary reasons why so many MIT undergraduates and graduates fail to develop the same level of proficiency in writing and speaking as they do in quantitative ability are readily identifiable. Most MIT undergraduates take few classes that require frequent writing and revising. Moreover, the opportunities in the curriculum for students to give formal or informal oral presentations are even more limited. When students are required to write or speak, they often receive little systematic instruction and minimal useful feedback. Finally, because the current curriculum does not reward students for mastering these abilities, students have powerful incentives to devote most of their limited time and attention elsewhere.

Limited Opportunities for Writing

Writing is a regular part of liberal arts education. Students in liberal arts programs usually write frequently in several courses each term, often composing ten or more substantial papers each year. Students in technical programs, however, usually encounter significantly fewer required writing assignments. In some cases, an engineering student will spend one or two years of undergraduate study without being required to write anything. Moreover, the few writing assignments they encounter in technical subjects are often single long exercises, such as lengthy design or laboratory reports that require complex combinations of cognitive and writing abilities. Writing, however, is, among other things, a skill. It is a muscle that needs to be developed through sustained and systematic practice. Asking a student who has not written in eighteen months to write a 5,000 word design document may very well be analogous to asking a sprinter who has not run at all in over a year to participate in a marathon. Instead of teaching students to write effectively, such approaches often leave students with the perception that writing is a painful, incomprehensible, and an always frustrating task, something to be survived and gotten through as quickly as possible. The end result is that some students give up on developing the communicative proficiencies that they, at least intellectually, know will be required of them after graduation. And even more sadly, these students fail to experience writing as a valuable method of enhancing and broadening their understanding of technical material.

Limited Opportunities for Instruction and Feedback

MIT undergraduates are required to take a minimum of three liberal arts classes as part of the eight humanities, arts, and social science courses required for graduation. Liberal arts faculty, however, differ widely in the emphasis they give to writing in these classes. Some instructors believe that because of the limited exposure most undergraduates will have to history, literature, and the arts, they should focus most of their efforts on introducing students to these essential aspects of world culture. Other liberal arts faculty, however, have developed writing assignments and feedback mechanisms for their classes that provide students with substantial instruction and experience in writing while simultaneously enhancing and refining their overall understanding of the course content.

In technical classes, faculty and teaching assistants often believe that instructing students in writing and speaking is largely the business of their liberal arts colleagues, and, consequently, focus their own comments on papers on technical matters. In addition, many technical faculty and TAs do not consider themselves competent to comment constructively on the form and style of student technical reports.

Lack of Value and Tangible Incentives

Robert Metcalfe, an MIT alumnus, developer of Ethernet, co-founder of 3Com, and now a columnist for *InfoWorld*, lectures each year to about 250 computer science majors on the importance of writing in engineering. He always begins his talk by recounting his experience in a class at MIT in the early 1960s taught by Jay W. Forrester, the inventor of core memory and one of the early giants of computer science. Metcalfe describes how Forrester spent considerable time commenting on each student's prose and required extensive and multiple revisions of each assignment. "Writing first became important to me," Metcalfe states, "because Jay Forrester treated it as important."[3]

Most students graduating from MIT, however, do not enjoy similar classroom experiences. Most technical faculty explicitly value abilities such as analytical thinking and creative problem solving. The basic unit of student work in most technical undergraduate programs is the problem set. These exercises rarely offer students opportunities to present information in a logical series of sentences and paragraphs, and, even when they do, instructors often focus on the technical solution and ignore any consideration of how the material is presented.

This lack of valuation and tangible incentives has, of course, a predictable effect on student behavior. An MIT aeronautical engineering faculty member once commented at a meeting, "Our undergraduates are extremely adept at minimizing any assignment's trajectory." Given the pace and pressure of most undergraduate technical programs, students in engineering classes maximize their performance by allocating almost all of their attention to an assignment's technical issues at the expense of any substantial effort to learn how to communicate the technical content appropriately. Ironically, this lack of focus on organizing and presenting ideas diminishes rather than enhances student comprehension of technical content and may very well increase student pace and pressure. Writing is a powerful tool for learning and understanding. By frequently expressing ideas in prose, students will almost certainly understand concepts more completely and accomplish tasks more efficiently.

Existing Programs

In addition to identifying the primary causes of the present current undergraduate program's inability to teach students to write and speak clearly and effectively, the subcommittee also identified several successful programs and resources already in place at MIT. This investigation had two general objectives. First, by examining each program's instructional practices, the subcommittee was able to identify pedagogical principles that should inform the design of the new curriculum. In addition, these programs provide models that have already proven their effectiveness within MIT's academic culture.

The Technical Writing Co-op

In the 1950s, MIT developed one of the first postwar writing-across-the-curriculum programs, the Undergraduate Technical Writing Cooperative (the Writing Co-op), which is still an active component of writing instruction at the Institute. For forty years, the Writing Co-op has been a flexible and adaptable mechanism for collaboration between writing program staff and technical faculty. Instructors from the School of Humanities and Social Science visit engineering classes, giving presentations about written and oral communication and providing feedback to students on particular assignments. The structure of individual co-ops differ significantly. Many co-ops include significant participation and instruction by writing program staff in specific formats for writing, speaking, or both. In other engineering classes, however, the co-op component may consist of one short lecture and the evaluation of a single piece of writing.

The Writing Practicum

The Writing Practicum evolved out of the co-op model with the primary objective of maintaining the co-op's flexibility and adaptability while providing a more structured and consistent context for the teaching of specific communication abilities. Initiated in 1993 as a pilot collaboration between the School of Engineering and the School of Humanities and Social Science, the practicum is a satellite class attached to an engineering course. Employing the content and context of the associated subject, a writing practicum provides students with substantial practice and instruction in writing, argumentation, revision, and presentation skills. These classes are not required, but offer six units of free elective credit (one-half the units of a regular MIT class).

As in writing co-ops, the exact relationship between each practicum and the associated technical course is highly flexible. Each practicum is a singular blend, combining coaching on some specific engineering assignments with a general exploration of the social, managerial, and human dimensions of the engineering process. When students participate in peer review, small group sessions, and oral presentations, they are developing not only communication skills but also their awareness of the relationships between the technical and managerial dimensions of engineering practice. In the practicum connected to the course Computer Systems Engineering, for example, students first revise, in groups, the ten-page design report each one wrote for the engineering class. Each student then writes a two-page memorandum, addressed to a manager lacking a technical background, recommending whether or not the design should be implemented. Significantly, in the past two years over half of these memoranda have presented sound arguments for why a student's project design is not feasible.

Another successful practicum has been developed in conjunction with the Experimental Projects Lab in the Department of Aeronautics and Astronautics.[4] The Experimental Projects Lab is similar to an undergraduate thesis in scope. Each team of two students chooses an original research project and is guided by a faculty advisor over the span of two semesters. The students participate in all aspects of experimental research, including project definition, a formal proposal, design of the experiment, construction of apparatus, completion of the experiment, and analysis and reporting of results. The communication practicum provides focused instruction in both written and oral communication skills involved in each of these activities.

These practica are taught primarily by graduate student teaching fellows drawn from a variety of technical and nontechnical disciplines. The fellows are supervised jointly by liberal arts and engineering faculty. After being selected, but before they begin to teach, the fellows participate in a semester-long series of workshops conducted by members of the liberal arts faculty. The training sessions provide the new teaching fellows with an overview of the nature of different types of effective technical communication, methods for providing students with effective and constructive feedback, general guidelines for developing and sequencing writing assignments, and specific techniques for encouraging effective peer review and collaborative writing. In addition, staff from MIT's English as a Second Language program provide strategies for instructing students who are nonnative or bilingual speakers of English.

A preliminary review of the program demonstrated both its success and its potential for further expansion.[5] That early review and subsequent reports from both faculty and students offer substantial evidence that these practica improved not only students' skills in technical writing, oral presentation, graphic design, peer review, and group writing, but also their overall understanding of the technical content of the engineering class. In its first year (1993–1994) there were practica in five engineering subjects. In the three following years (academic years 1994–1995, 1995–1996, and 1996–1997), over 100 students each year have enrolled in writing practica attached to eight different engineering classes in five engineering departments: Aeronautical and Astronautical Engineering, Civil Engineering, Chemical Engineering, Electrical Engineering and Computer Science, and Materials Science and Engineering.

The Integrated Studies Program (ISP)

ISP is an alternative freshman program that focuses on the role of technology in society and emphasizes learning by doing and self-discovery. Students investigate topics from a variety of viewpoints, and connections are made between material covered in the humanities, science, and engineering subjects. ISP incorporates sustained exercises in writing and speaking as part of its course entitled Technologies in a Historical Perspective. In addition to keeping journals, students write a description of their experience spinning wool, a "user's manual" for a nonlinguistic device to store data on strings, a group business plan, and a proposal for the design of a new automobile. Each group also gives an oral presentation on a group design assignment. As part of each assignment, students critique each other's work and write and revise several drafts before handing in the final document. This successful course provides a useful prototype for creating various types of communication-intensive freshman classes.

Technological Aids

MIT is continuing to develop effective and relatively inexpensive technological resources for the teaching of writing and speaking. In 1986, James Paradis and Edward Barrett developed the Networked Electronic Online System (NEOS), the first fully distributed computing environment for the teaching of writing, and the first university electronic classroom for teaching writing.[6] NEOS and the electronic classroom allow students to review each other's papers both inside and outside of the classroom, discuss a particular student paper displayed on a light valve projector, and access both specific course handouts and a limited but useful online style guide for technical writing.[7]

Barrett, Paradis, and Perelman in collaboration with Mayfield Publishing Company developed an extensive hypertext Electronic Handbook of Technical & Scientific Writing.[8] The Handbook covers most aspects of technical and scientific writing, including common document formats, documentation styles, and sections on English grammar, style, and usage designed specifically for nonnative speakers of English. Moreover, the Handbook's modular design allows easy development of course-specific versions containing examples and entries customized for specific writing assignments, class content, and professional styles.

MIT is also developing other hypertext and multimedia resources for the teaching of writing and speaking. Prototypes already have been developed for discipline-based, hyperlinked multimedia resources that include sample documents, commentary, videos, still images, and databases. Accessible through the Web, these documents provide students with specific report models (written and oral) and accompanying commentary.

Preliminary Design Phase

In response to the subcommittee's findings and conclusions outlined above, the Committee on the Undergraduate Program then charged the Committee on the Writing Requirement (CWR) to propose a new curriculum that would teach students to communicate effectively in a wide variety of contexts. The CUP instructed the CWR to formulate one or more formal proposals based on carefully developed design criteria and a thorough consideration of existing boundary conditions. These proposals would then serve as the starting point for further discussions with faculty, departments, and administrators, leading to a final curricular design that would be presented to the entire faculty for approval.

Design Criteria

The previous discovery process had already identified the basic elements necessary in the design of any undergraduate program that seeks to make proficiency in communication an integral and essential part of a technical education. Review of reports of similar efforts at other institutions have produced similar criteria.[9]

Substantial and Sustained Experiences in Both Speaking and Writing During Each Undergraduate Year

Sommers[10] and others have demonstrated the importance of sustained practice in developing fluency and competency in writing. Clearly, achieving proficiency in the various forms of oral presentation is an analogous process. Both activities are abilities rather than discrete bodies of knowledge. They are developed through practice and will atrophy through disuse. Individuals do not improve their writing, nor their skiing, nor their musical ability through one or two exercises a year. They improve through practice. Similarly, students will not learn to write well solely through one freshman expository writing class no matter how well it is taught. And whatever proficiencies they do develop will quickly fade unless they are reinforced through subsequent use. Writing and speaking, then, cannot be occasional undertakings, assigned at the beginning and, possibly, the end of a student's undergraduate career. They must be frequent and ordinary parts of an undergraduate's educational experience.

Useful Feedback and Instruction with Opportunities for Revision

Learning to write well requires not only practice but also effective instruction. Often, the best form of instruction is a well-crafted response to specific issues in a document. Such feedback does not just mark errors in grammar, spelling, and punctuation. It engages the writer in a dialogue about the issues raised in the essay and places matters of style within the context of the document's overall purpose.[11] Such feedback is most effective when it precedes and directs extensive revision of the original document. Sommers[12] and others have noted that the principal difference between the composing processes of experienced and inexperienced writers is that expert writers rework all elements of a document while novice writers mainly proofread and correct sentence-level problems. Effective feedback provides models of constructive responses to a document that students can use to help develop their ability to independently and comprehensively revise texts.

Integrated into All Parts of the Curriculum

Within the context of a technical university, instruction in professional communication is most effective when it is an integral part of professional technical education. To make instruction in writing and speaking solely the responsibility of liberal arts faculty ensures its marginalization. Furthermore, such a practice promotes the perception that proficiency in communication is completely separable from technical or even professional proficiency, a claim that is clearly not true in the workplace.

Component of Student Evaluation in All Parts of the Curriculum

Given the pace and pressure inherent in engineering education, students need to be motivated to learn to speak and write well. Effective communication in a technical curriculum must be valued and rewarded with incentives similar to those that exist in the workplace. Few managers would fail to note in a performance review an engineer's inability to write a comprehensible report. Engineering faculty should do the same.

Some engineering programs already incorporate clarity and effectiveness of expression into students' grades in some classes. There is a dramatic difference in attitude between students in writing practica attached to classes in which the writing quality constitutes a portion of a student's final grade and students in practica attached to classes that do not consider the quality of writing in evaluation. The first set of students largely view the writing instructor as a coach, an ally who can help then achieve their goal. The second group of students have the more traditional view of a writing teacher, as a gatekeeper to be gotten through (or around).

It will be impossible to make the ability to communicate course material clearly a significant component of a student's grade in all scientific and technical courses. Such a universal approach, however, is unnecessary. Our experience in the writing practica at MIT indicates that making the ability to communicate effectively a significant component of students' grades in only one or two classes sufficiently motivates them to improve their writing and speaking abilities substantially.

Boundary Conditions

In addition to describing the features necessary for successfully integrating communication instruction into MIT's undergraduate program, the committee also identified several restrictions that must be incorporated in any feasible design.

Overly Constrained Curriculum

MIT engineering students already have few opportunities during their four years to expand and enrich themselves through free electives. Engineering programs at MIT, as elsewhere, already mandate not only most of the classes students will take but also their sequence.

No Net Increase in Either Student or Faculty Workload

Writing is difficult and time-consuming work, both for students drafting papers and for instructors commenting on them. However, the current pace and pressure of both student and faculty workloads are already subjects of concern to the entire MIT community. The additional instruction must produce no significant increase in the overall demands placed on students and faculty.

Enhance Technical Education

Including instruction in writing and speaking as an integral part of engineering undergraduate programs has the potential of diminishing their overall excellence. As technical knowledge expands exponentially, engineering programs are already under considerable pressure to add more even more content and courses to their undergraduate programs. However, the experience of students, technical faculty, and instructors in the writing practica have demonstrated that assignments in writing and speaking can enhance and enrich student understanding of technical material. Rather than reducing the overall quality of technical education, assignments in writing and speaking must be designed to improve it.

Scalable Instructional Models

Some instructional models are extremely effective but impossible to expand to the scale necessary to accommodate large numbers of students. Eight to ten students meeting with one expert writing teacher twice a week in a writing workshop will almost certainly improve their writing skills. Such models, no matter how effective, are not, at most institutions, institutionally feasible. Effective methods of instruction must also be scalable. The writing practicum and the writing co-op already offer two such models and will serve as laboratories for the development of additional ones.

Modest Requirements for Additional Resources

In a time of fiscal restraint, instructional models must not only be scalable, they must be affordable. They cannot require significant new yearly expenditures.

Unresolved Issues

Discussions with departments highlighted several fundamental philosophical and design problems that need to be settled before the final formulation and implementation of the new curriculum: What are the most effective and educationally appropriate ways to integrate this additional instruction into all parts of the undergraduate program? What effect will the designation of certain subjects as communication-intensive have on student enrollment in other subjects? How can we best incorporate instruction in writing and speaking into the freshman year? Who should provide instruction and feedback to students, and how, if necessary, should these individuals be trained? Finally there is the fundamental issue of who is responsible for teaching students different types of communicative competencies. Is it solely the responsibility of the liberal arts faculty, or is it a shared obligation, with professional departments assuming a significant role in the teaching of professional communication?

Development Strategies and Procedures

Fortunately, discussions with faculty also revealed a consensus on the general objectives that were eventually embodied in the faculty motion on the communication requirement. Almost all faculty acknowledge that writing and speaking are essential skills and that MIT must do a better job in teaching them to our students. The devil, as usual, is in the details.

These discussions, however, also produced a consensus on a set of procedures and strategies to go forward in developing the new curriculum while recognizing the difficulty of resolving the specific points of disagreement.

Pilot Phase

The faculty have mandated a two-year period of curricular experiments designed to address the unresolved design issues and to inform the final formulation of the new requirement. These experiments will be coordinated by an Institute-wide faculty committee, but they will be developed at the grassroots of the institution by individual faculty and departments. Rather than seeking a single instructional model, these experiments will explore multiple design solutions that recognize and reflect differences in departmental and professional cultures.

In June 1996, MIT had already submitted a proposal to the Division of Undergraduate Education of the National Science Foundation requesting support for these curricular experiments. The NSF has funded the project, and the NSF program officer has provided valuable and useful advice and suggestions.

External Evaluation

During the discussions that preceded the awarding of the grant, the NSF requested that an external review panel evaluate the effectiveness of the curricular experiments. This suggestion was widely endorsed by the faculty as a way to ensure a valid and impartial assessment of each instructional model prior to the final formulation of the new curriculum.

Faculty Involvement and Faculty Development

The NSF program officer also emphasized the need to include formal structures for faculty development and involvement (long before similar concerns were raised by MIT faculty). Several different programs are already planned, including ongoing faculty colloquia on ways to integrate writing and speaking into undergraduate subjects. In addition to providing faculty with information and strategies that will help them integrate writing and speaking into their classes, we hope that these colloquia will develop and continually replenish a core group of faculty from all disciplines who will act as advisors and advocates for the initiative.

Technological Aids to Instruction

Finally, because writing instruction is inherently labor intensive, MIT will continue its efforts to use technology to make instruction more effective and more efficient. Computers can never replace teachers, but they can create virtual classrooms and writing centers that can expand and enhance a teacher's or tutor's effectiveness and accessibility.

Conclusion

MIT's experiences in developing a communication-intensive undergraduate curriculum may provide other technical and scientific universities with some helpful lessons for redesigning their own undergraduate programs to meet the new ABET criteria. First, MIT's experience with its own outcome-based writing requirement emphasizes the limitations of outcome-based measures in general, but especially in the evaluation of abilities, such as writing and speaking, that develop over time. Second, students clearly need better preparation in all forms of communication. Finally, the effort to provide this instruction within undergraduate technical programs will be difficult. It will be not only a complex and lengthy process in design, but also a delicate political and philosophical process in defining and redefining the roles and responsibilities of all faculty at science- and engineering-based universities.

Notes

Parts of this chapter are adapted from internal documents that were jointly authored with Professor Alan Lightman, Dean Kip V. Hodges, and Dean Rosalind H. Williams. I thank them for permitting their use.

* This committee became a division of ASEE in 1942 and merged with the Humanistic-Social Sciences Division to become the Liberal Studies Division in 1965. The name of the Liberal Studies Division was later changed to Liberal Education Division.

References

1. Committee on Educational Survey. 1949. Report of the Committee on Educational Survey to the Faculty of the Massachusetts Institute of Technology. The Technology Press, Cambridge, MA.
2. Zuckerman, Lawrence. 1996. Turn on a PC, tune in or drop out. But with passion. *New York Times* (28 January):3:12.
3. Metcalfe, Robert. 1995. Writing in the computer industry. Videotape of lecture. Massachusetts Institute of Technology, Cambridge, MA, 3 March.
4. Waitz, Ian. A. and Edward C. Barrett. 1996 Integrated teaching of experimental and communication skills to undergraduate Aerospace Engineering students. ASEE Meeting.
5. Williams, Rosalind. H., E. C. Barrett, and L. C. Perelman. 1994. The Writing Initiative: First year progress report. Massachusetts Institute of Technology, Cambridge, MA. 100 p.
6. Barrett, Edward C. and Paradis James G. 1988. Teaching writing in an on-line classroom. *Harvard Educational Review* 58:154–71. Committee on the Writing Requirement. 1997. Proposal for a new Undergraduate Communication Requirement. Massachusetts Institute of Technology, Cambridge, MA, 7 February. p. 12.
7. Barrett, Edward C. 1993. Collaboration in the electronic classroom. *Technology Review* 96 (2, February/March): 50–6.
8. Perelman, Leslie C., Edward C. Barrett, and Paradis James G. 1997. *Mayfield electronic handbook of technical and scientific writing.* Mayfield, Mountain View, CA. CD ROM for Windows and Macintosh.
9. Light, Robert J. 1992. Explorations with students and faculty about teaching, learning, and student life. The Harvard Assessment Seminars second report. Harvard University, Cambridge, MA.
10. Sommers, Nancy. 1994. A study of undergraduate writing at Harvard. Harvard University, Cambridge, MA.
11. Sommers, Nancy. 1982. Responding to student writing. *College Composition and Communication* 33 (2, May):148–56.
12. Sommers, Nancy. 1980. Revision strategies of student writers and experienced adult writers. *College Composition and Communication* 31 (4, Dec): 378–88.

Postscript

Author's Note: During the period between the writing and publication of this chapter, MIT successfully completed the development phase described in the text. On March 15, 2000, the MIT Faculty voted to establish a new communication-intensive curriculum integrating sequenced instruction and practice in writing and speaking in each undergraduate year and in all parts of the curriculum.

CHAPTER 6

JOHN BROWN

Refashioning the First-year Introductory Course on Communication Skills and Engineering Practice

Like most engineering colleges today, the School of Engineering and Applied Science (SEAS) at the University of Virginia confronts two novel curricular challenges. In response to the ABET 2000 proposals, we hope to reconfigure individual courses and thoroughly reshape the entire undergraduate curriculum. Yet such changes will occur in an environment facing tight constraints on resources given the present stringency in higher education funding across the country (particularly an issue at many state-supported institutions). Throughout all its units, the University of Virginia is seeking to boost institutional productivity in response to the combined challenges of fixed staffing and funding levels and anticipated enrollment growth. This chapter outlines one plan under consideration at the Division of Technology, Culture and Communication at SEAS to refashion its introductory writing and speaking class (TCC 101) to meet the opportunities posed by ABET within the constraints levied by the state legislature. Simply put, our goals are to:

- accomplish more teaching with fewer resources,
- improve the quality of our instruction in writing and public speaking,

- promote students' awareness of modern social, economic, and political issues relating to engineering practice,
- increase students' ethical awareness, and
- decrease the attrition rate of first-year engineering students (chiefly a problem of transfers into UVA's liberal arts college).

Fundamental curricular change would seem difficult to accomplish without new resources. But the ABET initiatives challenge us all to reconsider our means, ends, and outcomes. As this chapter will outline, we may have devised a way to literally do more with less.

For over twenty years, my department (known first as the Humanities Division and now the Division of Technology, Culture and Communication) has taught a basic introductory class in technical writing, required of almost all undergraduates in their first semester, although a significant minority are exempted with high school Advanced Placement credit.[1] In its present form, TCC 101 is a labor-intensive course. Taught only in the first (Fall) semester, it averages 380 students, divided into fourteen sections. Each section thus has one professor and twenty-seven students. The sections do intensive writing and speaking exercises within a curriculum whose intellectual content varies with the interests of each professor. All sections cover technical writing in such genres as technical descriptions, proposals, abstracts, and memoranda. The class includes exposure to other writing genres, and has significant public speaking requirements.

In its present form, the course is very popular with students, while the TCC faculty are mostly pleased with its structure and results. In particular, students enjoy the advantages of small classes and the chance to interact with professors rather than teaching assistants. The TCC 101 instructors themselves enjoy the satisfactions of teaching first-year students—a hardworking and enthusiastic group. So why change?

Simply put, our projected enrollment growth coupled with flat staffing levels will soon present overwhelming problems for the course in its present structure. When 101 began fifteen years ago, the student-teacher ratio was 23 to 1. Today we are at 27 to 1, with classes projected to reach over thirty students each within the next five years or so. One cannot teach communications and technical writing skills via lectures; these topics demand drills, repeated drafts, frequent student presentations, and detailed feedback from instructors. Intensive instruction in writing and public speaking simply becomes impossible with such large classes—no matter how good the teacher is.

The swelling class size also provides the impetus to reconsider a much older question "Is the school deploying its resources wisely when it uses Ph.D. faculty to teach composition and public speaking, without any resort to teaching assistants?" Using professors, rather than graduate teaching assistants, for instruction in composition classes reflects a high regard for teaching. But in this instance, the practice is downright expensive, especially given that trained TAs should be able to recognize the passive voice and poor topic sentences as well as a full professor can. Heightening the expense (at least from the perspective of the engineering school) is the fact that over 30 percent of our first-year students transfer out of the school sooner or later. If a less expensive but effective alternative is feasible, why employ professors to teach this group when nearly a third will leave the school anyway?

Our key problem is the enrollment growth. Given flat staffing, the swelling class size nearly demands a change in pedagogy. While the "no TAs" dictum for my department has been a popular marketing plank with deans and parents, can we afford this policy any more? And is this exclusive reliance on Ph.D. faculty justified on educational grounds?

Most likely, colleagues at other institutions share the problems I have sketched here. At Virginia I believe that the confluence of these difficulties argues for a through restructuring of our composition course. Equally, the timely appearance of the ABET 2000 goals provides a framework for change. Specifically these elements of the ABET 2000 criteria serve as developmental goals for a new version of TCC 101:

- an ability to function on multidisciplinary teams,
- an understanding of professional and ethical responsibility,
- an ability to communicate effectively, the broad education necessary to understand the impact of engineering solutions in a global/societal context,
- a recognition of the need for, and an ability to engage in, lifelong learning,
- the broad education necessary to understand the impact of engineering solutions in a global/societal context, and
- a knowledge of contemporary issues.

In its present form, TCC 101 does meet most or all of these goals. But with upwards of ten different professors teaching the various sections each year, the actual content and skills conveyed by the course has come to vary. So under the spur of the problems described earlier, the ABET goals provide a timely directive for a new beginning for a course that itself opens Virginia's undergraduate engineering curriculum.

First let me describe the new structure proposed for this course, then I will turn to its content. The plan calls for a single professor to teach the class with the assistance of six teaching fellows. The fellows will be education school graduate students in the Master of Arts in Teaching degree program. All 500 students in the entering class will take the course; none will receive advanced placement exemptions. The exemption has been difficult to justify in any event for a technical writing class as few students with AP credit received any instruction in technical communications in their high school English classes. Now that we are also seeking to provide a detailed introduction to engineering practice, the value of this class to all entering students is undeniable. Given their numbers, however, we will divide the first-year class equally between the fall and spring semesters.

The professor will give one lecture each week to 250 students, largely on topics that place engineering practice in its professional, ethical, political, economic, and social contexts (more below). That lecture will also serve as an intellectual framework upon which to hang communications instruction. Then the teaching fellows will take each section for two meetings a week. We plan to give every fellow two sections, each with twenty-one students. Thus the student-teacher ratio will fall by nearly one-third. Of itself, this should improve our ability to teach composition. The sections will do intensive work in technical writing genres and in public speaking (both formal speeches and extemporaneous presentations). Topics and issues raised in the professor's weekly general lecture will provide a common intellectual basis for many of these composition exercises.

The general lectures should further a number of essential goals. To date, our faculty have conceived of TCC 101 primarily as a skills course, although each instructor also has particular intellectual goals and substantive topics to convey. In its new incarnation, the course would still balance skills and knowledge, while invigorating both aspects. In particular its intellectual content should focus on providing a comprehensive and detailed introduction to the engineering profession.

That introduction must serve two ends, one narrowly pragmatic and the other broadly idealistic. While our first-year students come to us with great enthusiasm, they have little substantive knowledge about the disciplines of engineering, their characteristic concerns, or their problem-solving methodologies. Yet they must declare a major within their first year. Coupled with other classes in the first-year curriculum, this revised course will better provide this orientation to engineering practice and the issues of selecting a major. Beyond that pragmatic need, we also need new means to retain students' initial enthusiasm for engineering to achieve our idealistic goal. Too many of our students become dispirited with engineering. During their first three semesters most of their classes deal with theoretical and scientific foundations that seem far removed from real problems or design work. As a result, many transfer out of the school. Thus, in revising this class, we want to capitalize on their initial enthusiasm, for it is too valuable to squander.

Given this multiplicity of goals, we are renaming our composition class, calling it "The Worlds of Engineering Practice." While writing and speaking instruction will remain the course's dominant concerns, they will be subsumed within a larger intellectual content set by the professor in the weekly general lecture (supplemented by guest speakers as appropriate). I cannot describe the specific content of the lectures at this point, as the course is still on the drawing board. Furthermore my department includes faculty from a range of disciplines including English composition, history, sociology, literature, anthropology, and psychology, and staffing for 101 remains undecided at this point. Although all focus on technology studies, they would approach this course in different ways. Notwithstanding these differences, all will teach it as an introduction to engineering and its disciplines, the relations between technology and society, the nature of ethical theory and practice, and issues in communications.

My focus in teaching and research is the history of engineering and technology. I will describe here how ideas and perspectives from those fields could structure this new introduction to engineering practice. While specific course readings remain undecided, I would likely include such titles as Arnold Pacey's *The Maze of Ingenuity* or David Billington's *The Innovators*. To complement these historical studies, students would also subscribe to *Newsweek*. Incorporating a weekly newsmagazine into the class will allow the professor to inject a broad array of current issues relating to engineering practice into the course's lectures, tying the past to the present. This would also provide a vehicle for writing, researching, and speaking exercises in the sections on topics of clear relevance to the students' careers. While reading *Newsweek* will not, of itself, be a dominant element of the class, it will underscore engineers' need to have a broad familiarity with the world and contemporary issues.

The teaching fellows will use section time for specific instruction in composition and technical writing genres. This skills instruction could take place in the context of two larger group projects, both seeking to convey knowledge about engineering practice.

In my blueprint for the course, during its first six weeks students working in pairs will conduct the "Engineering Career Orientation Project" (ECOP), a project devised by my colleagues Mark Shields and Bryan Pfaffenberger. Fully described in another chapter in this volume, the ECOP incorporates a range of assignments on communications while it promotes students' knowledge of the engineering profession, its fields of study, and the career paths open to practitioners. Given that most entering students know comparatively little about engineering in general, prospective majors, or their career prospects, they find the ECOP an intensive and rewarding experience.

Knowledge gained through the ECOP helps students to declare a major during their first year, our chief pragmatic concern for the course. The second group project that I would incorporate into 101 would build on students' enthusiasm and idealism by providing a broader view of engineers' work and its place in the larger society. Tentatively called "Perspectives on Engineering" (POE), this team-oriented project seeks to accomplish a number of goals:

- give students a deeper understanding of their probable or possible major fields,
- trace the relations of those fields to other disciplines and professions,
- underscore the nontechnical and multifaceted elements that contribute to the success or failure of any given technology,
- provide a basis to discuss ethical considerations in engineering practice, and
- acquaint students with the real world challenges and accomplishments of engineers, both historic and contemporary.

The POE would structure the sections' work during the last nine weeks of each semester. Each discussion section will break down into five teams, with each team focusing on an engineering discipline (civil, mechanical, etc.). Working from a list provided by the professor, each group will select a triad of technologies to research and report on. The triads will include an example drawn from history, a problematic or failed case (either historic or modern), and a contemporary example of a technology—all in the same engineering field. Examples of triads might include: the DC 3, Comet jet, and Boeing 777 (aerospace), the Brooklyn Bridge, Tacoma-Narrows, and Channel-tunnel (civil), and IBM 360, IBM PC, and the World Wide Web (electrical engineering).

During the semester, setting up the POE will require a week, each triad element will have two weeks, and sections will use the final fortnight (weeks eight and nine) to draw the perspectives together.

One of the POE's chief intellectual goals is to underscore the varied elements that affect the success of any engineered product or technology. To demonstrate that point, for their first POE assignment each team member will concentrate on a particular perspective or field in researching each triad element. The perspectives will be scientific-technical, social-cultural, economic-business, legal-regulatory, and individual-biographical. At the end of week one in each triad case, the students will prepare memoranda to their teams (and their section leaders) outlining how issues within their specific fields affected the overall development of that case element in the triad.[2]

Following up on that research will be another individual writing assignment, due at the end of the second week of each triad element. In this three-to-four page (750–1,000 words) investigative report, each student will discuss the relationship of his or her field to the other perspectives and to the overall success or failure of the technology. Team members will be encouraged to discuss their approaches and findings collaboratively, although the actual report will be an individual product.

Each two-week triad segment of the POE will also have a team project. This will be an integrated verbal and visual description of the novel or defining design concepts of that technology. The goals of this assignment are to foster students' sense of visual thought in engineering, to require them to integrate different descriptive tools, and to hone their analytical skills in selecting the defining design concepts. Execution of the visual elements in the presentation will have to meet high standards, while the accompanying text would justify the selection of particular views and perspectives while also providing ancillary description.

With each two-week unit in a triad, students will be expected to show improvement and increasing sophistication in their individual and group assignments. After studying all three triad technologies, each team will have a group project due in week eight of the POE. In this ten-minute oral presentation, the teams will seek to integrate their POE research into an overall portrait of their discipline, its evolving concerns, its relation to other fields of engineering, and the ethical duties and dilemmas raised in their research. The POE will end with an individual writing assignment by each student—a personal evaluation of his or her effort, an assessment of the team's dynamics, and a reckoning of the project's overall success.

The POE project should bring the different disciplines and the actual work of engineering into sharper focus for students. The assignments should promote their writing and speaking skills, particularly since many tasks repeat with each triad unit. The POE's overall success (as well as that of the Engineering Careers Orientation Project), however, will depend on skillful instruction, provided both by the professor's weekly lectures and by the teaching fellows' work in the sections.

The success of this venture largely rests on the teaching fellows. Their somewhat unique title acknowledges that they will have a much broader role than the old-fashioned TA. In addition to running two discussion/recitation classes every week for each section, they will hold weekly office hours and attend the general lectures. They will also handle all the grading, within standards mandated by the professor. In the sections, the fellows will discuss proper communications techniques, lead workshops, direct discussions of readings, and oversee oral presentations. The job entails a fully rounded portfolio of teaching duties, hence the name "teaching fellows."

We are confident that we can secure qualified graduate students for the fellows' positions. The Master of Arts in Teaching program at Virginia is a nationally ranked course of study for teachers seeking graduate certification to teach in high schools and community colleges. Many come to the program with significant teaching experience, and a number of these MAT students choose to specialize in English education. They represent a notable resource that is underutilized at present because they lack sufficient teaching opportunities in the education school. Thus the MAT students would benefit professionally (and financially) from the opportunity to have significant teaching experience as part of their degree program.[3]

Given the quality of these MAT students (as well as their need for opportunities to teach in such a capacity), we expect to award the fellows' positions on a competitive basis. They will not have engineering or STS training *per se*, yet their selection will depend on having the requisite knowledge and experience in English composition. In this sketch of the course, the ECOP and POE structure much of the sections' work. Nonetheless blocks of time would remain available for basic instruction and exercises in composition, grammar, and public speaking. With close oversight and support by the coordinating professor, we are confident that the MAT students will do a fine job with the discussion sections.

To provide the necessary oversight, the 101 professor will have no other teaching commitments during the semester. In addition to the weekly lecture, the instructor will join section meetings in rotation. She or he will meet with all the teaching fellows every week to plot out the work of the discussion sections and to learn of any problems. By holding extensive office hours, the 101 professor will also field students' questions and problems directly. The professor will create and provide all handouts needed for the ECOP and the POE as well as providing handouts, transparencies, drills, and exercises for writing and speaking instruction. The job will involve far more than simply reading a lecture a week to a crowd of 250 students.

Those general lectures will provide the crucial foundation and thematic unity for all the elements of the course. The course I have described here might seem likely to disintegrate at any moment, with the ECOP, POE, skills instruction, readings in *Newsweek*, and course texts all apparently heading in different directions.[4] But the lectures should serve as the class's steady rest, introducing the disparate elements and drawing course themes together. In addition to introducing the ECOP, POE, and key genres of technical writing, those lectures could cover such topics as:

- the disciplinary character of engineering practice,
- the history of engineering,
- the relationship of engineering to business, especially in a corporate environment,
- ethical considerations in engineering,
- relations between government and engineering practice,
- social and cultural influences on engineering design,
- the varied consequences arising from technological innovations for engineers, consumers, the economy, and the society and culture.

Maintaining the interest of 250 students in a large lecture class will call for skillful and energetic lecturing—this is no place for a neophyte. But with sufficient preparation, oversight, and skill the lectures, projects, and section meetings should coalesce into a broad and valuable introduction to engineering practice and technical communications.

If successful, the combination of these elements should result in a course that increases the quantity and the quality of instruction, while releasing faculty for other research and teaching duties and lowering overall staffing costs. This innovative interdisciplinary liaison between the education school and the School of Engineering will offer unique benefits to all participants. This new version of TCC 101 will also promote students' abilities in six of the eleven target areas outlined in the ABET 2000 guidelines. Quite literally we can do more with less.

References

1. Initially this class was a two-semester sequence, required of all students without exception. Over the years, budget and curricular retrenchments similar to today's challenges first caused the course to shrink to one semester. Then students receiving high scores on the AP English exam were exempted from 101 altogether, although few had studied technical writing in high school. Today roughly 100 students (20 percent of the first-year class) are awarded AP credit in place of this class.
2. As they move on to the other two technologies in their triad, they will take different research fields, so that by the end of the POE they will have developed an understanding of technological change from three different perspectives.
3. We are exploring mechanisms whereby the education school may offer a graduate practicum (with instruction and credit) for those MAT candidates teaching in 101.
4. Actually, these different elements will not be disjointed in practice, because the content and timing of all these assignments and projects will originate with the professor. A single overarching syllabus will establish the requirements and due dates for the ECOP, the POE, course readings, essays, and final exam. Consideration of specific issues and articles in the *Newsweek* readings will be delayed a week after the issue date, allowing the professor and the section leaders to select relevant contemporary issues in advance and then weave these concerns into their lectures and discussions. In sum the different elements of the course will be arrayed like the rim of a wheel, with the professor as the hub, and the syllabus, handouts, and section leaders will serve as the spokes that tie the center to the students.

CHAPTER 7

CHARLES C. ADAMS

The Role of the Humanities in Distinguishing Science from Engineering Design in the Minds of Engineering Students

Introduction

Engineering problems differ from scientific problems. Therefore the proper identification, formulation, and solving of engineering problems require an understanding of the distinction between engineering and science. In this chapter it is argued that the humanities and social sciences (HSS) play a major role both in the distinction between engineering and science, and in clarifying that distinction in the minds of engineering students.

Modern engineering is a human cultural activity that involves an interplay between theory, experiment, and imagination, in which human beings form and transform nature, for practical ends and purposes, with the aid of tools and procedures.[1,2] Those "practical ends and purposes" involve human society in all its multifaceted complexity. Thus engineering design requires an integrated and holistic perspective on reality before engineering problems can be properly formulated and solved. The first two sections of this chapter discuss that distinction between engineering (or technology) and science, and examine the difficulties that ensue when that distinction is not made. To demonstrate the integrative character of engineering design, the next section looks at the kind of expertise required by a design team that sets out to solve the problem of transporting humans over long distances. Then four basic principles for guiding curricular and pedagogical reform are identified and discussed. Finally, a number of specific examples of integration of the HSS in the engineering classroom are considered. Throughout, an attempt is made to show that (1) the identification, formulation, and solving of engineering problems demand a holistic perspective that considers far more than just the narrow, "technical," dimensions of a problem, and (2) the HSS can play a major role in developing that holistic perspective in the minds of engineering students, even—and perhaps especially—in the engineering classroom.

The Distinction between Technology and Science

If engineering design requires an integrated and holistic perspective on reality, why is it that the tendency in modern technology seems to be in the opposite direction? That is, why is it easy for engineers to have their perspective so narrowed that important aspects of design problems are overlooked? Dams that disrupt the ecological balance of a region, VCRs that are too difficult for the average person to program, and the addressing of age-old medical problems with genetic engineering solutions that raise even greater ethical problems are just three kinds of design failures that result from what has been called the *narrowness tendency in design*.[3] Schuurman[4] blames this narrowness on the confusion of technology with science and the tendency for the methodology of science to be employed in situations where what is needed is not science but technology.

Invention, characterized by discontinuity, is central to technology. Imagination, as well as integration, play leading roles in engineering design. In contrast, abstraction and deductive logic, characterized by continuity, play leading roles in science. Schuurman distinguishes between the scientific method on the one hand, and the "technological scientific method" on the other hand:

> The method of science is that of analysis and abstraction. This means that the field of investigation is dissected: the particular, the concrete, the here and now, and the milieu are set aside in order to facilitate penetration to the laws and to the

order of the reality concerned. This scientific method leads to general or universal, lasting, and coherent knowledge of the law for a particular area of the reality being investigated.

The key point is that the scientific method is one of abstraction. It seeks general, not particular knowledge regarding the nature of reality. It ignores the peculiarities of the environment (milieu) and focuses on the one aspect of which knowledge is sought. This is almost the inverse of technology. In technology a solution is sought to a real problem. The problem is not abstract but holistic and multifaceted. The solution must work in that *real-world* environment and thus also must be holistic and multifaceted. The process of going from the problem to the solution, however, involves what Schuurman calls the *technological-scientific method* (in distinction from the scientific method) whereby a real-world solution is designed to solve the real-world problem:

> Now, the *technological-scientific method* is not to be equated with the scientific method, for while the latter seeks knowledge of the laws which hold for reality, the former is concerned with the formulation of the laws for both production and the product, that is, the laws for fashioning and for what is to be fashioned. The meaning of the technological-scientific method is technological in character, even though the method could not exist without a scientific foundation.

The fact that the technological-scientific "method could not exist without a scientific foundation" is important for describing how the scientific method and scientific knowledge are "reflected" in technology, both in the design stage and the final product. For example, the engineer analyzes the problem, breaking it into parts (abstraction) and seeking a generalized design solution for each part. A "generalized design solution" is a component part or function which has a degree of universality or standardization:

> Technological component functions, then, are attained through analysis. Abstraction sets each function off by itself, apart from others, so that the technological solution for a component function may be found in this way. The result consists of elemental building blocks which are neutral in their destinations and therefore universally applicable—for example, the screw, the rivet, and the welding. In this universal applicability, universal scientific knowledge is projected into technology as the result of analysis and abstraction. In other words, the universal utility of the solution to a component function is an analogy of scientific knowledge in technology.

In addition to universal utility, Schuurman sees other analogies of science in technology. For example, the desired durability of tools, in space and time, is seen as an analogy of durable knowledge of reality in science. Likewise, the integration of component parts and component functions in the production process is analogous to the coherence of scientific knowledge.

Notwithstanding these analogies, Schuurman insists on the distinction of the scientific method from the technological-scientific method, the distinction of science from technology. He points to the belief that technology is "applied science" as a clear example of that confusion, that failure to distinguish science and technology. Note, in the following quoted paragraphs, how that confusion leads to the "theoretification" of technology, a stifling of technological innovation.

> Those who conceive of technology as applied science are of course convinced that the so-called technological-scientific method should hold sway by itself. They make technological development a mirror of natural-scientific knowledge. This leads to the theoretification of technology, that is, to its being reduced to an exercise in theory, in the planning stage. The result is that human creativity as manifest in invention is precluded, and human freedom in technological forming is destroyed. The trait of continuity in natural-scientific knowledge as knowledge of the determined aspects is projected, as it were, into technology, rendering technological development a determined development.
>
> Technological development is stifled by the controlling influence of this *rationalism* in technology; the possibility of further disclosure is discouraged. As the designer's impulse toward theoretical control gains ground, labor declines in importance and status, meaningful initiatives are thwarted, the achievement of breakthroughs and new discoveries is rendered difficult or impossible, and the disclosure of meaning in technology is impeded. In other words, *the absolutization of technological-scientific thought resists the disclosure of meaning through technology.*[4]

Distinguishing Engineering from Science: The Two-cultures Problem

It may be argued that the two-cultures problem in undergraduate engineering (UE) education is an inherited one. That is, the *original* two-cultures problem, as articulated by C. P. Snow,[5] is between the culture of science and the culture of the humanities (literature in Snow's discussion). Because of the fundamental role of natural science in technology, and in engineering education in particular, the two-cultures chasm has split technology and the humanities just as it has science and the humanities. This is a point made in Monsma, where it is argued that scientism, that is, "regarding science as the standard to which reality should conform and by which various cultural manifestations are to be judged" (p. 98), is the source of technology's corruption.

In scientism there is an overriding proclivity toward substituting the analytical results, models, and abstractions of science for reality itself. Two characteristics of science have had a particularly strong impact on how modern technology is done, and when the practice of science degenerates into scientism, their negative impact on technology is hard to exaggerate. These characteristics are science's method of analysis and synthesis, and its abstract, theoretical, explanatory nature.

It is argued that although the method of analysis and synthesis is efficient and practical, it "exacts a price: the loss of a broader, holistic perspective."

> The danger is that no one sees the entire picture, no one asks how the technological procedures and objects being developed fit into society, how they affect the people who are to use them and the natural environment from which they come and in which they will operate.

Science's abstract, theoretical, explanatory nature contrasts with technology, which is "concrete, specific, and practical." Thus technology

> deals with real-life situations: with living persons and nations, and with concrete, material reality. Thus technology, responsibly done, must be holistic. It must ask about its effects on people, cultures, nations, and the natural environment; it must be based on normative principles. The hegemony of science in technology, however, discourages this sort of perspective. The mind-set and approach of modern science tells technologists that they need not concern themselves with the broad picture.

Lynn[6] makes this same point when he asserts that UE education "has moved monotonically to the development of courses and entire curricula that give primary (and in some cases exclusive) emphasis to analysis" (p. 150). He points to the fact that practically all engineering faculty have had doctoral level science, applied science, and engineering training, the goal of which is to support research undertakings. The nature of engineering research, particularly that funded by federal agencies and foundations, has been analytical. He also argues that faculty promotions have come to depend more on the success of the analytical content of their research than anything else. He goes on to make a case for the importance of design, as distinct from analysis, in engineering education.

Ladesic and Hazen[7] make the same point, but from the point of view of industry. They argue that UE education has "drifted away from an emphasis on empirically based systems design activities that accentuate engineering judgment and creativity" (p. 23) toward a textbook and computer-aided theoretical approach. They argue that this has resulted in a different kind of two-cultures problem, the culture of the classroom vs. the culture of the workplace, which is as problematic and intense as the traditional two-cultures problem.

96 | Liberal Education in Problem Formulation and Solution

Billington[8] contends that the work of engineers is misunderstood, that it involves more art than science, and that instead of being opposed to the liberal arts, it is one of them. Distinguishing engineering design from scientific investigation, he writes that "Engineers are about as dependent on modern scientific theories and discoveries as poets are on the hypotheses of modern linguists" (p. 87).

> Science is discovery, engineering is design. Scientists study the natural, engineers create the artificial. Scientists create general theories out of observed data; engineers make things, often using only very approximate theories.

Billington gives an example of how the hegemony of science in engineering has resulted in engineering failure. He points out how, early in the twentieth century, many American civil engineers embraced the mathematical "deflection theory" as a guide to designing bridges. This led to the design and construction of increasingly thin suspension bridge decks. In November of 1940, as a result of depending only on deflection theory (as well as ignoring dynamic effects), the Tacoma Narrows Bridge collapsed:

> A century before, the great engineers Thomas Telford (1757–1834) and John A. Roebling (1806–1869) had warned of the danger of thin decks. But *engineers who see their profession as an applied science* tend not to look back. Engineers as designers of large-scale works for society must.[8] (emphasis added)

Billington also points to the notion that there is "one best way" to solve an engineering problem as the confused influence of science in engineering. He contends that, like artistic expression, there are always an infinite number of legitimate paths to the desired end in engineering design. Likewise, he argues that a scientized technology "seems to dictate that artistic expression or personal taste are 'frills' that engineers must do without." This, he says, "is the logical outcome of the applied science view of technology."

Designing an Automobile

Consider the modern automobile, what it is, how it functions, and how it is designed. To fully understand an automobile, its function and design, one must have specialized knowledge in a wide variety of areas. Such expertise is impossible for any one person living today. Thus, instead of a single expert, a team of experts is required. To appreciate the range of expertise required of the team, it is helpful to consult a schema such as that shown in Table 7.1.

TABLE 7.1. Aspects of Reality

Aspect	Core Meaning
Fiduciary	Trust, faith, reliability: "the varying levels of reliability or trustworthiness a thing or person may have."[11] Manifest in religious faith, the trust that a child puts in a parent, or the reliability of a machine.
Ethical	Caring, love: as in parental love, love for one's country, caring for the environment (i.e., environmental ethics). Note that the term "ethical" is here not being used in the broad sense of morally right or wrong.
Juridical	Justice: fairness, "a well-balanced harmonization of a multiplicity of interests, warding off any excessive actualizing of special concerns detrimental to others."[12]
Economic	Stewardship: frugality and efficiency in managing, preserving, and caring for scarce resources.
Social	Social intercourse: the interacting of individual humans that characterizes the formation and functioning of groups.
Semantic	Symbolic meaning: signification; the making and interpreting of signs, lingual activity.
Aesthetic	Allusivity: nuancefulness, characterized by "a sheen of hinting ambiguity."[29]
Historical	Cultural-formative, technological: the human ability to form new things (artifacts, ideas, institutions, etc.) from what already exists.
Analytical	Logical distinction: cogitative analysis, the observation of diversity.
Sensory	Feeling, perception; the functioning of human and animal sense organs.
Biotic	Life.
Physical	Mass/energy.
Kinematic	Motion: continuous flowing.
Spatial	Continuous extension.
Quantitative	Discrete quantity: how much of something.

Source: Adapted from Kalsbeek;[9] Seerveld;[10] Clouser;[11] and, originally, Dooyeweerd.[12]

The field of study investigating the spatial aspect of reality includes geometry and topology and is considered a branch of mathematics. Expertise in these areas is required, for example, to understand, design, communicate, and properly fabricate the shape of the automobile body.

The kinematic aspect of the automobile includes not only the obvious motion of the car and its wheels as it moves along a road, but also the controlled vibrations of the interior compartment and the high frequency cyclical motion of the pistons, connecting rods, cams, gears, and other components of the drive train. A specialist in kinematic analysis and synthesis (a branch of mechanical engineering) is needed for this aspect of the automobile's design.

The physical aspect of the automobile is perhaps the one that is most apparent. The mass of the component parts and the system as a whole, the rate at which energy is used, the power produced by the engine, and the properties of the materials that make up the automobile's various subsystems are just a few of the physical attributes of the artifact that require the expertise of someone trained in physics. Because the physical aspect is so important in virtually all engineering design, the study of physics is an important part of any engineering curriculum.

The physical is the last subject function of the automobile. That is to say, the automobile is directly subject to physical law, as well as the laws for the kinematic, spatial, and quantitative aspects of reality. In the aspects beyond the physical, the automobile functions as an object. It is not directly subject to, for example, biotic laws, but nonetheless plays an important role with respect to the biotic functioning of those beings who are subject to biotic laws. The previous sentence may be repeated ten times, by simply substituting for the word "biotic," one at a time, the ten aspects found above the biotic in Table 7.1. The following eleven paragraphs offer brief descriptions of those object functions and point to the need for expertise, beyond that of mathematics and physical science, on the team mandated with designing an automobile.

There is a biotic function to an automobile: it must preserve the life of its occupants even while transporting those occupants at life-threatening speeds. Thus the chassis must have sufficient strength not only to contain a number of human occupants while traveling on the highway, but, as much as possible, to protect the lives of those occupants in the event of a collision.

This and the design of seatbelt and air-bag systems are only the most obvious examples of where expertise is required to understand the biotic response of human beings to kinematic and physical trauma. Persons riding in an automobile ought to be able to do so in relative comfort. Thus the sensory aspect must be considered when designing the artifact. Not only ought the seating be comfortable (sense of touch), but the experience ought to be visually and sonically comfortable. The design of the windows and the sound insulating quality of the interior are among the systems where the expertise of the physiologist and psychologist is required.

Because analysis is fundamental to understanding every one of its aspects, the automobile has an important analytical object function. But it is not only in the designing of the artifact that analysis plays a role. Driving an automobile requires a person to make decisions. Often those decisions become almost instinctive, as in the operation of the turn signals prior to making a right- or left-hand turn. But the automobile must be designed for situations where decisions can be made that are not of the mere stimulus-response variety but are, in fact, based on logical distinction. Whether or not a trained logician is required to provide the necessary expertise is not important. What is important is to recognize that no artifact, certainly not an automobile, can be designed without the design team having some expertise in logic.

As mentioned earlier, the historical or cultural-formative aspect is the founding function of an automobile. The expertise required here is engineering design in general. Part of that expertise, however, involves a knowledge of how the problem, or similar problems, were previously solved. Thus, the original designers of the automobile needed to understand how the problem of transportation had previously been confronted, i.e., they needed to know about the history of the bicycle and the horse-drawn carriage. A modern automotive design team needs to know the history of automotive design if it is to avoid "repeating the same mistakes" or "reinventing the wheel."

The aesthetic aspect of engineering design involves technological allusivity and the following definition of the latter is given: Technological allusivity in engineering design is achieved when the design successfully suggests a delightfully harmonious interaction, at the human-technical interface, whereby the product dissolves into an extension of the user.[14] It is sufficient here to simply call this "user-friendliness" and to suggest that it is the domain of the expert in ergonomics. But there are other ways in which the automobile functions aesthetically. The allusion of the tail fins on the 1959 Cadillac to rocket plumes, the allusion to wealth and prestige in the design of the Lexus, and the general allusion to personal independence in all automobiles when compared to mass transportation, are examples. While the design team may or may not require the services of a trained aesthetician, some expertise in aesthetics ought to characterize the members of the team.

How does an automobile communicate to its driver that fuel is running low, that it is traveling at excessive speed, that the cooling system is being overly stressed, or that some other kind of mechanical malfunction is occurring? How does the automobile communicate to those driving behind it that it is turning or stopping? How does the designer of the automobile communicate to the manufacturer how to build it? How does the manufacturer communicate to the potential customer the properties of the automobile that make it worth purchasing? To deal with these questions is to deal with the semantic aspect of the artifact. Perhaps a trained linguist is not required on the design team, but certainly some expertise in semantics is required of a variety of members of the team.

This section began by considering that an automobile is a socially qualified artifact and that the first kind of expertise needed to design an automobile is that of a sociologist, someone who has an understanding of the *need for,* and the *purpose of,* the artifact.

That an automobile has an economic aspect needs little argument to demonstrate. The purchase price, the fuel economy, and the recyclability of the parts are only the more obvious areas where expertise in economics is required by the design team.

The juridical aspect of the automobile is closely tied to the various laws that govern its design and use. There are laws that govern the safety of automobile occupants, govern the fuel economy of the vehicle, and govern the amount of pollutants it may expel into the atmosphere. There are also laws having to do with licensing, with insurance, and with traffic control. And, of course, there are laws regarding the operating of any business. Thus the design team requires some legal expertise. It should be added, however, that there are other questions of justice that the design team ought to deal with. These have to do with access to automotive transportation. For example, does the design allow for sufficient access by those who are physically handicapped, by those in deprived socioeconomic groups, and by those without significant technical training?

It may be argued that the properties of technological artifacts go beyond those of the physical and economic and include ethical properties. Ethical properties are those characteristics that incline the user in a direction that has to do with the love or caring for "the other." The "other" usually refers to other persons, but may also refer to things such as the environment, animals, or even artifacts themselves. There are many other ways in which the design of an automobile involves ethical concerns, and it soon becomes clear that ethical concerns are often interwoven with juridical, economic, and social concerns. Consider, however, just one ethical property that appears to be fundamental to the nature of the automobile. Many people report that they are at their misanthropic worst when behind the wheel of a car in a situation of heavy traffic. Somehow the design of the automobile is such that it alters the immediate relationship of the driver to the drivers in the surrounding traffic. The philosopher Marcel would say that the automobile facilitates the "spirit of abstraction"[15] such that the other drivers are seen abstractly as operators of machines, and not as fellow human beings. Dealing with this problem requires expertise, at least, in psychology, sociology, and ethics. Just as it may be necessary to include bio- or medical ethicists on a design team for a research hospital, so it may also be necessary to include some persons with expertise in ethics on the automotive design team.

Finally, an automobile functions objectively in the fiduciary aspect. The trust one puts in the braking system as one accelerates to high speeds on a highway is an example of that fiduciary aspect. Thus the design team needs to be concerned that the artifact it produces is worthy of that trust. In other words, the design team must be concerned for reliability and will likely include a person whose area of specialization is reliability.

The purpose of this lengthy exposition regarding the expertise required of an automotive design team is to demonstrate the necessarily holistic character of engineering design. There is no dualism between technical and HSS here. Fifteen aspects of the automobile and the automotive design process have been considered, some of which have been traditionally classified as "technical" and others of which have traditionally been classified as "HSS." The discussion shows, however, that they are all interrelated and of importance to the goal of designing an automobile.

Principles for Guiding Curricular and Pedagogical Reform

To overcome the two-cultures problem, clarify the distinction between science and engineering design for engineering students, and facilitate a holistic approach to engineering problem solving, engineering educators must be guided by four general principles: (1) holism (2) multidimensionality (3) the integrative character of design, and (4) harmonizing of abstraction and integration.

Holism: The Unity of Experience

Human beings experience reality in its wholeness. By means of abstraction it is possible, in one's mind, to break wholes down into parts. By abstraction it is possible to investigate the horizontal and vertical motions in the flight of a baseball. It is possible to consider a particular individual man as a father, a husband, a machine operator in a local factory, a member of a local board of education, a deacon in a local church congregation, or as an active member of a particular political party. A bottle of vinegar may be seen in terms of the chemical constituents of the vinegar: water and acetic acid. Acetic acid may be considered to be composed of hydrogen, carbon, and oxygen atoms that, in turn, are composed of protons, neutrons, and electrons. And reality itself may be abstracted into its irreducible aspects: quantity, space, motion, mass/energy, life, etc. The process of abstraction makes possible a deeper and more comprehensive understanding of the phenomena of experience, an understanding that enables human beings to alter and give form to their environment. But the process of abstraction also distorts the perception of reality by fragmenting experience.

Consider all the component parts of an internal combustion engine—pistons, connecting rods, filters, spark plugs, wires, etc.—laid out neatly on a large table. What exists on the table is incapable of providing mechanical energy to power an automobile, lawnmower, or other device, until it is assembled into a whole engine. Thus the internal combustion engine is more than the sum of its parts.[16] Analogously, the motion of the baseball is more than just the sum of its vertical and horizontal components.[17] The individual person is more than just the sum of his or her various roles. The vinegar is more than just the sum of its molecular or atomic parts. And the reality experienced by human beings is more than just the sum of its irreducible aspects.

Holism is the theory that reality is correctly seen in terms of interacting wholes that are more than the mere sum of interacting parts.[18] It suggests that there is a fundamental unity to human experience such that a true understanding of any part of reality requires the appreciation of the interacting relationships of that part to the whole of reality.

Multidimensionality: The Diversity of Experience

Human beings do not encounter reality monotonically, but rather as a rich diversity of varying experiences. The principle of holism or unity requires that those experiences be appreciated in their interrelatedness, but it does not require that they be reduced, monistically or dualistically, to only one or two kinds of experience. Thus the assertion by Thales that all things are water[19] or Descartes' mind/body dualism[20] are unhelpful distortions of experience.

The process of abstraction allows one to distinguish various aspects of human experience. Experience[21] demonstrates that those aspects may be categorized, and the number of categories reduced to a finite number that can be reduced no further. Dooyeweerd identifies that number as fifteen (see Table 7.1). By recognizing the irreducibility of these aspects one is recognizing the multidimensionality of reality and doing justice to the diversity of experience. By recognizing the interwoven character of the aspects in ordinary experience one is recognizing the unity of reality and doing justice to the holistic character of experience.

The principles of holism and multidimensionality are complementary. Absolutizing holism distorts experience by ignoring multidimensionality. Absolutizing diversity (that is, seeing each aspect as totally independent of another) distorts experience by ignoring its holistic character.

The Integrative Character of Engineering Design

As mentioned in the introduction to this chapter, modern engineering is a human cultural activity that involves an interplay between theory, experiment, and imagination, in which human beings form and transform nature, for practical ends and purposes, with the aid of tools and procedures. Engineering design is based on science but, unlike science, is characterized by integration. The engineering design process is one of understanding a problem, bringing together all relevant knowledge regarding the problem, imagining a solution to the problem, and specifying that solution in terms that can be communicated to those who will fabricate the solution. Design is thus integrative in three important senses. First, the designer must fully understand the problem by integrating all experience that is relevant to it. Second, the designer must properly integrate all applicable theoretical knowledge regarding the problem and its possible solution. Third, the proposed solution to the problem must be integrated into the milieu in which it will function.

Thus, as described above, engineering and technology ought to be carefully distinguished from natural science. Natural science is characterized by abstraction, which is, in a sense, the opposite of integration. This is valid because the purpose of science is to understand, or deepen understanding about reality. The purpose of engineering design, however, is to solve a practical problem by forming or transforming reality. That requires integration of previously abstract knowledge.

Natural science and modern technology (engineering) are thus linked in a kind of reciprocal relationship. To increase knowledge, science, based on empirical data, must engage in abstraction, producing a theoretical picture of reality. To solve *real-world* problems, technology must integrate that abstract knowledge provided by science. The results of technology provide new empirical data for further scientific exploration.

Harmonizing Abstraction and Integration

The last paragraph of the previous section describes the reciprocal relationship between abstraction and integration. It is in recognizing this relationship that these two opposing methods are harmonized. When it is not recognized, one ends up absolutizing one of the two methods and consequently either distorting or creating a distorted picture of reality. In the case where abstraction is absolutized, the abstract picture of reality is taken for reality itself. This partially explains the two-cultures problem. For example, the success of natural science produces a naturalist perspective on reality. When the abstract character of that naturalist perspective is ignored, the result is naturalism, the view that all reality can be explained in terms of the natural sciences. When engineering design is carried out from a naturalistic perspective, the results are distortions of reality: products that are created for an abstract world.

In the case where integration is absolutized, the higher (more complicated) aspects of reality are given favorable treatment and the lower (less complicated, those that require a higher degree of abstraction to appreciate) aspects are ignored. For example, in Romanticism, the aesthetic, juridical, ethical, and fiduciary aspects of experience are celebrated without adequate appreciation for the numerical, spatial, kinematic, and physical aspects on which those higher aspects are based. Consider, for example, the one-sided excesses of the Romantic poets or the "flower children" of the 1960s. The division of society into two camps, one with a perspective that takes the abstract picture painted by the natural sciences as true, the other with a perspective that vilifies, or at best ignores, that abstract picture and embraces a one-sided "integral" picture, is the two-cultures problem.

Avoiding these distortions requires a balanced, or harmonized view of abstraction and integration, of the natural sciences, and the HSS.

In the Classroom

The most effective vehicle for overcoming the two-cultures problem in UE education, and for enabling students to properly distinguish science from engineering design, is for UE professors to convey to their students a positive attitude with regard to the HSS and teach their particular subjects holistically. Here, consideration is given to how that positive attitude may be conveyed inside the classroom. What kind of incidental or planned activities will effectuate that conveyance?

Of primary importance is the concern that what the professor does be genuine and natural. By "genuine" is meant that an activity has behind it the professor's heartfelt enthusiasm. By natural is meant that an activity arise from the ordinary flow of events in a given classroom. These two requirements make it impossible to articulate any kind of predetermined "plan of action" that would be effective in all situations. Thus examples rather than prescriptions are offered. The requirements also mean that the professor must be committed to addressing the two-cultures problem and must have the necessary resources (for example, interests in a number of areas usually associated with the HSS) to do so. For that commitment to exist to any significant extent, some faculty development has to have already occurred. The resources, on the other hand, are already there. It is a two-cultures myth that engineers have no interests or abilities beyond those connected with their own discipline. The majority of UE educators, while spending a great deal of time on their areas of specialization, nonetheless have interests and abilities in areas that can easily be seen to relate to the HSS. The two-cultures problem is propagated by those professors not making connections for their students between those interests and abilities and their areas of engineering specialization, not by the absence of those interests and abilities.

Illustrative of an activity that can be effective in conveying a positive attitude toward the HSS to university engineering students is the overview with which most professors begin a course. An overview, by its very nature, tries to show connections between the technical course material to be studied and the concerns and issues of society in its day-to-day existence. One particular solar energy engineering course, for example, begins with a discussion of the historic "energy crises" that have occurred and discusses the reasons for studying and possibly working, as an engineer, in the area of solar energy. A specific point is made regarding the broad economic norm identified as "appropriate energy end use."[22] Here students are confronted with physical phenomena, economic norms, and political realities.

A particular fluid mechanics course begins with an overview of the subject that gives serious consideration to its history. A supplementary textbook[23] is used to deal with the history of fluid mechanics, going as far back as ancient Greece. That textbook, however, has an interesting and obvious philosophical slant to it. When that slant is pointed out early in the course, the students begin to see how even a technical course, such as fluid mechanics, is not free from the influence of political, social, and historical prejudices. Many fluid mechanics textbooks begin by admitting to the student that knowledge of the subject is far from complete. The course material on dimensional analysis and modeling is a way of dealing with that lack of knowledge. Yet students often take away from their study of fluid mechanics a sense that what *is* known is known with absolute certainty (that is, certainty of the Cartesian sort). For example, students easily accept the questionable idea that the Navier-Stokes equations[24] are "absolute truth," the problem being simply that we do not have the mathematical acumen necessary to solve them. The overview period at the beginning of a course provides opportunity to raise the philosophical issues associated with these physical concepts and to convey to the class the importance of understanding the history and philosophy of one's particular technical area of specialization.

The overview period (which may also occur at the end of a course) is an example of how a focus on holism can very naturally arise from the technical subject matter of a course. It is also easy to see, if an integrative overview is appropriate at the beginning and end of a whole course, how it may be appropriate for identifiable units of a course as well. But a UE course is more than just the abstract subject matter. It is an approximately fifteen-week sequence of events where a group of human beings (the students) with particular attributes (age, technical interests, technical level of competence, future expectations, etc.) encounter another human being (the UE professor) with his or her own particular attributes (personal history, professional history, style of teaching, etc.). It is natural for those particular attributes (either of the students or the professor) to be used to better accomplish the purposes of the course. Accordingly, two principles of effective pedagogy may be stated thus: (1) know your students (background knowledge, learning styles, cultural expectations, etc.) and use that knowledge to effectively construct the learning environment, and (2) know yourself (background knowledge, teaching style, social strengths and weaknesses, etc.) and use your strengths when constructing lesson plans. One of the greatest strengths that any UE professor has is experience, professional and otherwise. That experience ought to be exploited fully to teach effectively the subject matter *and* to fashion a holistic classroom environment. When done in that way, connections between the subject matter and other areas of life are clearly seen by the students and a positive attitude toward the HSS is conveyed. In the following paragraphs, several examples are given where the various experiences of professors are used for these purposes.

Viscosity, a property of fluids, is studied early in every introductory course in fluid mechanics. One professor took advantage of experience gained during his Ph.D. research to help his students see the connections between that topic and issues of economics, environmental ethics, and justice. His research had been to measure the viscosity of newly developed CFC alternatives as a function of temperature. Calling upon his experience enabled his students to see one particular interface of engineering and government (government regulation provided the impetus to move away from CFCs), the social impact of new technologies (automobile and window air conditioners are directly affected), and the trade-offs between environmental safety and economics.

Another professor's experience as an analytical heat transfer engineer in the aircraft industry provided him with both technical and HSS resources for teaching a UE course in heat transfer. The first turbine outer air seal on a jet aircraft engine must be designed to thermally respond in a manner that matches the thermal and mechanical movement of the first turbine blade tips. Too much gap between the blade tips and seal causes the engine fuel efficiency to suffer and can lead to a condition of engine stall. Too little gap presents the danger of interference and consequent catastrophic mechanical failure. The professor called on his experience in analyzing first turbine outer air seals to help teach the concept of transient thermal response, to provide a realistic exercise in the finite volume method of transient thermal analysis, and to get the students thinking about the relationship of ethics, economics, and industrial politics to engineering design. When analyzing an air seal for an engine used in an Air Force jet aircraft, the professor had considered test scenarios alternative to those described in the military specifications. He discovered that one particular combination of aircraft maneuvers would create an interference situation that would likely result in the engine exploding. He communicated this to the design engineer in charge of the air seal project, who, interestingly, had served as an Air Force pilot during the Korean conflict and confirmed the small but real probability of the problematic maneuver. A few days after this communication, the design engineer expressed, with consternation, that the "higher ups" had decided to ignore the problem because it was not on the list of performances specifications supplied by the Air Force. Relaying this experience to a UE heat transfer class provides an interesting case study as well as an open-ended problem in engineering ethics.

In an otherwise traditional fluid mechanics course, one professor divides the class into teams and assigns a pipe system design project. The pipe system supplies water to a village with a population of 2000 in a remote area of Brazil. Students are confronted with developing world economics, politics, and ergonomics, as well as traditional fluid mechanical design.

A UE professor agrees to do some consulting work for the theater arts department at his university by assessing the load handling capability of a makeshift balcony in one of the department's auxiliary theaters. The desire is to seat the audience in the balcony so that the whole of the floor area can be used in performing a particular play. The professor uses the consulting experience as a case study in his strength of materials course. He also uses the opportunity to teach the students about the structural implications of staging a play and the aesthetic concerns surrounding the design of a structure for such a purpose.

A professor teaches a senior level course in thermal-fluid systems design. To convey the social, economic, and political implications of energy consumption and distribution patterns, she relates to the class her experiences of the electrical energy "blackout" that occurred in the mid 1960s on the east coast of the U.S. and the two fossil fuel "energy crisis" periods that occurred during the 1970s. Having traveled across the country during the second of those periods, she relates the geographical differences in gasoline shortages as evidence for the complex social dimension of that particular technological problem.

In the process of teaching about four-bar mechanisms, a mechanical engineering professor calls upon his own interests and activities in woodworking and in music to present to the students a specific design problem. The problem is how to store a large collection of compact discs (CDs) in a cabinet (entertainment center) that is designed to hold audio-visual (AV) components. The AV design criteria include a cabinet depth (front to back) that far exceeds that needed by one row of CDs. Thus, either space is wasted behind that row, or more than one row of CDs are placed on a shelf, with the result that the back rows are hidden from view. The students are assigned the task of designing a mechanism that rotates shelves of CDs into view about a horizontal axis. The project integrates mechanical design principles with norms for economics (both money and space), communication (the need for a person to be able to identify all the CDs with equal clarity), and aesthetics (the design must conform to the style of the present cabinet).

A UE professor at a small college taught a course dealing with instrumentation and analysis regularly during the fall semester of each year. One of the topics covered near the end of the course was digital data gathering techniques, including analog to digital and digital to analog conversions. The professor had an interest in music and in music reproduction technology. During the early 1980s, when digital music reproduction in the form of the compact disc player was very new, he was particularly interested to read about debates concerning the relative quality of music reproduced digitally and music reproduced by analog techniques. He used those debates as a vehicle for enabling his students to see the relation of their technical studies to the psychological, economic, social, and aesthetic dimensions of music reproduction. Because the professor was one of the few people to own a compact disc player at the time, one class session, near the end of the semester, was held in his home (the class size usually numbered twelve or fewer). There students were able to experience digital sound reproduction and assess the differences between the new and the old sound reproduction technologies. In addition, because the professor had a wide range of interest in music, many of the students were exposed to particular forms of music for the first time and thereby gained an increased appreciation for this area of the arts.

The HVAC system serving the mechanical engineering department at one university provided examples of some of the difficulties of HVAC design. One professor used his experience with his own office environment to illustrate these difficulties to his HVAC class. The problems included lack of temperature control (excessive vertical as well as temporal temperature gradients) and lack of moisture control (excessive humidity leading to mold growth on books and wall areas). Going beyond the technical and ergonomic issues, the professor enabled the students to see the delicate relationship between the academic, administrative, and maintenance staffs of the university and how that relationship affected, and was affected by, the technical, HVAC, issue.

In a different HVAC related situation, a UE professor had designed and built a two-story, three-room addition onto his home. The heating and cooling load calculations made during the design served as a real-life example for his students. The pre- and post-addition utility bill data served as an ongoing example of HVAC performance analysis. But much more than HVAC loads are involved in the design, construction, and operation of even an addition to a building. In this particular case, the professor's interaction with the local building inspector; the construction workers who were hired for the foundation, framing, and masonry work; and the bank that financed the work, provided him with a wealth of anecdotal information regarding the interrelations of mechanical, economic, political, communication, and even ethical dimensions of the project. In addition, because the central room of the addition was designed for both AV entertainment and social gathering, there was an important aesthetic dimension that gave the project its overall direction. The professor was able to use all of this to help provide a more holistic context for his HVAC course.

The examples given in the paragraphs above are all closely related to technical topics that are taught in undergraduate mechanical engineering courses. It is possible, however, for UE professors to utilize their interests and experiences in the HSS, even without a connection to the immediate technical topic at hand, to provide a more holistic classroom environment. There may be, for example, a professor who is a bit of a Shakespeare buff and is ready to recite an appropriate line for every imaginable classroom situation. Many a professor has, on occasion, felt like quoting those famous words of Marullus, "You blocks, you stones, you worse than senseless things!" (*Julius Caesar*, I:i). Perhaps more appropriate, for those occasions when both the professor and the class feel the frustration that accompanies the slowness with which certain difficult concepts are grasped, might be Hamlet's anguished cry, "O, that this too too solid flesh would melt, thaw and resolve itself into a dew!" (Hamlet, I:ii). What occurs when a professor recites such a line is a kind of aesthetic commentary on the momentary situation. If done well (that is, if it is not forced), not only will it relieve tension, but, by example and in a positive context, it will help create a more holistic environment.

It may be suggested that UE professors ought to, on occasion, audit courses in the HSS and that doing so can have a very positive effect on the HSS students. In addition, it is quite common for auditing professors to glean insights into the points-of-view of students in general, thus better understanding their own engineering students. This opens avenues of communication that might otherwise be closed and demonstrates to the UE student the engineering professor's regard for the HSS.

The history and philosophy of science and technology are areas of interest shared by many UE professors. It is not uncommon to discover that, as a form of recreation, a professor reads biographies of technical persons or the histories of technical events. For example, there are numerous biographies of Albert Einstein[25-27] and respectable as well as readable histories of technologically important events such as the constructing of the first atomic bomb or the development of oil as a primary source of energy.[28, 29] The UE professor who is especially literate regarding areas such as these has a store of interesting information available that can be used to both enliven technical lectures and communicate a positive attitude regarding the HSS to students.

As a final example of activities within the classroom that promote a high regard for the HSS to UE students, consider the changes that are currently occurring in the UE professor's communication tools as a result of computer development. Chalk, blackboard, and textbooks have been primary among those tools since the early days of the profession. The overhead projector has made some inroads, but it is primarily a tool for formal rather than everyday classroom lectures. The current state of computer technology, however, is giving new shape to the overhead projector method of presentation and making the more formal presentation mode attractive for classroom lectures as well. Today a UE professor can prepare lectures on a personal computer using a presentation software package that allows for the easy incorporation of not only text, graphics, and equations, but motion graphics and sound as well. Doing so competently, however, means developing an appreciation for aesthetics. It also means that that appreciation is communicated to the UE student. Imagine preparing a lecture on beam deflection analysis using the multimedia capabilities of a computer. One might weave together music from Beethoven (the Fifth and Ninth symphonies may be appropriate), historic graphic material from Timoshenko's *History of Strength of Materials*,[30] and the basic textbook concepts and equations, to create a presentation that not only teaches the basics of beam deflection, but communicates to the UE student some understanding and appreciation for the HSS (particularly history, music, and aesthetics).

Conclusion

The humanities and social sciences play an important role in undergraduate engineering education. That role is not limited to providing "well-roundedness" in individual engineers, but is also a fundamental part of identifying, formulating, and solving engineering (design) problems. This is so because engineering problems are, by their very nature, holistic, in contradistinction to natural scientific problems, which are, by their nature, abstract. Thus the humanities and social science component of undergraduate engineering education ought not be limited to courses taken outside the school of engineering, but ought also be integral with engineering courses—particularly upper level design courses.

References

1. Monsma, S. V. (Ed.) 1986. *Responsible Technology*. Grand Rapids, MI: William B. Eerdmans Publishing Company.
2. Dordt College. 1993. *Self-Study Questionnaire for Review of Engineering Programs Using Engineering Topics Criteria*, A report submitted to the Engineering Accreditation Commission of ABET.
3. Van Poolen, L. J. 1987. Technological Design: A Philosophical Perspective, ASEE Annual Conference *Proceedings*, pp. 767—789.
4. Schuurman, E. 1980. *Technology and the Future*. Toronto. Wedge Publishing Foundation.
5. Snow, C. P. 1964. *The Two Cultures and A Second Look*. New York: Cambridge University Press.
6. Lynn, W. R. 1977. Engineering and Society Programs in Engineering Education, *Science*, Vol. 195, pp. 150–155.
7. Ladesic, J. G., and D. C. Hazen. 1995. A Course Correction for Engineering Education, *Aerospace America*, May, pp. 22–27.
8. Billington, D. P., 1986. In Defense of Engineers, *The Wilson Quarterly*, Vol. 10, No. 1, pp. 86–97.
9. Kalsbeek, L. 1975. *Contours of a Christian Philosophy: An Introduction to Herman Dooyeweerd's Thought*. Toronto: Wedge Publishing Foundation.
10. Seerveld. C. 1985. Dooyeweerd's Legacy for Aesthetics, in McIntire, C.T. (Ed.), 1985, *The Legacy of Herman Dooyeweerd*, University Press of America, Inc., Lanham, MD, pp. 41–79.
11. Clouser, R. 1991. *The Myth of Religious Neutrality*. Notre Dame, IN: University of Notre Dame Press.
12. Dooyeweerd, H. 1969. *A New Critique of Theoretical Thought*. Philadelphia: The Presbyterian and Reformed Publishing Company.
13. Flink, J. J. 1988. *The Automobile Age*. Cambridge, MA: The MIT Press. The many sociological "surprises" surrounding the automobile, e.g., the character of the social infrastructure required to support it, its socially formative effect on teenagers, etc., are evidence that a careful sociological analysis was missing during its earliest years. The first two chapters of Flink's *The Automobile Age* make very clear that the automobile is a socially qualified artifact.
14. Adams, C. C. 1995. Technological Allusivity: Appreciating and Teaching the Role of Aesthetics in Engineering Design, The Proceedings, IEEE/ASEE Annual Frontiers in Education Conference, pp. 3a51.
15. Marcel, G. 1952. *Men Against Humanity*. London: The Harvill Press Ltd.
16. By "sum of its parts" is meant, simply, the properties of the individual parts excluding the relationships between the individual parts that obtain when they are assembled to make

the whole and excluding the properties of the whole that do not obtain without the whole. Some examples of the latter, in the case of an internal combustion engine, are the shaft horsepower and torque that it is capable of delivering at a particular rpm, the economic value (dollars) of the whole as a replacement engine, and the reliability of the engine when placed in service.

17. At first glance this may not seem to be a good analogy. The vertical and horizontal components of the flight of a baseball are abstractions, they are kinematic subjects having no physical subject function. The parts of the engine are laid out on a table, thus they are physical subjects and seem not to be abstractions. But if one considers the design process, it is clear that at one point the parts were abstractions. The designer envisioned an engine that would have certain properties and then envisioned the parts that would be necessary to make that engine. Although the parts, when fabricated and laid out on the table are no longer abstractions in the conventional sense of the word, they are still parts of a whole that are separated from that whole. So they are a different kind of "abstraction." Thus the analogy holds. It is the commitment to a Cartesian mind/matter dualism that is at the root of our discomfort with the analogy.
18. Merriam-Webster. 1994. *Merriam-Webster's Collegiate Dictionary, Tenth Edition*. Springfield, MA: Merriam-Webster, Incorporated.
19. Copleston, Frederick, S.J. 1962. *A History of Philosophy, Vol. I: Greece and Rome*, New York: Doubleday (Image Book Edition, 1993).
20. Descartes, R. 1968. *Discourse on Method and the Meditations*. London. Penguin Books.
21. Consider biological taxonomy or educational taxonomy, to pick two very different areas of experience, as examples of categorization schemes. Dooyeweerd claims (see 1969, Chapter III) that all experience can be abstracted until one reaches a point where an aspect of experience simply cannot be reduced any further, i.e., it cannot be reduced to another aspect of experience. For example, one experiences the hardness of the bark of a tree and can explain it in terms of the physical properties of the bark. But the physical aspect of the tree cannot be explained in terms of, e.g., the spatial aspect of the tree. In other words, the physical aspect is irreducible to the spatial aspect.
22. By "appropriate energy end use" is meant that energy needs of a particular thermodynamic kind ought to be met with energy resources of the same thermodynamic kind. In other words, when there is a need for energy at high temperatures such as in refining particular metal ores (e.g., aluminum) or producing significant amounts of mechanical power (such as in mass transit vehicles), then it is legitimate to consider use of energy resources of a highly concentrated nature, for example fossil fuels. But where the energy need is at low temperatures (such as environmental space heating) then the appropriate energy resource is one that is diffuse, such as solar energy. Using fossil fuels to heat the space in a building is like using a 400 horsepower internal combustion engine to turn the crank on a pencil sharpener. It is a waste of the resource's potential.
23. Tokaty, G. A. 1971. *A History and Philosophy of Fluid Mechanics*. New York: Dover Publications, Inc.
24. The Navier-Stokes equations are simply the fluid mechanics equivalent of Newton's Second Law. They relate the forces acting on a fluid (gravity, viscous shear, etc.) and the properties of a fluid (density and viscosity) to the fluid's motion. In addition to the Navier-Stokes equations, there are two other sets of equations, continuity (equivalent to the conservation of mass) and energy (equivalent to the conservation of energy, or First Law of Thermodynamics) that together attempt to completely describe the physical state of a fluid.

25. Clark, W. 1971. *Einstein: The Life and Times*. New York: Avon Books.
26. Hoffmann, B. 1972. *Albert Einstein: Creator and Rebel*. New York: The Viking Press.
27. Pais, A. 1982. *Subtle is the Lord: The Science and the Life of Albert Einstein*. New York: Oxford University Press.
28. Rhodes, R. 1986. *The Making of the Atomic Bomb*. New York: Simon and Schuster.
29. Yergin, D. 1991. *The Prize: The Epic Quest for Oil, Money, and Power*. New York: Simon & Schuster.
30. Timoshenko, S.P. 1983. *History of Strength of Materials*. New York: Dover Publications, Inc.
31. Seerveld. C. 1980. *Rainbows for the Fallen World: Aesthetic Life and Artistic Task*. Toronto: Toronto Tuppence Press.

CHAPTER 8

HEINZ C. LUEGENBIEHL

DONALD L. DECKER

The Role of Values in Teaching Design

Some contend that engineering teachers neither do teach nor should teach values. We argue, however, that teachers convey values to students implicitly and explicitly, in all teaching environments, and particularly in the design classroom. We think it is worthwhile for teachers to decide consciously which values they intend to try to influence.

The Nature of Values

We all value specific things, such as our homes or our lives. We were taught values, such as friendship and loyalty, at home and in school. We know some things are more important than others. The term "values" is ambiguous, though; it does not mean the same to everyone. To focus on values in the design classroom, it is necessary to introduce and clarify a workable conception of the term. The list in Table 8.1 can be used to spotlight those central to the teaching of design.

Emmanuel Mesthene[1] has defined values as "those *conceptions of desirable states of affairs* that are utilized in selective conduct as *criteria* for preference or choice or as *justifications* for proposed or actual behavior." Given this definition, values may be of various types, including individual, societal, national, professional, and human. Their distinguishing features are that they are not absolute, but fit into a hierarchical framework, and they are perceived as relative ends to be achieved.

TABLE 8.1. Rankings Values from a 1985 ASEE Design Rap Session

Individual Values		Professional Values
Personal morality	More	Competence
Family	Important	Integrity
Initiative		Creativity
Friendship		Honesty
Flexibility		Conscientiousness
Hard work		Communication
Intelligence		Ability to analyze
Honor		Initiative
Quality of life	Less	Perseverance
Self-reliance	Important	Pride in work
Self-fulfillment		Professionalism

The fact that values are arranged in a hierarchical fashion indicates that, at times, one set of values may be emphasized at the expense of another.

Typically, a values clarification process involves choosing one goal to be achieved when several are possible. Values, which are held by people, must be distinguished from "value," which is possessed by things. An individual's computer has value, while its owner values privacy.

Appropriate ways to talk about values involve various terminologies: value commitments, value priorities, value conflicts, value subscription, value acquisition and abandonment, value redistribution, value emphasis and de-emphasis, and value standardization. Talk about "the values of a society" can then be interpreted as meaning that a particular society is committed to advocating acts that reveal and reinforce specific values in terms of a hierarchical perspective on values.

Several problems arise, however, when we attempt to apply this model. Values change more slowly than realities, and options change over time, leading to tension between traditional values and the need for social action based on knowledge. There is usually disagreement about which values are merely instrumental and which are basic, and, perhaps most important, there is little agreement about whether values simply reflect a particular society's attitudes or represent objective truths. The fundamental issue is whether a distinction can be drawn between values that are held and values worthy of being held.

Such weighty issues about the nature of values cannot be resolved here. It is important to be aware of them, however, for they influence our position in relation to the significance of values in teaching design. We will discuss how and why this is so.

Values and Engineering

Technology affects all of our values. Engineers, whether they know it or not, are powerful people, sometimes individually, but always collectively. It therefore seems worthwhile to reflect on what it is we transmit to our students in the teaching process. As educators of future professionals who will play a central role in determining the shape of our world, we have the responsibility to at least be aware of the values implicit in our teaching. Perhaps reexamining our own tendencies will lead to a clear and honest approach to teaching design. Not that we should indoctrinate our students, but unless we are aware of what we are teaching, we cannot even adequately aid them.

Engineering is a creative activity. Through engineering, things are brought into being. It involves a wrenching of nature. Through the design process, engineers put their mark on the world, which means that the world, to some degree, reflects their values. Through the process of making decisions, engineers make our world; their actions will either result in the world as we want it to be or in a misshapen image of that ideal. This means that the things created become an expression of values, even when we are unaware of the values exhibited.

Consider the accumulating effects of technological innovations. What engineers do has tremendous potential to foster or retard the prospect of certain values. Motives for technological change arouse fundamental questions. "Why is it necessary to develop atomic power plants? Why is it necessary to build large-scale supercomputers that 'think,' if only to monitor extremely complicated systems of our own design? What are our final ends, the good for human life, that underlie our philosophy of design?" Design should have some end or aim in mind. But what should that end be? The answer is determined by our values. But whose values, and why should one set count more than any other? The nature of technological design and the nature of technological development thus "ultimately forces us to face basic philosophical and spiritual questions that we can now ignore only at great peril." Teachers of engineering are also teachers of values.

Teaching and Values

Teachers are buffeted not only by a great variety of theoretical approaches to the teaching of values but by the conflicting demands of parents, religious leaders, and the community.

In traditional societies there is little conflict over values. Speaking with one voice, the various institutions of society reinforce each other's claims on the values of the individual, and divergence from established norms is rare. In a society

TABLE 8.2. A Compilation of Values

Four categories of values are highlighted in the following listing: individual, professional, societal, and human. The values in any category are not mutually exclusive, nor is the listing complete.

Individual Values

Curiosity	Intelligence	Self-advancement
Endurance	Leisure	Self-control
Family	Optimism	Self-fulfillment
Flexibility	Personal liberty	Self-reliance
Friendship	Personal morality	Self-respect
Hard work	Personal power	Self-worth
Honor	Personal security	Strength
Independence	Privacy	Success
Initiative	Property	Wealth
Intellectual simulation	Quality of life	Wit

Professional Values

Ability to analyze	Efficiency	Patience
Ability to synthesize	Fair play	Perseverance
Civic consciousness	Flexibility	Prestige
Collegiality	Forthrightness	Pride in work
Communication	Freedom of inquiry	Problem-solving ability
Compassion	Honesty	Professionalism
Competence	Idealism	Prudence
Conformity	Imagination	Rationality
Conscientiousness	Informedness	Realism
Cooperation	Initiative	Recognition of accomplishments
Courtesy	Innovativeness	
Curiosity	Integrity	Self-education
Decisiveness	Leadership	Selflessness
Devotion to principle	Literacy	Service to others
Duty	Loyalty	Tolerance
Economy	Obedience	Trustworthiness
Effectiveness	Openness	

Societal Values

Capitalism	Freedom of religion	Privacy
Centralization	Freedom of thought	Progress
Change	Governance by law	Public service
Competition	Individual rights	Social justice
Culture	Individualism	Societal harmony
Democracy	Liberty	Survival of society
Education	National pride	Tradition
Equality	National prosperity	
Equality of opportunity	Order	

TABLE 8.2. *Continued*

Human Values		
Autonomy	Health	Prevention of evil
Beauty	Hope	Progress
Beneficence	Human dignity	Promotion of goodness
Bravery	Humility	Prudence
Fairness	Idealism	Reason
Faith	Justice	Reverence for life
Freedom	Love	Self-sacrifice
Friendship	Morality	Truth
Happiness	Pleasure	

of rapid social and technological change the situation is completely different. Society's values are no longer homogeneous. Various institutions have different goals, depending on their role. The individual is bombarded by different messages, none of which necessarily provides a dominant focus

Research on values education in the 1960s and 1970s seemed to many educators to provide a fortuitous answer to this perplexing issue. The work on values clarification reconciled an emphasis in values with the inherent right of human beings to choose their own values. No longer was it the responsibility of teachers to inculcate a "correct" set of values; they could become mediators, clarifiers, unbiased discussion leaders who could help students "find themselves" without revealing their own biases. Most teachers could, in fact, abstract themselves from the issue altogether by seeing themselves simply as transmitters of knowledge.

Every teacher, however, at least in a small way, serves as a role model for students. In professional education, it is expected that teachers make students familiar with professional values. Given the nature of professional education, it seems clear that educators cannot simply avoid the issue. *They will influence the values students hold.* In light of a professional approach to education, it can even be argued that teachers have a positive obligation to influence students; otherwise, they are not concerned with their students' best interests and thus fail to fulfill the primary obligation of the professional to the client.

Educators cannot simply take a laissez-faire approach. They will, and must, transmit values, but should do so based on an adequate foundation and without imposing on the autonomy of student. To meet these two conditions, however, educators must first be aware of the values they consider central to the activity in which they are engaged. For future reflection, four categories appear in Table 8.2.

Conclusion

Engineering teachers may convey or teach values they do not intend to transmit. The process of discussing and ranking values can provide them with an opportunity to identify the values they want to encourage in students. By becoming explicitly aware of what values they and their colleagues hold, their teaching effectiveness in design and other engineering courses can be greatly enhanced.

References

1. Mesthene, Emmanuel. 1970. *Technological Change: Its Impact on Man and Society*. Cambridge, MA: Harvard University Press.
2. Dregson, A. R. The Scared and the Limits of the Technological Fix, *Zygon*, vol. 19, Sept. 1984, pp 265, 270.

CHAPTER 9

HEINZ C. LUEGENBIEHL

Engineering Ethics Education for the Twenty-first Century

Topics for Exploration

Engineering ethics education should be, like engineering ethics itself, a proactive activity. A major aim of acquainting students with ethical issues arising in the profession is to prepare them to avoid becoming entangled in ethical dilemmas. Analogously to defensive driving, awareness of the pitfalls likely to be encountered by inexperienced engineers will help them avoid circumstances potentially endangering the "safety, health, and welfare" of the public.

In the past, much of the focus in the teaching of engineering ethics has been placed on having engineers accept basic responsibilities regarding such issues as conflicts of interest, confidentiality, duties to fellow professionals and the profession, and responsibilities to the public. While these issues are, and will continue to be, central to engineering as a profession, it needs also to be recognized that the role of technology in the world, and the nature of the world itself, are undergoing rapid change. These changing circumstances bring with them the need to reexamine the engineering ethics curriculum. The following deals with a number of fundamental issues that ought to be considered for inclusion as central to the discussion of ethics, both in the classroom and in the codes of engineering ethics.

The issues of pressing concern for the future include the cross-cultural and multicultural dimensions of engineering practice; the increased reliance on technology to deal with technology, including especially the use of computers; the geometrically increasing complexity and scope of technology; increasing reliance on team-based project development; the move away from hardware to software or processes; questions of loyalty and gender in a changing business environment; new complexities regarding ownership of information; and the increasing trend toward the assignment of individual responsibility in professional contexts where such assignment may be difficult or misdirected. These issues, while peripheral in the past, will become a central focus of engineering practice in the future. Yet the traditional approach to engineering ethics takes little account of them. It is the responsibility of engineering educators to not only enable students to deal with traditional issues of engineering ethics but future ones as well. I will focus on these issues through the discussion of three interrelated topics and suggest a framework for future consideration of engineering ethics: the internationalization of engineering, cultures in the United States, and fundamental changes in the nature of technology.

The Current Emphasis in Engineering Ethics

An adequate examination of future trends in engineering ethics requires that the currently existing situation first be reviewed. The formal basis of engineering ethics in the United States was initiated with a fairly narrow range of concerns put forth in the first codes of ethics at the end of the nineteenth and the beginning of the twentieth century. These codes adopted a client-based model from the medical profession, which stressed the relationship of one individual to another in the context of the engineering profession. Subsequent recognition of the employed status of most engineers resulted in the introduction of codes that focused on three sets of duties for the engineer: to clients or employers, to the profession as a national entity, and to the public, interpreted as American society.

These obligations can be captured under a more general conception of loyalty.[1] Engineers owe loyalty to each of these groups and the primary aim of engineering ethics becomes the clarification of what these loyalties consist of in each instance. As a result, the major ethical discussion revolves around how conflicts of loyalty are to be resolved in general and in specific situations. For example, loyalty to employers is a primary duty of engineers. The code of ethics of the National Society of Professional Engineers (NSPE)[2] states: "Engineers shall act in professional matters for each employer or client as faithful agents or trustees" (Rule of Practice 4). Part of being a faithful agent is further specified as avoiding conflicts of interest. Ethical problems in interpreting this rule arise when engineers have, as they often do, both a past and a current employer. Such problems then need to be resolved by balancing engineers' rights to gainful employment with employers' right to protection of property. The NSPE code deals with the issue in the following manner: "Engineers shall not, without the consent of all interested parties, participate in or represent an adversary interest in connection with a specific project or proceeding in which the Engineer has gained particular specialized knowledge on behalf of a former client or employer" (Professional Obligation 4b).

Loyalty to clients or employers is most fruitfully interpreted in terms of an agency model. Engineers, in accepting a position, become agents or representatives of the employer. They thus become obligated to act in the best interests of the employer. This, in turn, provides the basis for engineers being required by the codes of engineering ethics to maintain confidentiality, to avoid conflicts-of-interest, to maintain competence, to be honest in their relationships with their employer, and to refrain from giving or taking bribes.

At the level of the profession, engineers are required to support the aims of the profession and to encourage other engineers to do so as well. This is founded on the service mission of occupational groups that have been designated professions. In turn, for the groups to do so, standards of association, civic service, furtherance of the profession's image, and the avoidance of unfair criticism of other engineers are established. The requirement for loyalty to the profession is justified on the basis of the claim that in the absence of control by the profession the group will not be able to guarantee the quality of its services. Thus, for instance, the NSPE code sets forth an obligation that under other circumstances would be considered self-serving: "Engineers shall endeavor to extend public knowledge and appreciation of engineering and its achievements and to protect the engineering profession from misrepresentation and misunderstanding" (Professional Obligation 2c).

Finally, and most important, engineers have duties to society as a whole. The rise of a profession can be viewed as the establishment of a contract between an occupational group and society, where the special expertise of the group is recognized in the form of it being named an exclusive service provider. In fulfilling their responsibilities to the public, the special obligation of engineers is to protect the safety, health, and welfare of the public. Potential conflicts can arise when this "paramount" responsibility must be balanced against obligations to employers or to the profession. Another potential conflict of loyalties thus exists. Codes have attempted to resolve such conflicts by appealing to the notion of "proper authority": "If their [engineers'] professional judgment is overruled under circumstances where the safety, health, property or welfare of the public are endangered, they shall notify their employer or client and such other authority as may be appropriate." (Rule of Practice 1a).[3]

In teaching potential engineers about their obligations as engineers and how to deal with conflicts among them, several background assumptions are typically made. Perhaps most significantly, it is assumed that engineers can use their own growth and education in society as the foundation for making ethical judgments.

That is, most students know right from wrong in ordinary settings even if they do not always act on their knowledge, but are so unfamiliar with the special ethical issues that arise in professional life that they are unprepared to make adequate ethical judgments in those contexts. These new circumstances are defined as "engineering or professional issues." Therefore, engineering ethics can be seen as the application of what is already known, as embedded in the European tradition of ethics. Such an interpretation places engineering ethics within the tradition of applied ethics as it is taught in many contemporary introductory ethics courses.

The other central assumption made is that engineering ethics is a form of professional ethics. An implied contract between society and an occupational group establishes a special relationship out of which arise obligations that ensure exemplary performance of the service function performed by the profession. This performance is compensated for by prestige and financial rewards. What needs to be noticed is that in this model obligations exist for professionals that do not exist for other citizens in society, such as the special obligation of engineers to protect the safety of society.[4]

The theoretical basis of engineering ethics as taught in the U.S. is thus generally a combination of applied ethics and professional ethics. The balance between the two often varies depending on who is doing the teaching. Engineering professors will more typically stress professionalism, while philosophy professors tend to rely on the European tradition of ethical theory as the basis for making judgments. The point to be made is that both emphases may be inadequate in light of changing trends in the world that potentially may influence the analysis of issues arising in engineering ethics.

It has been a long struggle to move engineering to professional status in the U.S. Just as real progress is being made, a new set of issues is arising that threaten the very status this struggle has been intended to achieve. The world is changing more rapidly than the underlying conception of what engineering is all about is able to absorb. No longer are employers or clients restricted to a national context. Engineers are increasingly working for multinational corporations or for foreign employers. No longer are engineers working primarily with others who are just like them. The workforce is becoming increasingly diverse. No longer are engineers able to take the traditional perspective on the products of their work. They must account for the scope, scale, and spread of technology, and their own reliance on technology. These changes have major implications for how and where engineers will practice and consequently on what is considered loyalty to an employer, to the profession, and to the public. A fundamental reexamination of the current foundation of engineering ethics will be required.

The Internalization of Engineering

During the last several decades the context in which the typical American engineer will operate during his or her lifetime has permanently changed. The world has seen an increasing trend toward multinational corporations, interaction of even local corporations with foreign corporations and customers, engineers working for either foreign or domestic corporations in foreign locations, and the increasingly rapid spread of technology throughout the world. While each of these factors in isolation has been present for some engineers for a long time, their cumulative impact as a totality is only now being felt. It is now appropriate to speak of a new paradigm of the internationalization of engineering practice.

Corporations have adapted to this changing scenario by increasing their emphasis on crosscultural training, for example, by bringing workers from one country to another for temporary training, by initiating fundamentals of training for survival in other cultures, by considering foreign language education a valuable part of a potential employee's resume, and by focusing on the understanding of a variety of customers and cultural contexts in the sales and marketing of products.

An international paradigm for engineering will of necessity affect the work of engineers. The question is whether they will also be prepared to deal with the resulting ethical implications of such changes. From the perspective of ethics the issues of particular importance include the definition of 'society,' environmental responsibility, obligations to consumers of technology, and ethical relativism as it is connected to cultural values.

Traditionally, "the public," in discussions of engineering ethics, has been viewed as American society. The professional status of engineers is taken as being granted in our society, as the name National Society of Professional Engineers indicates, by the nation. Questions of global impact, when they have been considered, have been seen as separable from the context of typical engineering decisions.[5] But few engineers today function in an exclusively national context. To limit their responsibilities to the American public would then imply that they have professional responsibilities and duties of loyalty to only a part of the actual public they serve, and that their concern for the public health, safety, and welfare is limited by national boundaries. This perplexity arises as a result of interpreting the special responsibilities of engineers in terms of the perspective of professional ethics, where professional status is granted by a contract with a particular society. On the other hand, if the perspective of applied ethics is used, then it becomes much more difficult to argue for special responsibilities of engineers. The seeming conflict can be reconciled if a perspective of global professionals is taken. Then responsibilities to the public will be all of those affected by engineering actions, no matter where they happen to live. Relationships among engineers throughout the globe would also be equalized.

A particularly significant example of such responsibility can be seen in the case of engineers' responsibility for the environment. This duty has in the past been interpreted in light of laws that govern a particular engineering activity. But in the contemporary world the environmental impact of engineering decisions is not restricted by national boundaries. Acid rain knows no borders. So the public that will be affected may well be in a distant part of the world. A classic case of a dual notion of responsibility is the chemical leak that occurred in the Union Carbide plant in Bhopal, India, in 1984. Different safety standards, different public alert and protection systems, and different operator qualifications were used for that plant than for a similar Union Carbide plant in West Virginia.[6] The apparent assumption was that responsibility for the environment and for the public in the two circumstances was different.

A similar issue arises in regards to the end users of engineering products. The legal system in the U.S. has forced engineers to be aware of and to understand the user of the product, as long as they are in the United States. When a product is sold in another country, however, even though the user may be harmed in the same way, differing standards of safety are applied. Further, engineers must recognize that users of their products will have different cultural and educational backgrounds. The typical way in which a product will be used will differ from circumstance to circumstance and American users by themselves cannot be defined as typical users. Therefore, responsibilities to end users, while formally the same, will actually differ based on the user of the product. Just as different standards are applied in the U.S. when a product is used by children as opposed to adults, so must the particular cultural background of users throughout the world be part of engineers' ethical deliberation.

Any discussion of the global responsibilities of engineers must also consider the problem of ethical relativism, that is, engineers will encounter a variety of conceptions of right and wrong in different parts of the world. What they have learned about what is right behavior in the U.S. may not be in accord with practice in another country. The teaching of engineering ethics in this regard has had an absolutist bend, given the framework for the teaching of ethics in the U.S. What is ethical for an engineer here, it is claimed, is ethical everywhere, and if practices are different somewhere else, then people in that society must be acting unethically. If our codes of engineering ethics reject attempts to "grease the wheel" in obtaining contracts, then this must be wrong everywhere, whether or not it is standard practice in that country. Alternatively, engineers are left simply with the idea that they should adapt to local customs. But this will be increasingly difficult as the very borders between engineers are broken down. Several computer software companies, for example, now operate in shifts, not in the same location, but based on the sun as it moves around the world.

Responsibilities to the public and standards of behavior toward fellow engineers must be rethought in light of the global environment in which engineers are now operating. They will need to interact with customers, the public, and other engineers on a global scale. Ethical demands suitable only for a national context will no longer be appropriate in the future.

Shifting Cultural Foundation in the United States

Not only are the stakeholders outside of the United States changing for engineers, a similar shift is occurring in the U.S. as well. Most noticeable for engineering is the changing makeup of the engineering workforce. Increasing numbers of women and members of diverse ethnic groups are becoming engineers and members of the workforce as a whole. There is a diversity of generational outlooks coming to the fore as well. And for all engineers, employer-employee relationships are changing with an increased emphasis on mobility in an age of corporate downsizing.

The implications of these factors for engineering ethics have largely been ignored. The codes of ethics, with the significant exception of the code of the Institute of Electrical and Electronics Engineers, have not changed to reflect this new professional environment. The teaching of engineering ethics needs to recognize that interaction between the genders and the generations requires new emphases. For example, the issue of sexual harassment needs to be specifically addressed. As well, it needs to be recognized that engineers will be interacting with other engineers with different background values, even if they did happen to grow up and receive their education in the same country. What is "conduct or practice which is likely to discredit the profession or deceive the public" (NSPE Professional Obligation 3) for engineers from one ethnic or generational group might well not be evaluated in the same way by those from another.

Even more fundamentally, perhaps, the changing nature of employment has major implications for the moral status of loyalty. Inherited conceptions of loyalty to an employer make sense only in a context where loyalty is a two-way street. They will thus ring false for many younger engineers. For engineers to be "faithful agents" requires that they perceive their employer to be worthy of trust. In an age of corporate downsizing this is no longer the case in some instances. Something new, therefore, needs to be substituted as a basis for action. I would suggest the notion of professional accountability for engineers, having them see their duties not focused around an employer, but rather around a set of values generated from the community of engineers. This would avoid the analysis of engineering activities focused on economic relationships. Rights and duties for engineers need to be established independently of their employer of the moment.

Fundamental Changes in the Nature of Technology

What is considered to be technology and, therefore, the domain of engineers has been undergoing a historical revolution. While all the traditional aspects of that technology are still in place, more and more demands on engineers exist in areas not covered by the traditional interpretation of what technology consists of. Furthermore, these new areas of technology have created novel concerns for the engineer both in terms of their own activities and in terms of the implications of their work on others. The major agent for this technological revolution has been the introduction of computers. Computers have brought with them a shift from hardware to software, have enabled a quicker spread of technology, have enlarged the scope and scale of technology, and have created distance between engineers and their products. But computers are not simply end pieces of use. Through computers, engineers are actually interacting with other human beings.

The foremost implication of this is that the appropriate model of the contemporary engineer is no longer that of an isolated individual (if that ever was the case in reality), but rather a team of engineers working together, perhaps even without personally knowing each other. But this means that ideas of individual accountability as they are taught in engineering ethics are no longer applicable to the working lives of many engineers. Nor can responsibility for actions be assigned in the same way when no one individual is responsible for the effects that are created through technological development.

A secondary issue related to this development is the concern with privacy and property. The engineering codes ignore what is becoming a fundamental concern in this regard. What are the rights to privacy of engineers? What rights to privacy do those who are affected by engineers' work have? What property rights does one have when the notion of property is becoming more and more elusive in an age of software? Engineering ethics education must address these issues in light of the new technologies and help students clarify their role in a system that is placing less and less emphasis on individual performance. It is evident from the current state of engineering education that teaming will be increasingly emphasized in the future. The Accreditation Board for Engineering and Technology (ABET) requires in its new *ABET 2000* criteria that students learn to function on "multi-disciplinary teams," yet students are often lost when it comes to the ethical demands implied by teamwork. Most interpretations of the rights and obligations of engineers still stress the idea of professional conduct as actions of one engineer serving one client. The new technologies make this an unrealistic ideal, and the discussion of ethics in engineering needs to be adjusted for the new environment of teaming and lack of face-to-face contact between engineers engaged on the same project.

Ethical Complexity

We are facing an age of increasing ethical demands being placed on engineers. The three factors discussed above indicate that the changes occurring in engineering are pulling engineers in different directions. They involve new conflicting demands to which the traditional solutions to engineering dilemmas cannot be applied in a straightforward fashion. Loyalty to employer, to a national profession, and to a particular society will no longer be a sufficient paradigm for engineering ethics in the future. The meaning of "employer," "profession," and "public" will vary with context. Thus, inherited conceptions of right and wrong will be insufficient for making appropriate ethical judgments. Yet we cannot ask engineers to adopt a form of ethical relativism, for this would mean having them shift the ethical basis for their actions on a daily or even hourly basis. The difficulty is that the factors discussed tend to make competing demands of engineers, especially when combined with traditional interpretations of engineering ethics, yet for engineers to serve a global society a universal perspective is needed. Such a perspective would take account of various values, but in the end find its foundation in the activity of engineering itself. New background conceptions of ethical behavior must be established for engineers. In my view, these ought to be based on a concept of professional conscience, on the idea that inherent in engineering are fundamental professional obligations that arise, not from a contract with a particular society, but from what engineers do.

The difficult part of such an undertaking will be to harmonize the competing demands of various constituencies, technologies, and values into one unified perspective so that engineers will have the guidance they need, because while each of the ideas discussed above raises ethical concerns other than those that have traditionally been emphasized, they also raise the possibility of ethical conflicts among themselves. Perhaps the way to proceed is to first examine the issues individually and then consider how problems among them can arise and how ethical guidelines might be constructed to deal with these. The very process of examination will surely show how difficult ethical considerations will be for engineers in the future and why, consequently, there needs to be an ever increasing emphasis on ethics education for engineers.

Conclusion

Given the competing demands and increasingly complex choices that will be faced by engineers in the future, how can engineering educators best prepare them? It seems clear from the preceding that neither education into nationally derived codes of ethics or a general reliance on a European model of ethical theory will be sufficient. Neither of these prepare engineering students for dealing with encounters with a variety of peoples of differing backgrounds and value assumptions. Yet it seems regressive to either demand no ethical reflection or simply an adaptation to local customs on the part of engineers, especially since the idea of local customs is an elusive one within the shifting trends of the utilization of technology.

It seems to me that the only viable solution to the problems raised is the establishment of a common framework for ethical decision-making for all engineers, whatever their nationality, their location, or their personal and group characteristics. This will require intense reexamination of what it means to be a human being in the world and what it means to be a professional engineer. Such an examination will not be easy and will, in fact, require compromise rather than simply abstract analysis. However, if it is not undertaken, we can expect increasing problems for both engineers and for the products they design. Whether the will exists among engineers to accomplish such a goal is another issue. Other professions, for instance medicine and nursing, have strived to accomplish this goal.

In effect, this chapter sets out the need for an international or global code of engineering ethics. Such a code can then be used as a teaching tool and a way of holding engineers accountable for their actions, whatever context they may be operating in. It would overcome both narrow perspectives of what constitutes culture and nationalistic biases. I do not kid myself. This will be difficult. In the past it has been difficult to even get American engineers to agree on a common framework for engineering ethics, based on a claimed difference in specialization. How much more difficult development of a common framework for all engineers will be can easily be imagined. But it seems to me an urgent and necessary task for engineering ethics at the dawn of the twenty-first century.

References

1. Baron, Marcia, *The Moral Status of Loyalty*, Dubuque, IA: Kendall/Hunt, 1984.
2. 1987 version of the "Code of Ethics for Engineers" as established by the National Society of Professional Engineers.
3. Herkert, Joseph, Codes of Ethics and the Moral Education of Engineer, *Business and Professional Ethics Journal*, Vol. 2, No. 4, 1983, pp. 41–61.
4. Herkert, Joseph, What Is Engineering Ethics?, *Science, Technology and Society*, Number 36, June 1983, pp. 1–6.
5. Martin, Mike W. and Schinzinger, Roland, *Ethics in Engineering*, 3rd edition, New York: McGraw-Hill, 1996.
6. Paul Shrivastava, *Bhopal—Anatomy of a Crisis*, Cambridge, MA: Ballinger, 1987.

CHAPTER 10

JOSEPH R. HERKERT

Integrating Engineering, Ethics, and Public Policy

Three Examples

Engineering ethics and engineering and public policy are both relatively new fields that have grown in stature and impact over the past two decades. Although each has a unique focus, in both cases engineers are challenged to move beyond traditional disciplinary boundaries and combine knowledge and methods from other disciplines with technical knowledge and methods. As would be expected, engineers have been involved in the development of both fields through research, education and the activities of professional engineering societies. In addition, applied ethicists, working on their own or in conjunction with engineers, have made significant contributions to the theory and practice of engineering ethics, as have social scientists in the area of engineering and public policy. Although a number of individual engineers, as well as the major professional societies, have been active in both, there have been few formal efforts to date to achieve greater integration and cross-fertilization of the two fields.

In this chapter I demonstrate the need for such integration by examining three current public policy issues with significant relevance to engineering: risk communication, product liability, and sustainable development. In each case I will argue that public policy positions typically advocated by engineers and engineering societies are often at odds with, or uninformed by, concepts that are fundamental to engineering ethics. This outcome, I will argue, is at least in part due to the conventional approach to engineering ethics, which takes as its focus the actions and motivations of individual engineers and professional societies, often to the neglect of broader questions of public policy. Consequently, the remedies I propose are for the most part aimed at encouraging more extensive treatment of public policy within the field of engineering ethics. While outside the scope of this chapter, related arguments could be made for elevating ethical analysis to a more prominent position in the field of engineering and public policy.

Following a discussion of critiques of the conventional approach to engineering ethics, as well as various schemes that have been proposed for classifying approaches to engineering ethics, I will focus on risk communication, product liability, and sustainable development as examples of the need for better integration of engineering ethics and public policy. I conclude with suggestions of how such integration might be better accomplished in research, education, and activities of professional engineering societies.

Engineering Roles and Ethical Responsibility

Langdon Winner, well known for his critical analyses of technological development, is equally critical of traditional approaches to engineering ethics.[1] Such approaches, Winner argues, focus almost entirely on specific case studies of ethical dilemmas to the exclusion of larger issues relating to the development of technology and the career choice of engineers:

> Ethical responsibility . . . involves more than leading a decent, honest, truthful life, as important as such lives certainly remain. And it involves something much more than making wise choices when such choices suddenly, unexpectedly present themselves. Our moral obligations must . . . include a willingness to engage others in the difficult work of defining the crucial choices that confront technological society and how to confront them intelligently.

Similar critiques of engineering ethics have been raised by others, including engineers such as Vanderburg[2] who draws a distinction between "microlevel" analysis of "individual technologies or practitioners" and "macrolevel" analysis of "technology as a whole."

A number of ethicists have attempted to categorize levels of ethical analysis relating to engineering (see Table 10.1). De George, for example, as reported by Roddis,[3] distinguishes between "ethics in engineering" and "ethics of engineering," the former referring to the actions of individuals and the latter to professional relationships and the responsibilities of the engineering profession to society. Ladd argues[4] that professional ethics can be delineated as "micro-ethics" or "macro-ethics" depending on whether the focus is on relationships between individual

TABLE 10.1. Engineering Ethics and Engineering Roles

	The Engineer's Roles		
Source	Personal	Professional	Public
De George, as reported in Roddis	*Ethics in engineering:* actions of individual engineers	*Ethics of engineering:* the role of engineers in industry and other organizations, professional engineering societies, and responsibilities of the profession	
Ladd		*Micro-ethics:* professional relationships between individual professionals and other individuals who are their clients, colleagues and employers	*Macro-ethics:* problems confronting members of a profession as a group in their relation to society
McClean	*Technical ethics:* technical decisions and judgments made by engineers	*Professional ethics:* interactions between engineers and other groups	*Social ethics:* technology policy decisions at the societal level

engineers and their clients, colleagues, and employers, or the collective social responsibility of the profession. Mclean,[5] an engineer, utilizes three categories: technical ethics, professional ethics, and social ethics dealing, respectively, with engineering design decisions, business and professional issues, and sociopolitical decisions concerning technology.

As indicated by the column headings in Table 10.1, combining these various schemes suggests a view that an individual engineer can confront ethical dilemmas in at least three distinct, albeit overlapping, roles that I will call the personal, professional, and public roles. By personal role, I mean the role of an engineer as an individual actor engaged in making technical decisions, whereas by professional role I refer to ethical rights and responsibilities that arise in the relationships entered into with others of the same profession, clients (including the public), or employers. The public role entails the involvement and responsibilities of engineers, individually or collectively, in relation to questions of broad technology policy.

While some work in engineering ethics has been focused on the public role of engineers, Winner's critique is essentially correct. The primary focus on engineering ethics, whether by ethicists, engineers, or professional engineering societies, has been on the personal and professional roles of engineers. Examples of this focus are case studies such as the DC-10 case, Challenger disaster, and Goodrich air brake case, where individual decisions by engineers and managers are called into question. Similarly, engineering codes of ethics, while stressing moral responsibilities to the public, tend to be aimed at the duties and rights of individual engineers in carrying out such responsibilities. In the traditional approach to engineering ethics, public policy is often only considered tangentially, as in the case of legislation to protect whistleblowers, or the ethical responsibilities posed by environment, health, and safety regulations.

The public role of engineers has been a central focus of efforts falling under the banner of Engineering and Public Policy (EPP). EPP approaches, however, tend to be long on policy and economic analysis and short on ethical analysis, at least of the sort that would connect back to the engineer's personal and professional roles. As will become apparent from the activities of professional engineering societies and the beliefs of individual engineers that will be examined in subsequent sections of this chapter, the connections between ethics and policy are often made in a shallow manner by such groups and individuals or not made at all.

This state of affairs is problematic for a number of reasons. From a societal viewpoint, we need policies that are ethical and ethical viewpoints that are sensitive to social problems and issues. From the individual's viewpoint, engineers need ways of dealing in a consistent and holistic manner with ethical issues that arise in their various roles. In the absence of integration of ethical considerations from their personal and professional roles with issues that may arise in their public roles, engineers might become confused or complacent regarding the importance of ethics and its connection with public policy.

In order to illustrate these problems, in the following sections I consider three major public policy issues that have caught the attention of engineers and professional engineering societies-risk communication, product liability, and sustainable development. In each case I highlight some ethical concerns related to these issues and provide examples of how engineers or professional engineering societies have either ignored such ethical questions, or avoided facing them head-on in their considerations of public policy.

Risk Communication

Risk communication serves as a rich example of the interrelation of engineering ethics and public policy. Questions related to risk assessment and risk management are arguably the most prominent issues encountered in engineering and public policy. During the past few decades engineers have been embroiled in numerous controversies surrounding technological and environmental risks, including nuclear power, toxic chemicals, and transportation safety.

The duality of risk and safety, and the clear ethical issues involved in determining acceptable risk, are central to most treatments of engineering ethics.[6] Less obvious to engineers are the ethical issues involved in determining or measuring risk[7] and the ethical issues raised by differences in risk perception between technical experts and ordinary citizens.[8]

As I have argued elsewhere, such perceptual differences, which have been widely studied and characterized by behavioral and social scientists, reflect real limitations to expertise, resulting in an ethical obligation for engineers to listen to and attempt to understand the concerns of nonexperts, and to incorporate such concerns into their consideration of risk. Others have made similar arguments in favor of incorporating lay risk perception in risk assessment activities based upon the principle of informed consent.[9]

The implication of these arguments is that, in order to succeed, efforts at risk communication should be two-way, with risk information being exchanged between experts and nonexperts, as opposed to the traditional, one-way model of risk communication in which the expert merely informs or educates the lay public as to the facts of the situation. As psychologist Paul Slovic has argued:

> there is wisdom as well as error in public attitudes and perceptions. Lay people sometimes lack certain information about hazards. However, their basic conceptualization of risk is much richer than that of the experts and reflects legitimate concerns that are typically omitted from expert risk assessments. As a result, risk communication and risk management efforts are destined to fail unless they are structured as a two-way process. Each side, expert and public, has something to contribute. Each side must respect the insights and intelligence of the other.[10]

This argument also extends to the importance of considering both quantitative methods and information, favored by engineers and other experts, and qualitative methods and information, utilized by some social scientists and the lay public. Fischhoff and Merz[11] have noted for example:

> Quantitative understanding is essential if people are to realize what risks they are taking, decide whether those risks are justified by the accompanying benefits, and confer informed consent for bearing them. Qualitative understanding is essential to using products in ways that achieve minimal risk levels, to recognizing when things go wrong, and to responding to surprises.

Even if one holds to the traditional model of risk communication, however, in which the expert is the primary communicator and the nonexpert the primary receiver of risk information, there are still ethical questions that need reckoning with, including issues relating to the purpose of the risk communication, the motivation of the risk communicator, and whether the risk message is overt or covert.[12]

While researchers in the field of engineering and public policy often incorporate such considerations in their work, many engineers and engineering societies that address risk issues cling to the traditional model of risk communication, belittle and devalue lay perceptions of risk, and characterize any approaches to risk other than their own as irrational and irrelevant to public policy. For example, a leader of the Committee on Man and Radiation (COMAR) of the Institute of

Electrical and Electronics Engineers (IEEE) maintains that all views on the health impacts of electromagnetic radiation other than that of COMAR are simply incorrect, and spawned by "electrophobia," an irrational fear of electrotechnology. His indictment goes beyond members of the lay public, extending even to other IEEE members and staff who do not accept COMAR's position.[13] Similarly, a civil engineering educator recently argued that:

> extreme environmentalism, akin to a religion, is causing the expenditure of massive amounts of limited resources of both money and attention to solve relatively unimportant problems. The public perception of environmental problems is far removed from that of the scientific consensus. Environmental policy is now based more on emotion and debate than on facts and rational calculation. A moderate, rather than fearful, reaction to environmental concerns is urged. Engineers should promote quantitative solutions to environmental problems, and should appreciate the economics of pollution control and risk reduction.[14]

In addition to skirting serious ethical questions, such posturing on the part of engineers and engineering societies is often counterproductive, contributing to a negative public image of engineers[15] and obscuring important risk information that might fall outside of the bounds of expert risk models.

Even engineers with a strong concern for public safety in their personal and professional roles can retreat to the traditional model of risk communication in their public role. Consequently, a good place to begin persuading engineers to take public risk perception more seriously would be through development of a framework for engineering ethics that is conducive to better integration of the personal, professional, and public roles.

Product Liability

The recent interest in tort liability reform is a particularly vivid example of how issues of risk can potentially cause conflicts between public policy and engineering ethics. The 104th Congress passed legislation that would severely limit the effect of product liability litigation by placing a cap on punitive damages and enacting stricter requirements for holding manufacturers liable. Though President Clinton, as expected, vetoed the bill, product liability is almost certain to be a recurring issue in future congressional sessions.[16]

The proponents of product liability reform have argued that the current system unjustly rewards plaintiffs and stifles technological innovation, resulting in a lack of competitiveness on the part of U.S. manufacturers and decreased product safety. On the other hand, supporters of the current system point out that while there might be a need for minor refinements, the system generally works as intended in discouraging the manufacture of defective products and compensating people injured by such defects.[17]

Some observers view the debate over product liability reform as a classic industry versus consumer confrontation. A *New York Times* editorial,[18] for example, described the proposed legislation as "The Anti-Consumer Act of 1996." Engineers and engineering societies, on the other hand, have tended to side with the proponents of product liability reform. For example, a vice-president of engineering of a major U.S. automobile company has argued that product liability restricts engineering practice by inhibiting innovation, discouraging critical evaluation of safety features, and preventing implementation of new or improved designs.[19]

The position statement on product liability tort reform of the United States Activities Board of the IEEE,[20] issued in 1993, calls for stringent limits on product liability, including holding the manufacturer blameless when existing standards are met, adequate warnings are provided, or the product is misused or altered by the user. Other engineering societies, such as the American Society of Mechanical Engineers,[21] have also actively supported product liability reform.

While many engineers are supportive of changes in the product liability system, few have considered the effect that decreasing the impact of product liability would have on engineering ethics, most notably on engineers seeking to call to the attention of their employers design defects that might jeopardize safety. Indeed it is dismaying how little attention the engineering community has paid to the ethical implications of product liability. For example, a major 1994 study of product liability and innovation by the National Academy of Engineering, while considering such issues as corporate practice, insurance, regulation, and the role of scientific and technical information in the courtroom, contained no ethical analysis at all, with the exception of a single chapter by Fischhoff and Merz which, in addressing public risk perceptions, raised the issues discussed in the preceding section of this chapter.

Even the ethics literature is equivocal on the issue of product liability.[22] In De George's well-known essay on engineering responsibility in the Pinto case,[23] for example, he advocates stronger regulation and fines and imprisonment for corporate officials when market mechanisms fail to achieve desired levels of safety, giving only passing notice to the role of product liability litigation in this regard.

While the evidence appears to be mixed concerning whether or not product liability rewards result in improvements in product safety, the role of product liability litigation in creating an environment wherein engineers with safety concerns are given a hearing by their managers is certainly worthy of consideration. As Ladd[24] and others have argued, corporations are not moral agents. The sole goal of a corporation is to generate profits. In order to influence a corporation's behavior, then, we need to make it in their economic interest to do the right thing. Product liability litigation would seem one mechanism for achieving this goal. At the very least, the connection between the threat of product liability suits and the ability of engineers to raise safety concerns is a question that ought to be scrutinized by those concerned about engineering ethics.

Another aspect of product liability reform that merits further attention is the effect that sweeping changes might have in weakening the legal doctrine of informed consent, which evolved from tort law, and its role in providing support for the ethical principle of informed consent, which, as noted previously, is often the basis for arguments concerning engineering responsibility for understanding and accommodating the public's views on risk.

Finally, it should be noted that the overriding concerns here are not limited to product liability but also to other forms of tort liability of interest to the engineering profession, including malpractice issues and liabilities resulting from environmental risks and accidents in public facilities. As liability issues grow in importance in the public policy arena, engineers will need ways of reconciling their public positions with their personal and professional ethics.

Sustainable Development

The concept of sustainable development has become a major public policy issue worldwide, including within the engineering community.[25] Following the Brundtland Commission report in 1987, which defined sustainable development as "development that meets the needs of the present without compromising the ability of future generations to meet their own needs,"[26] the concept attracted considerable attention within the international community and on national agendas. The 1992 UN Conference on Environment and Development in Rio de Janeiro produced a blueprint for global sustainable development, Agenda 21, which resulted in the establishment of dozens of national commissions, including the US President's Council on Sustainable Development.[27] The engineering community has also reacted, with the establishment of the World Engineering Partnership for Sustainable Development (WEPSD) in 1992,[28] and issuance of position statements by committees formed within the traditional engineering organizations, including the American Society of Civil Engineers (ASCE)[29] and IEEE.[30]

Sustainable development theory emerged from the field of ecological economics. According to this theory, sustainable development involves achieving objectives in three realms: ecological, economic, and social (see Figure 10.1). The ecological objective involves maintaining a sustainable scale of energy and material flows through the environment such that carrying capacity of the biosphere is not eroded. The economic objective seeks to provide an efficient allocation of resources in conformance with consumer preferences and the ability to pay.

```
                    Social Objective
                          △
                         ╱ ╲
                        ╱   ╲
                       ╱     ╲
                      ╱       ╲
                     ╱_____╲
         Ecological Objective    Economic Objective
```

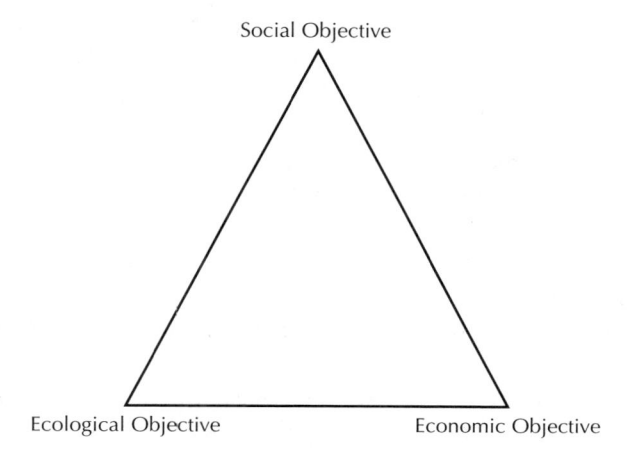

FIGURE 10.1. Objectives of a sustainable society

The social objective aims at a just distribution of resources among people, including future generations. The overall objective of a sustainable society is the achievement of sustainability in economic, ecological, and social systems.[31]

The success of public policy to promote sustainable development is dependent upon achieving all three objectives of a sustainable society. However, questions of just distribution and other questions of equity (such as risk distribution) are often left off the table when engineers (and others) consider sustainable development policies and issues. Indeed, almost all the effort of engineers and engineering organizations on the issue of sustainable development is focused on the need to strike a balance between economic development and environmental protection. While these efforts are commendable, they are limited by their failure to come to grips with the third essential element of sustainable development—the social objective. In other words, as Beder[32] warns, sustainable development requires more than the mere application of technological fixes to environmental problems.

The social dimension of the sustainable development problem is either ignored entirely or relegated to a secondary status by many of the engineers and engineering groups active in the promotion of sustainable development. It is common, for example, for engineers writing on the topic to refer to development that is "environmentally and economically sustainable"[33] or to view sustainable development as an attempt to balance environmental quality and economic growth.[34] Social objectives, if made explicit at all, are often only considered to the extent that they have an impact on the economic or ecological objectives, as in the following statement by a prominent member of the WEPSD:

Factors in the project that pertain to economics, culture, environment, or technology are all interdependent. This interdependence requires that the projects be examined in terms of their impacts on the ecosystem.

The IEEE Environment, Health, and Safety Committee (EHSC), charged with tending to sustainable development, has downplayed the concept explicitly because it is "somewhat value-laden . . . implying for some people, for example, redistribution of wealth or a need to restrict current consumption." The EHSC has instead embraced the concept of "industrial ecology," which limits its focus to "industrial and economic systems and their linkages with fundamental natural systems."

Most disturbing are the not-so-hidden messages that suggest the motivation behind the involvement of some engineers and engineering groups in sustainable development. For example, risk and liability concerns have been frequently cited by these groups as impediments to innovation for sustainable development,[35] suggesting that, for some, sustainable development is merely another vehicle to promote the narrow engineering views discussed in the previous two sections of this chapter. Engineers who promote sustainable development often ridicule other stakeholders, such as environmental groups and ordinary citizens, and government regulations are often vilified by them.[33] A founder of the WEPSD, in articulating what is clearly a technocratic vision of sustainable development, even goes so far as to portray engineers as the best arbiters of *all* knowledge that needs to be brought to bear on the problem:

> It is important for engineers to provide the implementing interface between science, society, and the decision-making bodies (public and private) to ensure that the most appropriate knowledge in natural, technological, and social sciences is implemented to meet the needs of both present and future generations.

There have been some efforts aimed at incorporating environmental and social equity concepts into engineering ethics. Here again, however, social concerns have been secondary to environmental issues. For example, while a proposal from the World Federation of Engineering Organizations, one of the founding organizations of the WEPSD, promotes social and cultural sensitivity, it does so within the framework of a "Code of Environmental Ethics for Engineers."[36] Similarly, a proposed addition to the ASCE Code of Ethics called for engineers to "sustain the world's resources and protect the natural and cultural environment," where the phrase "cultural environment" apparently refers to the built environment. Even the inclusion of this limited notion of sustainable development in the ASCE code generated considerable controversy for several years.[37] Although subsequent revisions of the ASCE code, adopted in 1996, make explicit the engineer's commitment to the environment and introduce the concept of sustainable development, the language utilized strongly implies that the principles of sustainable development extend only to the protection of the environment.

As in the cases of risk communication and product liability, it would seem that engineering ethics would have much to contribute to the understanding of the engineer's public role with respect to sustainable development, both in terms of the social objective of sustainable development, and in better integrating environmental ethics and engineering.

Integrating Ethics and Public Policy

The cases of risk communication, product liability, and sustainable development highlight the importance of integrated approaches to engineering ethics and public policy, including the need for innovative pedagogies in engineering ethics, increased attention on the part of professional societies to consistency in their treatments of ethical and public policy issues, and increased understanding on the part of policy analysts and policymakers of the fundamental importance of professional ethics.

An example of an integrated approach to engineering ethics and public policy are the preventive approaches to engineering and ethics advocated by Vanderburg, Harris,[38] and others. By focusing on proactive measures to avoid narrowly conceived engineering designs and circumstances that precipitate ethical dilemmas, preventive approaches circumvent the problem of reducing engineering ethics to a series of individual cases while at the same time making engineering students and practitioners aware of the broader environmental and societal contexts in which engineers operate.

Educational innovation aimed at stronger integration of engineering ethics and public policy can occur at both the levels of curricular reform and classroom methodology. At the curricular level, Devon,[39] for example, advocates an integrated sequence of design courses that stress the interdisciplinary nature of engineering "in terms of the technological, ecological and social systems within which it operates." In a similar vein, Mclean, noting that "engineering is *all about* ethical questions," argues for thorough integration of ethics and the engineering design curriculum.

Curricular innovation can also entail stronger integration of engineering education with studies in the humanities and social sciences. The Benjamin Franklin Scholars Program at North Carolina State University, for example, is a five-year, dual-degree program in which engineering students earn a conventional B.S. in engineering or computer science along with a conventional or self-designed multidisciplinary degree in the humanities and social sciences. The scholars' programs are tied together by a core of three multidisciplinary courses that progress from history, philosophy, and sociology of technology, through an ethical analysis of the idea of progress, to a capstone course involving technology assessment and public policy.[40] In addition, students who pursue a degree in multidisciplinary studies (MDS) for their second major often choose to combine studies of ethics and public policy. For example, Franklin Scholars have designed MDS majors in "Environmental Politics and Ethics," "Ethics and Technical Management" and "Issues in Medicine and Bio-technology."

In addition to traditional topics and case studies in such areas as whistleblowing, conflict-of-interest, and trade secrecy, engineering ethics courses should include substantial emphasis on public policy issues that have significant relevance to engineering. In addition to the examples highlighted in this chapter, public policy issues that might be considered include such concerns as privacy and freedom of expression on the Internet, the role of technology in healthcare, and the effects on engineers and engineering of corporate restructuring and downsizing. Many of the cases typically used in engineering ethics courses are amenable to discussion of public policy issues (for example, the Ford Pinto case), while less common cases that have significant public policy implications, such as the Bjork-Shiley heart valve case,[41] should be developed and utilized to a greater extent.

Collaborative learning pedagogies,[42] where students work together in semi-autonomous pairs or small groups, are especially well suited to the integration of ethics and public policy. In addition to improving student performance and enthusiasm for learning and developing student skills in such areas as communication, the exercise of judgment, and teamwork, collaborative learning provides opportunities for engineering students to explore ethical and policy problems from the vantage point of the personal, professional, and public roles, and to develop strategies for reconciling these various roles. Such activities as role playing, interactive exercises, and group presentation of case studies are particularly effective. For example, playing the role of a juror or a plaintiff in a product liability case, group response to risk perception exercises, and research and presentation of engineering ethics cases with public policy implications expose students to viewpoints and information they might not otherwise consider, as well as give them practice at dealing with people from other backgrounds and with other viewpoints.

Professional engineering societies need to engage in efforts to consolidate their public policy and ethics activities. One approach to this problem would be to revise their codes of ethics to incorporate a broader appreciation of the public role of engineers, as a complement to the public responsibility of engineers in their personal and professional roles, a concept which is already present in most of the codes. The IEEE Code,[43] for example, in addition to the standard admonition for engineers to uphold the public safety, health, and welfare, has a provision that, by calling on engineers to "improve the understanding of technology, its appropriate application, and potential consequences," requires moral responsibility in the engineer's public role. Taking this notion even further, Vesilind has proposed that the ASCE code be replaced with a "Code of Professional Responsibility" that would require engineers to consider public viewpoints on such issues as risk and include the public in decisions concerning their projects.

The professional societies should also ensure closer linkage between their bodies that make policy recommendations and their bodies that oversee ethics concerns. IEEE, for example, has been considering mechanisms by which policy statements could undergo rigorous technical review by the appropriate experts within the organization.[44] Why not also subject policy recommendations to ethical review by the organization?

While the focus here has been on recasting engineering ethics so as to better incorporate public policy considerations, the field of engineering and public policy would likewise be strengthened by more serious attentiveness to ethical issues, particularly microethical concerns that often escape the attention of policy analysts and policymakers. It is possible and desirable, for example, to include considerations of public viewpoints and perceptions in the conduct of risk assessments,[45] and to include social indicators among the sustainability metrics utilized in decision models that evaluate technology choice for sustainable development.[46] Courses in engineering and public policy, many of which already deal with macroethical questions such as intergenerational equity, should also focus attention on how public policy affects the personal and professional roles of engineers.

In this chapter, I have made a case for the centrality of ethical considerations in engineering and public policy, using the examples of risk communication, product liability, and sustainable development, and argued for stronger integration of engineering ethics and public policy in research, education, and the activities of professional engineering societies. The need for this integrated approach evolves from recognition that the social implications of technology permeate the three spheres in which the engineer operates: personal, professional, and public. Limiting the focus of engineering ethics to the personal and professional spheres, and public policy to the public sphere (the conventional approaches), can leave the engineer (and the engineering student) with a disjointed sense of his or her role in addressing important issues of ethics and public policy, and can result in public policy positions by engineers and professional engineering societies that are lacking in ethical foundations.

References

1. Winner, Langdon. 1990. Engineering Ethics and Political Imagination, in Durbin, P., ed., *Broad and narrow interpretations of philosophy of technology: Philosophy and Technology*, 7.

2. Vanderburg, Willem H. 1995. Preventive Engineering: Strategy for Dealing with Negative Social and Environmental Implications of Technology. *Journal of Professional Issues in Engineering Education and Practice* 121: 155–160.
3. Roddis, W. M. Kim 1993. Structural Failures and Engineering Ethics. *Journal of Structural Engineering* 119: 1539–1555.
4. Ladd, John. 1980. The Quest for a Code of Professional Ethics: An Intellectual and Moral Confusion, in Chalk, R., Frankel, M. S. and Chafer, S. B., eds., *AAAS Professional Ethics Project: professional ethics activities in the scientific and engineering societies*. AAAS, Washington, DC: 154–159.
5. Mclean, Gerard F. 1993. Integrating Ethics and Design. *IEEE Technology and Society* 12 (3): 19–30.
6. Flores, Albert., ed. 1989. *Ethics and risk management in engineering*. Lanham, MD, University Press of America.
7. Mayo, Deborah.G., and Rachelle D. Hollander, eds. 1991. *Acceptable Evidence: Science and Values in Risk Management*. New York. Oxford University Press.
8. Herkert, Joseph R. 1994. Ethical Risk Assessment: Valuing Public Perceptions. *IEEE Technology and Society* 13 (1): 4–10.
9. Martin, Mike W. and R. Schinzinger. 1989. *Ethics in Engineering*, 2nd. ed. McGraw-Hill, New York.
10. Slovic, Paul. 1987. Perception of Risk. *Science* 236: 280–285.
11. Fischhoff, Baruch and Merz, J.an F. 1994. The Inconvenient Public: Behavioral Research Approaches to Reducing Product Liability Risks, in Hunziker, J.R. and Jones, T.O., eds., *Product liability and innovation*. National Academy Press, Washington, DC: 159–189.
12. Morgan, M.Granger and Lave, Lester. 1990. Ethical Considerations in Risk Communication Practice and Research. *Risk Analysis* 10: 355–358.
13. Osepchuk, John M. 1996. COMAR After 25 Years: Still a Challenge! *IEEE Engineering in Medicine and Biology* 15 (3): 120–125.
14. Chadderton, Ronald A. 1994. Should Engineers Counteract Environmental Extremism? *Journal of Professional Issues in Engineering Education and Practice* 121: 79–84.
15. Vesilind, P.Aarne. 1993. Why Do Engineers Wear Black Hats? *Journal of Professional Issues in Engineering Education and Practice* 119: 1–7.
16. Lewis, Neil A. 1996. Clinton Vetoes Bill to Curb Awards in Product Liability Suits. *New York Times* (May 3).
17. Hunziker, Janet R. and Trevor O. Jones, eds. 1994. *ProductL Liability and Innovation*. National Academy Press, Washington, DC.
18. *New York Times* (editorial), 1996. The Anti-Consumer Act of 1996. *New York Times* (March 21).
19. Castaing, Francois J. 1994. The Effects of Product Liability on Automotive Engineering Practice, in: Hunziker, Janet R. and Jones, Trevor O., eds., *Product liability and innovation*. National Academy Press, Washington, DC: 77–81.
20. IEEE United States Activities Board, 1993. Product Liability Tort Reform (position statement). Washington, DC. [Available from World Wide Web site, http://www.ieee.org/usab/DOCUMENTS/FORUM/LIBRARY/POSITIONS/tort.html.
21. ASME International, 1996. Letter to the House and Senate Conferees on Product Liability Bills. [Available from World Wide Web site, http://www.asme.org/gric/96-01.html.]
22. Rabins, Michael J., Edward Harris, and Michael Pritchard, 1992. Engineering Design: Literature on Social Responsibility Versus Legal Liability. [Available from World Wide Web site, http://ethics.tamu.edu/ethics/essays/design.htm.]

23. De George, Richard T. 1981. Ethical responsibilities of engineers in large organizations: the Pinto case. *Business and Professional Ethics Journal* 1: 1–14.
24. Ladd, J. (1982) Collective and individual moral responsibility in Engineering: Some Questions. *IEEE Technology and Society* 1 (2): 3–10.
25. Herkert, Joseph R. 1997. Sustainable Development and Engineering: Ethical and Public Policy Implications, in *Proceedings, 1997 International Symposium on technology and Society: Technology and Society at a Time of Sweeping Change*. IEEE Society on Social Implications of Technology: 175–180.
26. World Commission on Environment and Development, 1987. *Our Commo Future* (The Brundtland Report). Oxford. Oxford University Press.
27. President's Council on Sustainable Development Home Page (not dated) [Available from World Wide Web Site, http://www.whitehouse.gov/WH/EOP/pcsd/index.html.]
28. Carroll, William J. 1993. World Engineering Partnership for Sustainable Development. *Journal of Professional Issues in Engineering Education and Practice* 119: 238–240.
29. Baetz, Brian W. and R.M. Korol. 1995. Evaluating Technical Alternatives on Basis of Sustainability. *Journal of Professional Issues in Engineering Education and Practice* 121: 102–107.
30. IEEE TAB Environment Health and Safety Committee, 1995. *White Paper on Sustainable Development and Industrial Ecology*. [Available from World Wide Web site, http://www.ieee.org/ehs/ehswp.html.]
31. Farrell, Alex. 1996. Sustainability and the Design of Knowledge Tools. *IEEE Technology and Society* 15 (4): 11–20.
32. Beder, Sharon 1994. The Role of Technology in Sustainable Development. *IEEE Technology and Society* 13 (4): 14–19.
33. Hatch, Henry J. 1993 Relevant Engineering in 21st Century. *Journal of Professional Issues in Engineering Education and Practice* 119: 216–219.
34. Prendergast, John. 1993 Engineering Sustainable Development. *Civil Engineering* (October): 39–42.
35. Civil Engineering Research Foundation, 1996. *Engineering and Construction for Sustainable Development in the 21st Century: Assessing Global Research Needs*. Washington, DC.
36. World Federation of Engineering Organizations (not dated) *The WFEO Code of Environmental Ethics for Engineers*. [Available from World Wide Website, http://sunsite.anu.edu.au/feiseap/policies.htm#environmental ethics].
37. Vesilind, P.Aarne. 1995. Evolution of the American Society of Civil Engineers Code of Ethics *Journal of Professional Issues in Engineering Education and Practice* 121: 4–10.
38. Harris, Charles E., Jr. 1995. Explaining Disasters: the Case for Preventive Ethics. *IEEE Technology and Society* 14 (2): 22–27.
39. Devon, Richard F. 1991. Sustainable Technology and the Social System. *IEEE Technology and Society* 10 (4): 9–13.
40. Porter, Richard L. and Joseph R. Herkert, 1996. Engineering and Humanities: Bridging the Gap, in: *Proceedings of the 1996 Frontiers in Education Conference*. American Society for Engineering Education and Institute of Electrical and Electronics Engineers (available on the World Wide Web at URL http://www.caeme.elen.utah.edu/fie/procdngs/se8c4/papers5/96407.htm).
41. Fielder, John. 1995. Defects and Deceptions-the Bjork-Shiley Heart Valve. *IEEE Technology and Society* 14 (3): 17–22.
42. Herkert, Joseph R. 1997. Cooperative Learning in Engineering Ethics. Presented at Mini-Conference on Practicing and Teaching Ethics in Engineering and Computing,

Sixth Annual Meeting of the Association for Practical and Professional Ethics, Washington, DC.
43. Institute of Electrical and Electronics Engineers. 1990. *IEEE Code of Ethics*.
44. IEEE Ad Hoc Committee on Technology Policy Development. 1993. *Technical [Available from World Wide Web site, http://www4.ncsu.edu/unity/users/j/jherkert/ethics.html].information for the public welfare—fifth draft working paper*.
45. Shrader-Frechette, Kristen S. 1995. Evaluating the Expertise of Experts. *Risk: health, safety & environment*. 6:115.
46. Herkert, Joseph R., Alex Farrell.and James J. Winebrake. 1996. Technology Choice for Sustainable Development. *IEEE Technology and Society* 15 (2): 12–20.

*An earlier version was presented at the 1996 American Society for Engineering Education Annual Conference, Engineering and Public Policy Division, June 23–26, Washington, DC.

CHAPTER 11

MICHAEL E. GORMAN

JULIE M. STOCKER

MATTHEW M. MEHALIK

Using Detailed, Multimedia Cases to Teach Engineering Ethics

ABET has decided to switch to outcome-based assessment of engineering programs rather than lists of required courses (see the last page of this paper for resources on outcome-based assessment). Thus, programs will be evaluated according to their contribution to the skills that ABET has decided must be demonstrated. While a progressive move in some ways, it raises serious questions for the teaching of ethics. What kind of outcome-based measure can be used to assess whether we are producing ethical engineers from our undergraduate programs, programs that must demonstrate an "emphasis on effective communication and professional and ethical responsibility, awareness of the global societal context of engineering, and knowledge of contemporary issues"[1]

Adapting Mortimer Adler's Paideia Proposal[2] to ethics, we would argue that we seek outcomes in three areas:

1. Knowledge: This educational goal corresponds to what? Every student ought to know about ethical theories. For example, Harris, Pritchard, & Rabins[3] teach students to distinguish between a utilitarian approach and one based on

respect for persons (RP). The assessment of knowledge is straightforward. A standard test or quiz will tell us whether students can list facts about each of several ethical theories and whether they know the codes of their professions.

2. Skills: This goal corresponds to what students will also need to know about how to apply ethical theory and professional codes to practical problems—they will have to be able to engage in moral reasoning. Clever essay problems on tests might be able to assess aspects of this reasoning, but the best method is to use cases, an approach adopted in textbooks like Harris, et al., and Martin & Schinzinger.[4]

3. Wisdom: There are two aspects to this goal: when and why. Students ought to know when to apply a particular ethical framework to a problem, and also to be able to ask the deepest question: why are we doing this project or design? Will it make the world a better place? Ultimately, we hope we can turn students into virtuous practitioners. Gioia,[5] in his discussion of the Ford Pinto case, makes the distinction between ethical decisions, which accord with accepted professional standards and codes of ethics and moral decisions, which stem from a higher conviction about what is "right" (Gioia, 1992). Note that this higher conviction can only emerge after students have mastered ethical decision making. The key final step here is to grasp the spirit of the codes and be able to apply them to new situations for which they were not specifically designed. Moral imagination is a crucial part of this process.

Moral imagination is "an ability to imaginatively discern various possibilities for acting within a given situation and to envision the potential help and harm that are likely to result from a given action."[6] Moral imagination includes "our ability to elaborate and appraise different courses of action which are only partially determined by the given content of moral rules"[7] as well as an ability to understand the assumptions and conceptual framework in which a problem is embedded. Moral imagination is more than just a skill; it requires students to be able to switch frames, to really understand how a problem might appear from another point of view.

In his provocative novel *Ishmael*, Daniel Quinn[8] tries to create such a frame shift, arguing that the Taker myth that permeates our culture should be replaced by one based on Leaver values; the former sees Nature as something to be controlled, dominated, and manipulated, whereas the latter sees Nature as something to be left alone, whenever possible. In terms of moral theory, Quinn believes in creating virtuous people by altering the myth most of us live by. Once we have internalized a new myth, we will know how to share resources, not just with other human beings but also with other species. What Quinn is outlining is a process of moral imagination: recognizing that one's cultural worldview is a myth, trying out another view, and seeing how problems look from that perspective.

But one should not hold the new view dogmatically. Moral imagination is a tool for combating dogma, for recognizing that there are different ethical perspectives that can be applied to a problem. The hope is that by exercising moral imagination, practitioners will become reflective, considering alternative views and arriving at decisions that are better than one could develop from only one frame of reference. As Caroline Whitbeck argues, "[s]olving actual moral problems is not simply a matter of choosing the "best" of several possible responses. It is also a matter of devising possible responses."[9]

Donald Norman[10] distinguishes between experiential and reflective cognition. The former is exemplified by the expert in a domain working on a familiar problem. The "obvious" solution emerges from her experience with previous similar cases. The latter is exemplified by the expert moving into a new domain, where her previous experience does not immediately suggest a solution. This kind of experiential/reflective distinction can also be applied to ethics. Hopefully, students will become skilled at certain kinds of ethical reasoning, so that certain moral decisions will be virtual "no-brainers." But they will also need to be able to recognize situations where they have to reflect and apply moral imagination.

Recent research in cognitive psychology supports the view that experts learn from cases.[11] The case-study approach is being used increasingly to teach engineering design,[12] engineering ethics,[13] and environmental engineering. However, few cases studies exist that incorporate these three topics. Thus students tend to compartmentalize their thinking about design as if one could design something without considering its ethical and environmental implications. Many of the existing cases that encourage inclusion of environmental concerns in design development fail to account for the fact that any engineering design must be practical, that is, it must be capable of being produced given present technology, and it must be a viable product in the marketplace.

In introducing cases into the engineering curriculum, in many instances students are presented with short hypothetical scenarios or truncated vignettes from real events. These short cases are useful in pinpointing ethical issues. The danger of only using short cases is that it might encourage students to attack the particular issue, while neglecting the actual context and practical constraints in which any decision process is imbedded. It has been shown that longer, real-life cases that describe actual ethical and engineering dilemmas are also very effective pedagogical tools.[14] If so, then there is a role for longer, more fine-grained cases that combine design and ethical decision making. Ideally, these cases should be based on real-life events and should include multiple decision points for which there is no one simple "right" answer.

At the University of Virginia, we have created a website, *http://cti.itc.virginia. edu/~meg3c/ethics/home.html*, that contains a series of such case studies. This website was awarded first prize in a competition sponsored by MIT's Ethics Center for Engineering and Science. The site features primarily cases that have been researched and written by students, with faculty support from the School of Engineering and Applied Science and the Darden School of Graduate Business Administration. The website consists of a collection of cases that focus on ethical considerations in the early stages of the invention and design process, rather than as aftermath of a completed design. Because of the growing use of cases in engineering courses, and because it is difficult to separate out design issues from those in ethics and in the environment, we are developing cases that encourage students to think imaginatively about design in light of the increasing concern for the environment and other issues that will be challenging to them in their work as engineers. We hope to produce engineers who will be better able to make ethical decisions about creating and marketing new technologies.[15]

The first set of cases we are developing illustrate organizations that make sound engineering design decisions based on the best knowledge available; that think carefully about safety, the public good, and the environment; yet find themselves in trouble (Dow Corning [A] and [B]). What they lack is moral imagination: the ability to disengage themselves from their engineering/scientific point of view, to be aware of the ways in which other people frame and structure their experiences, and to understand and evaluate their activities through perspectives that are different from, even alien to, their own.

A second set of cases are being built around the challenge of trying to meet a rule-based Kantian imperative that argues that risk should be virtually eliminated. Engineering students are typically trained in a more utilitarian perspective that engages in a cost/benefit analysis of design-environmental costs, risks, and outcomes (DesignTex [A] and [B] and Rohner Textil cases). These cases will encourage students to exercise moral imagination by trying to meet challenges such as "Can one design a product that is environmentally sustainable and viable in the marketplace?" "Can one take into account and/or avoid social and political risks that are inherent in any product design?"

The goal of the project is to develop and disseminate cases and supporting materials that teach students to exercise good judgment and moral imagination, that help them learn that design always entails an ethical perspective, and that demonstrate that environmental design is both challenging and viable. These materials have been or will be tested in the classroom and should have appeal in a variety of disciplines including engineering, technology, environmental studies, and ethics. They will be published in the Darden Graduate School of Business Case Bibliography and eventually in a textbook.

This presentation will highlight two of the cases, both of which have been used in fourth-year courses required of all engineering students at the University of Virginia (Technology, Culture and Communication 401 or 402). The fourth-year course includes the kind of culminating design or research experience called for in the new ABET criteria, and also has a strong ethics and communications component. The first of the sequence of cases described below is intended for use early in this fourth-year sequence, when students are still thinking about what topic they want to pursue for their senior thesis; the second sequence of cases is intended for use in the second semester, when students are thinking about the kinds of problems that might emerge later in a research or design project. These two case-sequences are part of a larger set of cases that include other ones we have developed as well as more traditional cases like the Challenger and Bhopal. We also hope the case sequences described here can be used in other courses; indeed, we have tried them with great success in a course on technological thinking given to first-year honors students in engineering.

Toward a Sustainable Tomorrow: The Design of an Environmentally Intelligent Fabric

The new ABET criteria explicitly include environmental sustainability among the important considerations that students should consider when they invent and design. This case shows students how sustainability can be used to shape the design process. All of our cases have more than one decision point; the first decision is labeled A, and then succeeding letters of the alphabet are used for later decisions that the students must make based on the outcome of earlier ones.

Summaries

DesignTex (A): Susan Lyons, a vice president at DesignTex, a firm that develops high-end custom fabric collections, wants to create an environmentally responsible fabric that would provide a model for sustainable design. Ms. Lyons consulted with William McDonough, Dean of the School of Architecture at the University of Virginia and a noted designer of environmentally sustainable buildings. McDonough argues that designers and inventors need to think about cost, performance, and aesthetics, but adds two additional constraints: Will the design process and eventual product be ecologically intelligent? Will they meet the criteria of social justice? McDonough's stated ideal is that "No environmental risk is acceptable." Engineers tend to talk about "acceptable" and "unacceptable" risk so that McDonough's statement challenges a traditional paradigm and seems outrageous as a criterion for a practical design. Students have to engage in moral imagination if they are going to comprehend McDonough's point of view.

In this case, Susan Lyons took up his challenge to see whether or not DesignTex could design a fabric that would meet McDonough's ideal. The case follows the development of a new furniture fabric and puts students in Susan Lyon's shoes both before and after she consults with William McDonough, asking students to decide what it means to be environmentally sustainable and whether McDonough's principles go too far i.e., whether it is really necessary or feasible to redesign the chemical protocols by which all fabrics are manufactured in order to produce a completely compostable product that emerges from an absolutely clean manufacturing process.

DesignTex (B): This case describes the successful creation of a fabric that met McDonough's specifications and asks students who should be required to pay the additional cost of implementing McDonough's protocol.

Evaluation of the DesignTex Cases in Actual Engineering Classes

We believe the DesignTex cases offer valuable additions to engineering curricula. They offer students concrete examples of the importance of ethics in engineering. Students enjoy the chance to learn such lessons through "hands-on" experience. These cases encourage engineering students to consider ethics as a component of design, rather than as an afterthought. They force the students to steep themselves in that quagmire called ethics, an area not often emphasized in engineering classes, but a very real consideration of practicing engineers. The cases successfully required the students to consider ethical, practical, and economic components of a product in making decisions about its design and production. Rather than bore these students with moral platitudes and pre-packaged lectures, the DesignTex Cases require the students to actively consider the factors at play in integrating ethics into the engineering profession and the business world. They not only introduce the students to a real-world concept, but foster a greater empathy for the role of managers as decision makers (roles that most engineers move into as their careers progress). Students appeared to enjoy this change and acknowledged the need for such considerations now and in the future. When discussing the cases, students revealed a growing awareness of the tension between modern business/engineering design and the extent to which one infuses ethical considerations into these areas.[16]

Survey Results

About fifty fourth-year students from two sections of a fourth-year engineering course with an ethics component responded to surveys on the DesignTex cases. Students filled out the surveys on different days, and the number of students attending each day varied slightly. The questions immediately precede the summary tables.

The results below reveal students' tendency to compartmentalize ethics away from other topics. For example, they recognized the environmental nature of the case issues (as shown below), but did not recognize these issues as ethical ones, revealed in the lower score on whether the case "Raised important ethical issues." Overall, however, the students found the DesignTex case and the class structure around it interesting and relevant to their engineering education. While this is an encouraging observation for the future of engineering, it needs continual reemphasis through cases similar to DesignTex (Russell & Stocker, 1996).

Summary Tables of Selected Survey Questions

In the following, students rated their level of agreement in a scale from 5 (Strongly Agree) . . . 1 (Strongly Disagree).
This case . . .

a. Raised important ethical issues.
   Response Frequency 1-4, 2-14, 3-19, 4-10, 5-3, N/A-1, Total 51
b. Raised important engineering design issues.
   Response Frequency 1-2, 2- , 3-8, 4-24, 5-10, N/A-0, Total 51
c. Raised important environmental issues.
   Response Frequency 1-1, 2-3, 3-, 4-21, 5-19, N/A-0, Total 51
d. Contained interesting subject matter.
   Response Frequency, 1-4, 2-6, 3-18, 4-18, 5-5, N/A-0, Total 51

In the following, students responded on the following scale: 5 (Very Helpful) . . . 1 (Completely Unhelpful).
How valuable did you find . . .

a. The World Wide Web for the presentation and organization of the case?
   Response Frequency 1-0, 2-2, 3-11, 4-17, 5-19, N/A-2, Total 51

These results suggest that a few students still see a dissonance between engineering ethics, design, and environmental issues: forty students thought the case raised important environmental issues, thirty-four thought it raised important design issues, and only thirteen thought it raised important ethical issues. But much of this, we suspect, has to do with the fact that the ethical issues are embedded in the case, and not always explicitly labeled as such. In other words, when we teach this case, we need to do more to link it with students, knowledge about ethics, derived from other readings in the course. The students were very enthusiastic about the Web presentation.

Rohner Textil (A-D): We have also prepared a detailed set of cases based on the decisions made by Albin Kaelin, director of the Rohner Textil mill who had to produce the fabric. Kaelin had to build a complex network including spinning and twisting mills, dye companies, potential customers, Swiss regulators, and his own parent company. This set of cases follows his attempts to keep the network together and focused on the same set of environmental and ethical goals. For example, one dilemma focuses on what happens when the fabric fails a robotics test imposed by Steelcase, a major potential customer; in order to meet the new standards, a chemical will have to be added to the manufacturing process. Another dilemma focuses on employee autonomy: a dyemaster substitutes a dye without adhering strictly to the EPEA standards used by the mill. These cases involve the sorts of daily decisions an engineer-manager will have to make.

Science or Superstition? The Dow Corning Breast Implant Case

Engineering students frequently assume that it is sufficient to present good data in order to make a point, that the data will speak for itself. The Dow Corning case illustrates what happens when a company takes this stance.

Summaries

Dow Corning Corporation (A): There are several cases already written on the breast implant litigation situation. However none of the case writers have been allowed access to Dow Corning's records or key personnel. We have been offered this access. In particular, we explore the set of original design decisions that led to the creation, manufacture, and marketing of implants, and what was entailed in their subsequent redesign and alleged improvement. We asked the students to decide whether to put this "new, improved" implant on the market.

Dow Corning Corporation (B) (in progress): This is a follow-up to the (A) case, with emphasis on Dow Corning's reaction to increased regulation and changing federal and scientific community standards for testing of medical products, such as implants, as well as the first lawsuits against Dow Corning's implants.

Usually, at the end of the class session, we bring the students up to date on the status of the litigation. If the independent epidemiological data continue to be consistent with a multitude of present findings, it will turn out, just as Dow Corning has claimed, that the causal relationship between breast implants and disease is, at worst, a very weak correlation. "The best evidence now is a relative risk of 1.0, indicating no contribution of implants to the disease" (Angell, 1996, p. 197). Yet despite what appears from a scientific and engineering point of view to be "good scientific evidence," and despite the fact that Dow Corning scientists and managers believed they acted virtuously according to professional scientific and engineering standards, this company has been sued for over 4 billion dollars in a class action suit involving 440,000 women, and it is currently in Chapter 11 bankruptcy. These cases should be good illustrations of the importance of design decisions for the development and future of a product. They also raise questions about moral imagination, because they exemplify what can happen legally as well as morally when one has a product such as a breast implant and one does not take into account the perspective of women who received the implant.

Evaluation of the Dow Corning Cases in Actual Engineering Classes: When piloted in two freshman-level classes, the students reacted to Case A with a "no-brainer type" decision to market the implant, as originally anticipated by the authors. However, in some senior-level classes, there was difficulty overcoming the piecemeal knowledge the students had of the breast implant litigation, since it has been so well publicized. Because the students already know the outcome (bankruptcy for Dow Corning), they will often request further testing, relying on science to save the day. Typically, the students resist having to make a judgment call, since it forces them to share in the responsibility for the product and the amount of testing done to ensure its readiness for market.

Still, overall, both the first and fourth-year students recognized the inherent difficulty in deciding when a product has been tested enough to go to market. This case especially highlights the need for moral imagination, since the Dow Corning scientists, and thus the students using this case, must break out of their engineering mold and look at it from the perspectives of the managers, corporate lawyers, marketing personnel, and the women receiving the implants.

Case B continues this facilitation of moral imagination in the students. It presents the second phase of the history of Dow Corning's breast implant, from the time the new-and-improved version went to market to the first legal decisions against it. This case shows what happens when the scientists at Dow Corning failed to use their moral imaginations. As far as they were concerned, their science demonstrated the safety of the implants, but they failed to see the situation from the view of a woman with implants who has started hearing frightening news about them causing disease, although she feels fine at the time. In the end, students acknowledged the difficulties of relying solely on science in defending a product, since society overall, and especially the courts, treat evidence differently than is done in science.

Example of Survey Results: About forty fourth-year students from a fourth-year engineering course with an ethics component filled out the survey about Case A, and about thirty-five for Case B. Students filled out the surveys on different days, and the number of students attending each day varied slightly. The questions immediately precede the summary tables.

In contrast to DesignTex, students appeared to find both engineering and ethical issues in the Dow Corning cases at about the same rate, which indicates good balance between the ethics and the science of the cases. Overall, the students found both cases and the class structures around them interesting and relevant to their engineering education.

Case A -Dow Corning: In the following, students rated their level of agreement in a scale from 5 (Strongly Agree) . . . 1 (Strongly Disagree). This case . . .

    a. Raised important ethical issues.
       Response Frequency 1-0, 2-3, 3-9, 4-20, 5-9, N/A-0, Total-41
    b. Raised important engineering design issues.
       Response Frequency 1-0, 2-2, 3-14, 4-16, 5-9, N/A-0, Total 41
    c. Raised important environmental issues.
       Response Frequency 1-5, 2-12, 3-12, 4-6, 5-3, N/A-0, Total 38
    d. Contained interesting subject matter.
       Response Frequency 1-0, 2-3, 3-12, 4-15, 5-10, N/A-0, Total 40

In the following, students responded on the following scale: 5 (Very Helpful) . . . 1 (Completely Unhelpful).
How valuable did you find . . .

    a. The World Wide Web for the presentation and organization of the case?
       Response Frequency 1-0, 2-2, 3-8, 4-17, 5-9, N/A-0, Total 36

Case B -Dow Corning: In the following, students rated their level of agreement in a scale from 5 (Strongly Agree) . . . 1 (Strongly Disagree). This case . . .

    a. Raised important ethical issues.
       Response Frequency 1-0, 2-5, 3-12, 4-8, 5-11, N/A-0, Total 36
    b. Raised important engineering design issues.
       Response Frequency 1-0, 2-3, 3-9, 4-16, 5-8, N/A-0, Total 36
    c. Raised important environmental issues.
       Response Frequency 1-1, 2-16, 3-10, 4-4, 5-1, N/A-0, Total 32
    d. Contained interesting subject matter.
       Response Frequency 1-1, 2-1, 3-10, 4-16, 5-8, N/A-0, Total 36

In the following, students responded on the following scale: 5 (Very Helpful) . . . 1 (Completely Unhelpful).
How valuable did you find . . .

    a. The World Wide Web for the presentation and organization of the case?
       Response Frequency 1-0, 2-0, 3-10, 4-18, 5-6, N/A-0, Total 34

The pattern of results from Case A and Case B are very similar. Thus, we know that the cases work well together, and that they successfully balance the science with the ethics component, allowing the student to delve into both areas simultaneously.

Business and engineering classes will use the Dow Corning cases to prepare students for real-world problems, where an algorithm is not always available for solution. The case study approach is a widely accepted method for encouraging students to wrestle with problems in engineering, especially ethical issues. Engineers do not create products or processes just for use in the lab or a corporation. These processes are part of a larger system, society, of which we are all stakeholders. Challenging engineering students to think about such problems today will better prepare them for the working world they will inevitably enter. Along with the two cases, there is a teaching note for Case A that supplies a detailed teaching plan for the professor, facilitating use of the case teaching method in the classroom. These materials demonstrate the relevance of topics such as Dow Corning and their breast implant to the Engineering fields.

Conclusions

We hope this chapter demonstrates that cases can be a powerful method for introducing students to moral imagination and ethical reasoning in a way that shows a clear link to the design process. These cases lead to at least two outcomes:

1. Students learn to exercise moral imagination by: (a) adopting the point of view of a participant, like a small textile manufacturer in the DesignTex cases or a Dow Corning implant chemist, guessing what decision the person will make and defending it; (b) making and defending their own decisions about the case, which might be different from the participant's.
2. Students also learn to engage in moral reasoning as they make and defend their decisions. For example, can they identify how utilitarian and RP perspectives might lead to different decisions in the DesignTex case? Or, in the Dow Corning case, can the students support their decision whether to market the implant, by examining the different segments of society interested in the implant, and the degree of risk acceptable in different cases? We use role-playing to facilitate this kind of imagination and reasoning.

We are now at the stage where we are trying to design ways of measuring these outcomes. Our current evaluation surveys address student satisfaction, but not student ability to engage in moral imagination and reasoning. For example, in a recent section of the fourth-year thesis course, eight out of nine students indicated that they found the cases "helpful," but that very general response does not help us assess what students have learned from them that they will apply in their professional careers.

We have students write journals and essays, and need to be able to extract information from these that will help us do a better job of assessing whether they are engaging in moral imagination and reasoning. One way to accomplish this goal would be to require students to keep a portfolio of their work during the semester, in which they comment occasionally on what they are learning (Edgerton, Hutchings, & Quinlan, 1991)

One good demonstration of learning from these cases would be to see if the students gained enough wisdom to apply what they had learned to their own design projects. For example, as part of the fourth-year course, students have to propose an undergraduate thesis topic and carry it to completion. We are trying to give them the DesignTex and Rohner Textil cases before their project proposals, to see if it influences their choice of topic and approach, and the Dow Corning cases later, to see if it influences how they think about the impact of their project on society. At least one student in Gorman's current fourth-year class has decided to do her thesis on junk science, inspired by these cases, and is also seeking employment opportunities in sustainable development.

But for too many of the students, the cases and their projects are still at a distance from one another. That would be fine if these students applied the lessons later, in their careers. The ASEE needs to consider such long-term outcomes of learning-for example, student reports of actual ethical dilemmas they encountered after they graduated, and how they resolved them.

Acknowledgments

We would like to thank the NSF Science, Technology and Society Program Grant SBR#9319983 and the Darden Graduate School of Business Administration for their support in our studies of ethics, invention, design, and the case study approach. We would also like to thank Pat Werhane, the Ruffin Professor of Business Ethics at the University of Virginia, for her assistance in our efforts.

Please note that the opinions expressed in this work are solely those of the authors.

References

1. Luegenbiehl, Heinz. 1996. Message from LED Division Chair Heinz Luegenbiehl. ASEE Liberal Education Division April 1996 Newsletter, School of Engineering & Mathematics, Lake Superior State University.

2. Adler, Mortimer J. 1982. *The Paideia Proposal: An Educational Manifesto*. New York: Macmillan.
3. Harris, C. E., Pritchard, M. S., and Rabins, M. J. 1995. *Engineering Ethics: Concepts and Cases*. Belmont, CA: Wadsworth.
4. Martin, M. W., and Schinzinger, R. 1989. *Ethics in Engineering* (2nd ed.). New York: McGraw Hill.
5. Gioia, D. 1992. Pinto Fires and Personal Ethics. *Journal of Business Ethics* 11:384-385.
6. Johnson, M. *The Moral Imagination*, Chicago: University of Chicago Press. 1993, 202.
7. Larmore, C. Moral Judgement, *Review of Metaphysics*, 35, 275-296. 1981, 284.
8. Quinn, Daniel, 1992, *Ishmael: An Adventure of the Mind and Spirit*, New York: Bantam/Turner.
9. Whitbeck, Carolyn. 1996. Ethics as Design: Doing Justice to Moral Problems. Hastings Center Report, 26(3):9-16.
10. Norman, Donald A. 1993. *Things That Make Us Smart: Defending Human Attributes in the Age of the Machine*. New York: Addison Wesley.
11. Kolodner, J. 1993. *Case-based reasoning*. San Mateo, CA: Morgan Kaufmann Publishers, Inc.
12. Fitzgerald, N. 1995. Teaching with Cases. ASEE PRISM, March 16-20, 1995.
13. Panitz, Beth. 1995. Ethics Instruction: An Undergraduate Essential. ASEE PRISM, October 21-25, 1995.
14. Rest, J, Barnett, R., Bebeau, M., Deemer, D. Getz, I., Moon, Y., Spickelmeier. J,. Thoma, S., & Volker, J., *Moral Developent: Advances in Research and Theory*, New York: Praeger Press. 1988.
15. Mehalik, Matthew M. and Stocker, Julie M. 1996. Site Context: Summary .http://cti.itc.virginia.edu/~meg3c/id/ethics/ethics frame.html.
16. Russell, Edward P. III and Stocker, Julie M. 1996. Hands-On Ethics: Experiences with Cases in the Classroom. 1996 ASEE Conference Proceedings.

Outcome Assessment Resources

http://www.abet.ba.md.us/EAC/eac2000.html
http://www.asee.org/asee/announce/frameworkee/
Ferrier, Michelle Barrett. 1994. In Search of Effective Quality Assessment. ASEE PRISM, September 1994.

CHAPTER 12

EDWARD WENK, JR.

Teaching Engineering as a Social Science

Today's public engages in a love affair with technology, yet it consistently ignores the engineering at technology's core. This paradox is reinforced by the relatively few engineers in leadership positions. Corporations, which used to have many engineers on their boards of directors, today are composed mainly of M.B.A.s and lawyers. Few engineers hold public office or even run for office. Engineers seldom break into headlines except when serious accidents are attributed to faulty design.

While there are many theories on this lack of visibility, from inadequate public relations to inadequate public schools, we may have overlooked the real problem: Perhaps people aren't looking at engineers because engineers aren't looking at people.

If engineering is to be practiced as a profession, and not just a technical craft, engineers must learn to harmonize natural sciences with human values and social organization. To do this we must begin to look at engineering as a social science and to teach, practice, and present engineering in this context.

To many in the profession, looking at teaching engineering as a social science is anathema. But consider the multiple and profound connections of engineering to people.

Technology in Everyday Life

The work of engineers touches almost everyone every day through food production, housing, transportation, communications, military security, energy supply, water supply, waste disposal, environmental management, health care, even education and entertainment. Technology is more than hardware and silicon chips.

In propelling change and altering our belief systems and culture, technology has joined religion, tradition, and family in the scope of its influence. Its enhancements of human muscle and human mind are self-evident. But technology is also a social amplifier. It stretches the range, volume, and speed of communications. It inflates appetites for consumer goods and creature comforts. It tends to concentrate wealth and power, and to increase the disparity of rich and poor. In the competition for scarce resources, it breeds conflicts.

In social-psychological terms, it alters our perceptions of space. Events anywhere on the globe now have immediate repercussions everywhere, with a portfolio of tragedies that ignite feelings of helplessness. Technology has also skewed our perception of time, nourishing a desire for speed and instant gratification and ignoring longer-term impacts.

Engineering and Government

All technologies generate unintended consequences. Many are dangerous enough to life, health, property, and environment that the public has demanded protection by the government.

Although legitimate debates erupt on the size of government, its cardinal role is demonstrated in an election year when every faction seeks control. No wonder vested interests lobby aggressively and make political campaign contributions.

Whatever that struggle, engineers have generally opted out. Engineers tend to believe that the best government is the least government, which is consistent with goals of economy and efficiency that steer many engineering decisions without regard for social issues and consequences.

Problems at the Undergraduate Level

By both inclination and preparation, many engineers approach the real world as though it were uninhabited. Undergraduates who choose an engineering career often see it as escape from blue-collar family legacies by obtaining the social prestige that comes with belonging to a profession. Others love machines. Few, however, are attracted to engineering because of an interest in people or a commitment to public service. On the contrary, most are uncomfortable with the ambiguities of human behavior, its absence of predictable cause and effect, its lack of control, and with the demands for direct encounters with the public.

Part of this discomfort originates in engineering departments, which are often isolated from arts, humanities, and social sciences classrooms by campus geography as well as by disparate bodies of scholarly knowledge and cultures. Although most engineering departments require students to take some nontechnical courses, students often select these on the basis of hearsay, academic ease, or course instruction, not in terms of preparation for life or for citizenship.

Faculty attitudes don't help. Many faculty members enter teaching immediately after obtaining their doctorates, their intellect sharply honed by a research specialty. Then they continue in that groove because of standard academic reward systems for tenure and promotion. Many never enter a professional practice that entails the human equation.

We can't expect instant changes in engineering education. A start, however, would be to recognize that engineering is more than manipulation of intricate signs and symbols. The social context is not someone else's business. Adopting this mindset requires a change in attitudes. Consider these axioms:

- Technology is not just hardware; it is a social process.
- All technologies generate side effects that engineers should try to anticipate and to protect against
- The most strenuous challenge lies in synthesis of technical, social, economic, environmental, political, and legal processes.
- For engineers to fulfill a noblesse oblige to society, the objectivity must not be defined by conditions of employment, as, for example, in dealing with tradeoffs by an employer of safety for cost.

In a complex, interdependent, and sometimes chaotic world, engineering practice must continue to excel in problem solving and creative synthesis. But today we should also emphasize social responsibility and commitment to social progress. With so many initiatives having potentially unintended consequences, engineers need to examine how to serve as counselors to the public in answering questions of "What if?" They would thus add sensitive, future-oriented guidance to the extraordinary power of technology to serve important social purposes.

In academic preparation, most engineering students miss exposure to the principles of social and economic justice and human rights, and to the importance of biological, emotional, and spiritual needs. They miss Shakespeare's illumination of human nature—the lust for power and wealth and its corrosive effects on the psyche, and the role of character in shaping ethics that influence professional practice. And they miss models of moral vision to face future temptations.

Engineering's social detachment is also marked by a lack of teaching about the safety margins that accommodate uncertainties in engineering theories, design assumptions, product use and abuse, and so on. These safety margins shape practice with social responsibility to minimize potential harm to people or property. Our students can learn important lessons from the history of safety margins, especially of failures, yet most use safety protocols without knowledge of that history and without an understanding of risk and its abatement. Can we expect a railroad systems designer obsessed with safety signals to understand that sleep deprivation is even more likely to cause accidents? No, not if the systems designer lacks knowledge of this relatively common problem.

Safety margins are a protection against some unintended consequences. Unless engineers appreciate human participation in technology and the role of human character in performance, they are unable to deal with demons that undermine the intended benefits.

Case Studies in Socio-Technology

Working for the legislative and executive branches of U.S. government since the 1950s, I have had a ringside seat from which to view many of the events and trends that come from the connections between engineering and people. Following are a few of those cases.

Submarine Design

The first nuclear submarine, *USS Nautilus*, was taken on its deep submergence trial February 28, 1955. The subs' power plant had been successfully tested in a full-scale mock-up and in a shallow dive, but the hull had not been subject to the intense hydrostatic pressure at operating depth. The hull was unprecedented in diameter, in materials, and in special joints connecting cylinders of different diameter. Although it was designed with complex shell theory and confirmed by laboratory tests of scale models, proof of performance was still necessary at sea.

During the trial, the sub was taken stepwise to its operating depth while evaluating strains. I had been responsible for the design equations, for the model tests, and for supervising the test at sea, so it was gratifying to find the hull performed as predicted.

While the nuclear power plant and novel hull were significant engineering achievements, the most important development occurred much earlier on the floor of the U.S. Congress. That was where the concept of nuclear propulsion was sold to a Congressional committee by Admiral Hyman Rickover, an electrical engineer. Previously rejected by a conservative Navy, passage of the proposal took an electrical engineer who understood how Constitutional power was shared and how to exercise the right of petition. By this initiative, Rickover opened the door to civilian nuclear power that accounts for 20 percent of our electrical generation, perhaps 50 percent in France. If he had failed, and if the *Nautilus* pressure hull had failed, nuclear power would have been set back by a decade.

Space Telecommunications

Immediately after the 1957 Soviet surprise of Sputnik, engineers and scientists recognized that global orbits required all nations to reserve special radio channels for telecommunications with spacecraft. Implementation required the sanctity of a treaty, preparation of which demanded more than the talents of radio specialists; it engaged politicians, space lawyers, and foreign policy analysts. As science and technology advisor to Congress, I evaluated the treaty draft for technical validity and for consistency with U.S. foreign policy.

The treaty recognized that the airwaves were a common property resource, and that the virtuosity of communications engineering was limited without an administrative protocol to safeguard integrity of transmissions. This case demonstrated that all technological systems have three major components—hardware or communications equipment; software or operating instructions (in terms of frequency assignments); and peopleware, the organizations that write and implement the instructions.

National Policy for the Oceans

Another case concerned a national priority to explore the oceans and to identify U.S. rights and responsibilities in the exploitation and conservation of ocean resources. This issue, surfacing in 1966, was driven by new technological capabilities for fishing, offshore oil development, mining of mineral nodules on the ocean floor, and maritime shipment of oil in supertankers that if spilled could contaminate valuable inshore waters. Also at issue was the safety of those who sailed and fished.

This issue had a significant history. During the late 1950s, the U.S. government was downsizing oceanographic research that initially had been sponsored during World War II. This was done without strong objection, partly because marine issues lacked coherent policy or high-level policy leadership and strong constituent advocacy.

Oceanographers, however, wanting to sustain levels of research funding, prompted a study by the National Academy of Sciences (NAS). Using the report's findings, which documented the importance of oceanographic research, NAS lobbied Congress with great success, triggering a flurry of bills dramatized by such titles as "National Oceanographic Program."

But what was overlooked was the ultimate purpose of such research—to serve human needs and wants, to synchronize independent activities of major agencies, to encourage public/private partnerships, and to provide political leadership. During the 1960s, in the role of Congressional advisor, I proposed a broad "strategy and coordination machinery" centered in the Office of the President, the nation's systems manager. The result was the Marine Resources and Engineering Development Act, passed by Congress and signed into law by President Johnson in 1966.

The shift in bill title reveals the transformation from ocean sciences to socially relevant technology, with engineering playing a key role. The legislation thus embraced the potential of marine resources and the steps for both development and protection. By emphasizing policy, ocean activities were elevated to a higher national priority.

Exxon Valdez

Just after midnight on March 24, 1989, the tanker Exxon Valdez, loaded with 50 million gallons of Alaska crude oil, fetched up on Bligh Reef in Prince William Sound and spilled its guts. For five hours, oil surged from the torn bottom at an incredible rate of 1,000 gallons per second. Attention quickly focused on the enormity of environmental damage and on blunders of the ship operators. The captain had a history of alcohol abuse, but was in his cabin at impact. There was much finger-pointing as people questioned how the accident could happen during a routine run on a clear night. Answers were sought by the National Transportation Safety Board and by a state of Alaska commission to which I was appointed. That blame game still continues in the courts.

The commission was instructed to clarify what happened, why, and how to keep it from happening again. But even the commission was not immune to the political blame game. While I wanted to look beyond the ship's bridge and search for other, perhaps more systemic problems, the commission chair blocked me from raising those issues. Despite my repeated requests for time at the regularly scheduled sessions, I was not allowed to speak. The chair, a former official having tanker safety responsibilities in Alaska, had a different agenda and would only let the commission focus largely on cleanup rather than prevention. Fortunately, I did get to have my say by signing up as a witness and using that forum to express my views and concerns.

The Exxon Valdez proved to be an archetype of avoidable risk. Whatever the weakness in the engineered hardware, the accident was largely due to internal cultures of large corporations obsessed with the bottom line and determined to get their way, a U.S. Coast Guard vulnerable to political tampering and unable to realize its own ethic, a shipping system infected with a virus of tradition, and a cast of characters lulled into complacency that defeated efforts at prevention.

Lessons

These examples of technological delivery systems have unexpected commonalities. Space telecommunications and sea preservation and exploitation were well beyond the purview of just those engineers and scientists working on the projects; they involved national policy and required interaction between engineers, scientists, users, and policymakers. The Exxon Valdez disaster showed what happens when these groups do not work together. No matter how conscientious a ship designer is about safety, it is necessary to anticipate the weaknesses of fallibility and the darker side of self-centered, short-term ambition.

Recommendations

Many will argue that the engineering curriculum is so overloaded that the only source of socio-technical enrichment is a fifth year. Assuming that step is unrealistic, what can we do?

- The hodgepodge of nonengineering courses could be structured to provide an integrated foundation in liberal arts.
- Teaching at the upper division could be problem- rather than discipline-oriented, with examples from practice that integrate nontechnical parameters.
- Teaching could employ the case method often used in law, architecture, and business.
- Students could be encouraged to learn about the world around them by reading good newspapers and nonengineering journals.
- Engineering students could be encouraged to join such extracurricular activities as debating or political clubs that engage students from across the campus.

As we strengthen engineering's potential to contribute to society, we can market this attribute to women and minority students who often seek socially minded careers and believe that engineering is exclusively a technical pursuit.

For practitioners of the future, something radically new needs to be offered in schools of engineering. Otherwise, engineers will continue to be left out.

CHAPTER 13

CRAIG GUNN

Orienting Engineering Students to Contemporary Issues through a Broader Perspective

Introduction

"Variety's the spice of life, that gives it its flavor." These lines in "The Task II" by William Cowper (English poet 1731–1800) reflect an attitude that must be fostered in the minds of engineers. No man is an island, and no field of study can divorce itself from the activities, interests, and positive reinforcement of divergent areas of instruction. Students who become embroiled in the quest for a degree in engineering can quickly close the doors to many of the more liberal pursuits and the world in which the pursuits exist. There is a feeling that any time spent on "non-engineering" pursuits is not beneficial to a career. Efforts, therefore, must be made as with writing across the curriculum to include in the educational structure of every engineer ample connections to one of those seemingly liberal pursuits: the contemporary world. These ties can be implemented by working in two areas: insertion of courses into the existing curriculum or extracurricular, activities. The logistics of inserting additional courses into an engineer's program requires a great deal of effort. This in itself is a wide-ranging project and deserves many papers of its own. The other more personal activity can involve faculty/staff members who are interested in sponsoring activities that foster and promote a broader perspective than just engineering. This chapter proposes methods by which the engineering student can obtain both an engineering degree and a contemporary perspective through faculty/staff involvement. It is also concerned with barriers that need to be removed before improvements can be implemented.

Contemporary Viewpoint

The 1992 movie *Sneakers* portrayed a variety of engineers who found little in life to be of value besides engineering. In the 1993 movie *Falling Down*, Michael Douglas showed the American public the stereotypical engineer with pocket protector, slightly unbalanced look, and a very narrow view of life. Movies made for both the big screen and television seem to capture the picture of the humorless engineer who finds pleasure only in regimen and coldness. Even Halloween costumes preparing small children to be engineers can easily be created with long-sleeve, white shirts and white tape on broken glasses. With this shortsighted view of the engineering profession comes a belief on the part of even engineers themselves that reality is what they see and hear in the media. A view that the engineers' world contains only technical content with no connection to contemporary society follows easily from the pictures that are created in the public eye.

When the scientific revolution took place in the 1700s, the movement to divorce the scientific world from the arts was loudly applauded by liberal studies activists and scientific proponents alike. Scientists were trying to carve their place in the world and, therefore, had to focus on only scientific issues. Science would take center stage and the liberal arts would have to follow as a distant second. Complaints over the past decades have reflected upon this separation of disciplines. We have been deluged with "right brain, left brain" concerns, with cries that "engineers *cannot* communicate," and with the whole smokescreen of what is perceived to be the "stereotypical engineer." We can decry these attitudes, but many are ingrained into the society in which we live and work. While the public's perception of engineers may require a vast reeducation, it is with the engineers that the first steps must be taken to change the incorrect vision.

This overly simplified misconception allows many engineers to go through their college or university careers unaware that many of their peers studying fluids, circuits, controls, composites, or calculus actually do have vast experience in the liberal arts and the real world. These talents and interests lie hidden while technical courses are taken and technical knowledge is gained. It is important to the rounded education of engineers that an effort be made to bring to light the broader interests of engineers. Age-old stereotypes that influence the way that engineers perceive themselves need to be investigated, modified, and in many cases destroyed.

The comments made here will not deal with the vast misconceptions that are prevalent in the world outside of engineering. To address that issue would take much greater space than can be allowed here. The area that will be investigated approaches the issue from the inside by targeting the engineering faculty, support staff, and students. When these groups truly realize what is involved in the makeup of a complete engineer, the issue of separation of engineers and humanists will fade away. It is important to realize that the ideas discussed below will seem obvious, but for many the most obvious ideas are those that are most easily ignored. Yes, everyone needs to be cognizant of the contemporary world, but is any amount of appreciable time provided to actually integrate these needs into the engineering curriculum.

No visible change can take place in the attitudes of people if interest and a willingness to discuss are not present. In order for the attitudes of the world to change, engineers must first believe that there is no wall between the sciences and the issues raised through a liberal education. As engineers grow in their realization that the humanities do hold an important place in their lives, the word will reach out to the masses of people outside the engineering disciplines. Interested parties in every engineering department should be functioning as catalysts for discussion on contemporary issues.

It is interesting to note that one finds a very interesting situation when one approaches contemporary issues and their connection to engineering. The stereotypical vision of the engineer stated above is rampant in both the engineering community and the world outside engineering. It is enlightening to note, though, that the stereotype is only a superficial belief among the vast majority of engineers. It appears to be necessary for many to foster this belief in order to keep the engineering area pure from contemporary issues where people are unable to provide distinct and singular answers to questions. The importance of the technical education is somehow enhanced by the fact that it is not connected to the issues associated with the ever-changing world. When one sits down with engineers in a non-threatening environment where true feelings can be expressed, a different impression about liberal education is expressed. Students begin to speak about the instruments that they play and the particular level of competence that they have achieved in the musical world. When discovered working on the computer, students will blush when they speak of the poetry that they write, the most recent play that they have penned, or the latest collection of short stories that they have produced. In many cases, these admissions seem to be something that should be left hidden in a veil of secrecy. Students never seem ready to admit to a sharing of liberal ideas, needs, and wants. This demonstrates the need to revitalize the education of future engineers. Efforts must be made on the part of professors in engineering to share with their students the interests that they too have in the world around them. Humanities courses offer the content, but engineering faculty members offer the pulsing blood that will allow their students to realize that there is a connection with their technical studies.

Methods

Simple allusions to books read outside of the science disciplines, music listened to or played, and interesting compositions created can pique the interest of individuals who have felt that they alone create poetry as engineers, enjoy Bach cantatas, or relish the intricacies of avant-garde versions of contemporary plays. As the number of engineers openly sharing their interests in the humanities increases, a greater variety of liberal pursuits will surface. Discussion can then lead to a visible increase in the number of individuals who no longer feel that what they find fascinating in the humanities is "an unengineering activity." Faculty members who spend a few minutes at the end of class inquiring into the most recently read fiction, plays attended, or music listened to will start the process of linking the interest in engineering to interests in the humanities. Rejection of a one-dimensional stereotype begins the process of attitude change. The less eager students who are uncomfortable with revealing their interests in the world around them will gain from open discussion and, hopefully, feel the desire to become part of the movement to join engineering to the contemporary world. As more and more engineers are made aware of the numbers of humanist engineers, the stereotypes will begin to change.

Beyond the simple process of making students and faculty alike aware of the interests of other engineers, there are a multitude of activities that can be individually tailored to an engineering department. Activities can encompass sponsored trips to cultural events, simple evenings sharing excerpts from favorite works of literature, or a carefully planned evening of musical entertainment performed strictly by engineers.

The unique experience of travelling with a group of engineers to a weekend of Shakespearean plays is hard to comprehend for those who do not understand the closeness of engineering to the intricate use of language in these plays. Stimulating conversation concerning conciseness, clarity, and detail coupled with audience issues, the development of characters, the technical nature of plot formation, and the connection of classic plays to modern life make for a broad avenue upon which to connect the two areas. When the group travels, it is even more interesting if the makeup of the group includes humanists, who can also gain from the interaction with engineers. Anyone returning from a theater weekend and in-depth conversation will have a new perspective on engineers and their multifaceted interests.

Many departments have also come to realize that their faculty and students, while having strong engineering aptitudes, possess numerous talents in music. Collecting information on these talents is an eye-opening experience when one discovers that a large percentage of the group is familiar with at least one musical instrument. This interest can lead, as it inevitably does, to evening gatherings and formal productions sponsored by engineering departments and colleges. Again the diversity of interest away from the traditional views of engineers helps to unite the humanities and sciences.

Any ventures that enhance the education of engineers are a valuable activity. Their importance lies in the issue of providing engineers with outlets for their difficult schedules; times when they can enjoy reading fiction, immerse themselves in good drama, or sit down to create a piece of text that does not have to relate to a technical issue. Contests have been created to allow engineers to show their photographic talent; their ability to construct short stories, poetry, and plays: along with the standard essays on issues relating to engineers. One formidable project calls for teams of students to edit and rewrite chapters in published texts that are deemed inferior, so as to become involved with the text and with the mechanisms required to build the improved text.

All the above activities may be of interest to those who believe that the world is a vital part of an engineer's education. The problem exists when we approach the engineering student body and inform it that the humanities should be investigated, utilized, and made a part of an engineer's life. Here the necessity is to make a case for the unification of the technical and liberal sides to education. The intricacy of musical pieces, the knowledge that can be gained from working carefully through a text, or the meticulous effort that needs to taken in painting a work of art are activities that complement the direction that engineers commonly follow in gaining their knowledge. Engineers are meticulous. They strive for clarity and conciseness, and their research requires exacting pursuit until every contingency is addressed. The creative mind presents these same requirements. As engineers delve into the world around them, they will find connections between that world and engineering. The catalyst that can begin this process is as close as the department in which they study. Conversation leads to discussion. Discussion can open wide the doors. Discussion being the catalyst, engineering faculty can direct the conversation on what effect engineering has on the world. Does improved technology really benefit humans or does it raise issues that may never have been considered? If a dam is built on a river that creates a huge reservoir, which then floods a town containing items of cultural importance that cannot be removed, is the engineer to be concerned? If improved medical equipment allows many more infants in third world countries to survive birth only to die of starvation, should engineers be bothered? Is there an issue when you build a new airport that destroys an endangered species? What is the position of the engineer in the controversial areas of cloning, abortion, and assisted suicide? Areas outside of mainstream engineering that have consequences for engineers can be discussed in every engineering course. Communication, ethics, and trust all can be tried to issues relating engineering to contemporary issues. It is these conversations that can draw faculty and students alike into the broad perspective of education, an education in and of the world we inhabit.

Conclusions

Listening, thinking, writing, and speaking about contemporary issues are activities that can easily be sponsored and perpetuated in engineering departments. Students made aware of the need to investigate their connections to the contemporary world will bring to engineering a whole new breadth and depth of study. Faculty and staff members should utilize any means possible to draw connections between engineering and the world around engineers. At the simplest level it is an awareness issue. As more individuals are made aware of the mutual interests of their colleagues, the separation of the liberal arts and the sciences will shrink. With this shrinkage will come a greater awareness on the part of the rest of the world and with it a firmer commitment to see the whole person and not the stereotype.

References

1. Bordogna, Lawrence et al. Manufacturing and Engineers' Education, *Issues in Science and Technology*, 7, no. 1 (fall 1990): 20(3).
2. Booth, William. Curriculum Sparks Debate at MIT, *Science*, 236 (1987): 1515(2).
3. Filho, Matthew. Humanist Education for the Lives of Today's Engineers, *IEEE Communications*, 30, no. 11(1992): 72 . 4.
4. Florman, Samuel. Learning Liberally, *Prism*, 3 no. 3 (1993):18(3).
5. Kirkely, Lawrence. Our Industry Could Lead a Liberal Arts Renaissance, *Datamation*, 29, no.3 (1993): 29.
6. Kranzberg, Melvin. Engineering the Whole Engineer, *Prism*, 3, no. 3 (1993): 26(6).

CHAPTER 14

JOHN KRUPCZAK

Reaching Out across Campus

Engineers as Champions of Technological Literacy

Introduction

Recent developments illustrate that engineers can play a unique and central role in the promotion of science and technology education for nonscience and nonengineering students. On several campuses, engineering faculty have pioneered new approaches to teaching science and technology to majors in nontechnical fields. These efforts are helping to reach a long-neglected population of students and are establishing more reciprocal relationships between engineering programs and their larger campus-wide communities. Coincidentally, courses for nonengineers are shedding new light on the liberal education of engineering students. It has been found that some engineering students can benefit from a hands-on familiarity with technological devices and an understanding of the cultural context of technological development as addressed in some courses intended for nonengineers. The ABET Engineering Criteria 2000 facilitates widespread adoption of these innovative new developments by giving engineering departments greater scope in the definition of their program constituencies and objectives.

A Call to Action

American programs of higher education in the sciences, mathematics, engineering, and technology provide high-quality professional training for those students who major in these areas. However, the quality and scope of education in science and technology for nonscience and nonengineering majors have created a situation in which "roughly 95% of the American Public is consistently found to be scientifically illiterate by any rational standard."[1] While precise definitions of scientific and technological literacy are open to debate, colleges and universities are nonetheless being called upon to ensure that "All students have access to supportive, excellent undergraduate education in science, mathematics, engineering and technology, and all students learn these subjects by direct experience."[2] Science and engineering educators must maintain effective programs for educating scientists and engineers while establishing new efforts to define and meet the growing need for science and technology education for all students through high-quality, hands-on educational programs.

Elements in this recent call to action are also recognizing that the commitment to teaching nonscience and nonengineering students must extend beyond the interest of individual faculty members. In studies of the problem of widespread scientific and technological illiteracy, the National Research Council emphasized that institutional factors must be addressed to provide the necessary incentives for sustained improvements in nonmajors science and technological education.[3,4] The NRC further recognized that accomplishing this goal will require changes in the priorities and missions of many postsecondary institutions.

Historical Background

Concern for the scientific and technological knowledge of nonengineering and nonscience students is not a new phenomenon. In the past twenty years, there has been a series of studies and recommendations on the issue. For example, in 1982 the National Research Council published *"Science for the Non-Specialist,"* a review of the quality of college science education for nonscience and nonengineering majors.[5] This review documented the lack of attention paid to the science education of students in nonscience and nontechnical majors at the college level. The report also recognized that nonspecialists have backgrounds and interests requiring science courses of a different character than those for science majors.

Sloan Foundation New Liberal Arts Program

In 1982 the Alfred P. Sloan Foundation launched the New Liberal Arts Program with the intention of improving the quality of education that undergraduates receive in the areas of technology and quantitative reasoning.[6] The Sloan Foundation sponsored development of a variety of courses on technological topics for nonscience majors. The results of the work led to a number of books, monographs, and course syllabi.[7] In the area of technology, a wide variety of materials and courses were attempted. Generally, three types of courses were developed at participating institutions. In some cases, existing mathematics and science courses were modified to include more technological topics. Some of the courses that were developed were based on specialized technological topics such as forensics or reproductive technologies. A third category of course merged technology with other disciplines such as music and electroacoustics or the history and philosophy of technology. While some engineering faculty did participate in the program, the majority of the technology courses developed under the New Liberal Arts Program originated in science departments.

The New Liberal Arts Program broke new ground in establishing technology as the intellectual peer of science at the college level.[8] These new initiatives brought many nonscience or nonengineering students into meaningful contact with technological devices in ways that would not otherwise have been possible. It also prompted science faculty to more closely examine the relationship between the sciences and their engineering applications.

An important outcome of the New Liberal Arts Program was the identification of the technological applications of science as a topic of high interest to the nonscientist and nonengineer. Participating faculty observed that students from nontechnical majors appear especially interested in learning science when motivated by the context of familiar or intriguing technological devices. Technology was identified as an important factor in science education, but exactly how best to employ technological topics in science education could not yet be isolated. While an array of new technologically oriented course and curriculum materials for nonscience and nonengineering majors was developed, consensus on the most effective means to promote scientific and technological literacy was yet to be achieved.

Recent Efforts from Science Departments

Building on the result of the New Liberal Arts Program, some recent innovations from physicists have reformulated traditional science courses to emphasize technological applications. Dudley and Bold[9] have developed an introductory physics course for science majors that is organized around technological applications of physics. Other efforts targeting general students are exemplified by physicists Bloomfield[10] and Watson.[11] Bloomfield has created a textbook that explains the physics behind familiar technological devices. Watson has prepared a series of internet-accessible lessons targeting nonscience students explaining the scientific concepts used in high technology. What distinguishes these new efforts is an attempt to produce broad technologically structured science survey courses for nonscience and nonengineering majors as compared to the more specialized and topical focus of the earlier New Liberal Arts initiatives.

Consciously or unconsciously, these most recent efforts from science departments to reach out to nonmajors incorporate a way of thinking that is more akin to engineering than traditional models of science teaching. In previous attempts to reach nonscience majors, physics texts focused on the scientific understanding of the natural world rather than human-made technology. Representative examples include: *Physics for the Inquiring Mind* (Rogers)[12] and *Physics for Poets* (March).[13] These earlier works reflect an effort to explain to the nonscientist a physicist's understanding of the natural world. These books are organized according to the traditional sub-fields of physics. The more recent works by Bloomfield and Watson show a shift of perspective toward applications and an explanation of the human-made environment—in other words, the products of the engineering disciplines.

Pre-college Technology Education

As some college science departments have become more engineering-like in their latest courses for nonscience majors, a parallel movement in K-12 education has created formal "Technology Education" programs, which are in some cases essentially pre-engineering courses for the general K-12 school population.

Recent changes in K-12 curriculum have resulted in a revision of former industrial arts education into more broadly defined technology education. Technology education seeks to incorporate into the general education program a hands-on curriculum that includes a study of the structure of our designed environment, manufacturing and production processes, transformation of energy, communication and materials technology, technological problem solving, and design.[14] While this movement has some roots in vocational education, its goals are not the training of a subset of students in specialized skills as preparation for direct employment. Instead, K-12 technology education is envisioned as general education for all students about the technological world in which they live.

At the cutting edge of technology education are programs that emphasize the solution of design problems by methods that include the use of mathematics and knowledge from science.[15] Were these programs not identified as taking place in middle schools,[16] engineering faculty would have difficulty distinguishing the objectives from those found in many introductions to engineering college classes. Curriculum innovators at the K-12 level are now using the term "Pre-engineering" in developing integrated mathematics, science, and hands-on technological activities.[17] These programs are targeted for general distribution among the K-12 population, not just to students who anticipate careers in engineering.

Innovations from the Engineering Community

These movements by university science departments and K-12 technology educators to improve scientific and technological literacy converge in an educational domain in which schools of engineering were previously the lone occupant. Respect for technological topics as intellectually challenging, the study of science with an emphasis on applications, the use of mathematics in the context of solving technical problems, and direct contact with the workings of the designed world are the essential characteristics of engineering endeavors. As other educators have adopted perspectives borrowed from engineering, the opportunity exists for engineering educators to begin to envision their constituents as all students and not just those in engineering schools.

An additional timely development is the institution of the ABET Engineering Criteria 2000. In the past, the proscriptive nature of the accreditation requirements would have made it difficult to justify any long-term effort by an engineering program beyond that required for teaching engineering students. The ABET Engineering Criteria 2000 gives engineering programs a new freedom and responsibility to define their own program constituents, goals, and objectives. Engineering programs can expand and broaden their educational vision to include all students, not just engineers, and utilize self-determination of the ABET 2000 criteria to sanction such visions.

As engineering programs begin to offer courses designed for nonengineering majors, liberal education in the context of engineering education takes on a broader meaning. The liberal education of engineers has been treated as a service to be performed by other departments. The departments teaching liberal arts have had the responsibility of broadening or rounding out the education of the engineer. In return, engineering programs offered very little for broadening or rounding out the educational experience of the liberal arts students. It would seem a more equitable relationship within university faculty for engineering programs to contribute in a direct way to the liberal education of nonengineers as the engineers programs have so long expected of the liberal arts faculty.

Several representative samples and a more detailed case study of the new courses developed by engineering departments are included as examples and models for those contemplating similar initiatives. Representative courses developed by engineers include "Introduction to Physical Devices and Systems," a hands-on collaborative laboratory course developed by the College of Engineering at the University of California at Davis; "Science at Work," developed at Mission College; "Exploring Technology," taught at Lake Superior State University; and "Science and Technology of Everyday Life," taught at Hope College. These courses exemplify teaching nonengineering students about science in the context of technological applications. All incorporate hands-on laboratory experiences to familiarize students with the workings of technology. In each case, the course innovations were initiated by engineering programs as a contribution to the liberal arts educational offerings of their respective campuses. A review of additional information on technological literacy courses taught by engineers has been complied by Byars.[18]

It is not possible at this time to attribute any newly offered technological literacy courses to the existence of the ABET Engineering Criteria 2000. As all of the courses to be described were initiated prior to the adoption of the ABET Engineering Criteria 2000, conclusions cannot be made regarding the influence of the EC 2000 on the initiation of these courses. However, in the case of Hope College, the campuswide interest in a technological literacy course has prompted a revision of the departmental mission statement to include the education of nonengineers as part of the ABET accreditation process.

Introduction to Physical Devices and Systems,
University of California, Davis

A collaboration from the College of Engineering at the University of California, Davis, created a set of projects that emphasize hands-on exploration of how things work.[19] A primary goal is to increase the familiarity of students with the hardware aspects of technology. Several laboratory projects were developed, including introduction to hand tools, a mechanical dissection project, building a multimeter, exposure to two-stroke engines, use of various electronic sensors, and the design of a house. An evaluation of this course by Bland[20] identified the following activities as especially useful to the target group: the opportunity to take apart, reconstruct, and build devices; specific emphasis on the links between abstract and concrete concepts; and laboratory activities that include approaches from multiple disciplines.

Exploring Technology, Lake Superior State University

Mahajan, McDonald, and their colleges in the School of Engineering at Lake Superior State University in Sault Ste Marie, Michigan, developed a course[21] entitled "Exploring Technology." Their audience is freshman from both engineering and nonengineering majors. In this course, students build and test electronic circuits, design and build a small mechanical part, operate an automated manufacturing system, and program robots. The course is intended to be a survey of technology satisfying the natural science elective for nonscience and nonengineering majors. To take advantage of individual areas of expertise, the course is team-taught by six instructors.

Student assignments in the "Exploring Technology" course included work with computers, maintenance of a journal, and problem-solving assignments. The relationships between engineering and other fields is highlighted. The nonengineering students responded enthusiastically to the course. After taking the course, students felt a sense of familiarity and confidence with modern technology. The nonengineers showed an understanding and appreciation of how engineering affects their lives. Students cited the experiential activities as both novel and enlightening.

Science at Work, Mission College

Disney and coworkers at Mission College in Santa Clara, California, have created a technical literacy course for nonengineering students.[22] The course uses examples of engineering systems from the local community such as highway bridges with which students are familiar. Basic science is introduced as needed to aid in understanding the engineering application. Hands-on exercises have been developed including a suspension bridge, a telecommunications laboratory, and a microcontroller project. The course activities are portable and have found circulation within the Community Colleges of California system.

Case Study

Science and Technology of Everyday Life, Hope College

The "Science and Technology of Everyday Life," taught at Hope College, will be described as a more detailed example of a college-level course in scientific and technological literacy that is being offered by an engineering program.[23] This case study will illustrate a distinguishing feature of technological literacy courses offered by engineering programs—an investigation of scientific principles in the context of modern technological devices. The case study also emphasizes hands-on contact with workings of modern technology, another hallmark of technological literacy efforts of engineers.

The "Science and Technology of Everyday Life" is intended for students from nontechnical majors and includes students from business, history, fine arts, and elementary education. First offered in the spring 1995 semester, the objective of the course is to develop a familiarity with both the engineering aspects of how various technological devices work, and an understanding of the basic scientific principles underlying their operation. The course focuses on the wide variety of technology used in everyday life to help in engaging the student's interest.

The course format is three hours of lecture and three hours of laboratory per week over a fifteen-week semester. Laboratories involve experiential learning activities such as disassembling a car engine, building a simple electronic music keyboard, and making a hologram. Enrollment is fifty students each semester. The lecture portion of the course is taught in a single section by one member of the engineering faculty. There are two laboratory sections of twenty five students each. Each laboratory section is run by one faculty member assisted by two undergraduate teaching assistants. The books used as texts are David Macaulay's *The Way Things Work* and Bloomfield's *How Things Work: The Physics of Everyday Life*."[24] A supplemental reading list compiled from a variety of sources is also used.

Course Topics

The course topics were selected to represent the technologies most frequently encountered in everyday life and were based partly on results of surveys of student interests. Topics covered include the automobile, radio and television, computers, and medical imaging. A more detailed list of the course material, divided into subtopics, is listed in Table 14.1.

The first topic discussed is the automobile. The automobile is described as a group of interconnected systems, each performing a particular function. The systems studied are the engine, fuel and air, exhaust, cooling, lubrication, power train, brakes, steering, suspension, and electrical. The components associated with each system and their interrelations are described. Common problems are discussed. In this portion of the course, an emphasis is placed upon helping students with little technical background to become accustomed to the process of thinking through cause-and-effect relationships. Such a facility is essential to understanding any technological system. In the laboratory, students disassemble and reassemble four-cylinder automobile engines. This activity brings nonengineering students into a degree of contact with technology that they are unlikely to have experienced previously.

The next segment of the course begins to discuss technological devices that are based on applications of electricity and magnetism. Topics covered include the photocopier, batteries, the electric motor, electric power plants, and house wiring. The principles of science addressed are electrostatics; the concepts of current, voltage, and resistance; basic electrical circuits, joule heating, and electromagnetic induction. Attention is next directed to light and electromagnetic waves. Devices studied include incandescent and fluorescent lighting, lasers, lenses and eyeglasses, and photographic film and processing. The electromagnetic frequency spectrum is described along with the fundamentals of wave propagation. From the topics described above, the course emphasis on the technological applications of science is evident.

The concepts developed in electricity and magnetism and electromagnetic waves now make it possible to move into telecommunications equipment. Here

TABLE 14.1. Course Topics

Automobile	*Sound and Music*
Automobile Subsystems	Sound, Microphones, Transistor Amplifier, Speakers
Internal Combustion Engine: 4-Stroke Cycle	Magnetic Tape Recorder
Fuel-Air System, Exhaust, Cooling, Ignition	Audio Compact Disc
Manual and Automatic Transmissions	
Brake System	*Telecommunications*
Steering and Suspension	Telephone
Electrical System	Radio
	Television
Electrical and Magnetic Devices	
Static Electricity, Photocopier, Lightning	*Medical Imaging*
Current Electricity, Batteries, Photo-voltaic Cells	X-rays, Ultrasound Scanner, Magnetic Resonance Imaging
Household Wiring, Electrical Safety, Ohm's law	
Magnetism, Permanent and Electromagnets, Motors	
Electromagnetic Induction, Generators, Power Plants	*Computers*
	Binary Code, Adders, Calculator, Microprocessor
Light and Photography	Computer Memory, Disk Drives, CD-ROM
Incandescent, Fluorescent Lights, Lasers	Mouse, Printers, Internet
Lenses, Mirrors, The Eye, Eyeglasses, Color Perception	*Miscellaneous Topics*
Black and White Photography, The Pinhole Camera	Refrigerator and Air Conditioner
Color Photography, The Single-Lens Reflex Camera	Microwave Oven

the topics covered are the telephone, radio, and television. The key step of using an electromagnetic microphone to produce a voltage whose amplitude is proportional to the sound wave intensity is discussed. The concept of encoding information in either amplitude or frequency modulation on a carrier wave is studied. Creation of a color TV picture by superposition of electron beams for the red, green, and blue signals is explained. As the course progresses, discussion of those scientific principles usually deemed as important for liberal arts students takes place; however, the order and degree of emphasis may vary from a more traditional course for nonscience majors.

The next segment focuses on audio equipment. Transistor amplifiers are explained along with electromagnetic speakers. The concept of recording both analog and digital signals is developed. Devices studied include magnetic tape recorders and the audio compact disk. The concept of the digital representation of the audio signal is explained. Issues of sampling frequency and the need for error correction are studied.

Having established a foundation in basic electricity and magnetism along with an understanding of both sound and electromagnetic waves, a variety of miscellaneous topics can be addressed. The common medical imaging devices of x-rays, ultrasound, and magnetic resonance imaging are studied. Microwave ovens, refrigerators, and air conditioners are also examined.

The last section of the course is about computers. The topic of computers is so broad that the treatment must be highly selective. Emphasis is placed on understanding basic aspects of computer hardware rather than developing experience with using any particular type of software. Binary numbers and ASCII code are explained. Boolean functions are introduced to the point that it is possible to describe the functioning of a four-bit full adder. The remainder of the time is devoted to describing the functioning and interrelation of computer components such as the processor, memory, storage devices, and printers. The descriptions are aimed at making it possible for students to interpret the information given in personal computer advertisements. An overview of the Internet and computer networks is included.

This review of specific topics illustrates the range of technological devices that can be included in a one-semester course for liberal arts students. A wide selection of the products of the engineering disciplines form the topics of study. All topics are drawn from the context of the technology with which students are accustomed to encountering in their daily lives.

Course Laboratories

The laboratory projects were developed to provide hands-on contact with modern technological devices and to show the application of a particular scientific principle in some area of technology. The projects used are summarized in Table 14.2. Each project serves two functions: to illustrate a scientific understanding of some aspect of nature and to demonstrate how that understanding is used in a technological device. The projects are designed to be accessible to college students with no prior technical training or expertise. The laboratory projects aim to further the understanding of technology by introducing the nonengineering students to using tools, building things, and making things work. The laboratories expose students

TABLE 14.2. Laboratory Projects

Laboratory	Technological Activity	Science Topic
Automobile Engine	Disassemble and reassemble nonworking engine	Combustion
Simple Machines	Work with levers, pulleys, inclined planes	Conservation of Energy
Electricity	Basic Voltage, Current, Resistance Experiments	Voltage, Current, Resistance
Power Plant Tour	Inspection of major plant components	Electromagnet Induction
Electric Motor	Assemble a DC motor from basic components	Electromagnetism
Pinhole Camera	Make camera, expose and process film	Light Propagation
Lasers and Optics	Assemble diode laser, make telescope, microscope	Light Refraction
Simple Radio	Build simple crystal radio	Electromagnet Waves
Music Keyboard	Draw and etch circuit board, solder components	Resistance, Capacitance
Telephone	Disassemble telephone, study microphone	Electromagnet Induction
Holography	Expose and process hologram	Electromagnetic Waves

to the workings of actual technological devices of which they were previously only consumers.

The majority of projects are designed to result in tangible evidence of accomplishment on the part of the student. Half of the projects result in a working device or other product that the students take home. Such projects extend the impact of the course to include peers and family members. Other projects focus on an uncommonly high degree of contact with the workings of technology, such as construction of a diode laser light source or disassembly of an automobile engine. Most laboratories are relatively inexpensive and do not require any highly specialized equipment or facilities. The projects are able to be modified to suit different institutional circumstances.

The Internal Combustion Engine

In this laboratory, students disassemble and then reassemble real four-cylinder automobile engines. Major engine components such at the water pump, alternator, and starter are identified and studied. The relationship between the engine valves, camshaft, pistons, and crankshaft is examined. The engines are not intended to run, so the requirements for assembly are not especially stringent. Despite the nonfunctionality, the intimate contact with this technological device provides a unique learning experience for the nonengineering and nonscience student. Small model airplane engines are also disassembled to explain how the basic principle of internal combustion can be scaled down in size.

Simple Machines

In this laboratory, simple machines such as inclined planes, levers, and pulleys are built and studied as examples of some of humankind's oldest technological devices. The simple machines are the basic building blocks of more complicated equipment. Students measure the relationship between force and distance and calculate work done. Conservation of mechanical energy is demonstrated. The experiments make use of readily available equipment such as ten-speed bicycles, clamps, meter sticks, and spring scales. The project includes a discussion of metric system measurements. The scientific principle illustrated is conservation of energy in the context of simple mechanical devices.

Electricity

This laboratory familiarizes students with some aspects of electricity. Activities include the use of digital and analog multi-meters, measurement of battery voltages, construction of a copper-zinc-lemon battery, creation of simple circuits with a light bulb, measuring the output of a solar cell, and operation of light emitting diodes. Students wire a light socket and plug to make a lamp. This laboratory serves to develop an understanding of the key electrical concepts of current, voltage, and resistance.

Tour of Local Power Plant

A field-trip is taken to a local coal-fired electrical power plant. The tour is conducted by plant engineers and includes inspection of the coal pulverizers, boilers, burners, precipitators, turbines, and generators. Questions and discussions emphasize the application of the interrelation of electricity and magnetism in the plant generator. During the tour, students are able to see for themselves how the various components of the power plant work together to produce electricity. Through this laboratory, students see how electromagnetic induction is used to generate the electricity for consumer use.

Building an Electric Motor

Though often hidden from view, the electric motor is an extremely common and essential technological device. In this laboratory, students build a simple DC electric motor from basic components. Construction requires that students wind the field and armature coils and assemble the component parts. Students keep the completed motors. The project shows an application of electromagnetism in the form of a versatile technological device.

Pinhole Camera

In this project, students construct a simple pinhole camera using a box or container they bring to the laboratory. The pinhole camera project is used to illustrate some of the characteristics of light propagation and the basics of photographic processing. The cameras are used to take photographs. Students process the negatives and positives in the darkroom.

Light, Lasers, and Optics

Through a series of mini-projects, basic concepts of geometrical optics and laser light are introduced. Students measure the focal length of lenses and construct a simple telescope and microscope. Diode lasers, Helium-neon lasers, and fiber-optic cable are studied. An intensity-modulated laser is used to transmit sound.

Building a Simple Radio

Each student builds a simple AM "crystal" radio. To facilitate the understanding of the basic principles at work, the radio design is extremely simple, utilizing only a coil wound around a cardboard tube, a germanium diode, an earphone, and an antenna. The radios are kept by the students. The radio facilitates an understanding of electromagnetic waves and the important principle of information encoding in the area of telecommunications.

Constructing an Electronic Keyboard

A simple one-octave electronic music keyboard is constructed from elementary components and a single integrated circuit. Students make a printed circuit board on which the components are soldered. The completed keyboards are kept by the students. This project serves to provide students with a hands-on experience in electronics. Most of the student effort is devoted to constructing the device and getting it to work. The underlying function of key electrical components is examined and the concept of sound pitch and frequency is explained.

The Telephone

Students disassemble and reassemble a telephone. The internal electrical and mechanical components are identified and explained. Electromagnetic microphones are removed from the telephones and used to construct a simple person-to-person telephone line. This project demonstrates another application of electromagnetic induction but in a technological context which is entirely different from the power plant.

Making a Hologram

In the hologram project, each student makes a white light reflection hologram using a single-beam apparatus. Students bring in small objects as subjects for the hologram. Completed holograms are taken home. Holograms serve as an example of the application of interference and diffraction effects in light propagation. Understanding holography helps to develop an appreciation of the wave nature of light.

Assignments and Assessment

The student assignments for the Hope College "Science and Technology of Everyday Life" course consist of papers, quizzes, laboratory reports, homework assignments, and participation in class activities. Quizzes emphasize knowledge of basic concepts, explanations of underlying science, and familiarity with relevant terminology. Students write three papers describing the workings of technological devices. These papers must include a description of how the device in question works, the underlying science involved, a brief history of its development, and the student's assessment of the impact of the device on our culture and society. A longer final term paper is included. The primary purpose of the papers is to develop the ability of the student to be self-educating in scientific and technical topics. For the final paper, students must select a technological issue not previously covered in class to demonstrate a capacity for independent thinking and synthesis of information.

Observations about Teaching Nonengineers

While some of the laboratory activities may seem familiar to those with engineering backgrounds, nonengineering and nonscience students have rarely attempted any of these projects before. Further, while many versions of some of these laboratory projects can be found both in the literature and from commercial sources, effective use with students of limited technical background requires that the procedures be optimized for a target audience. Several characteristics must be designed into the projects. Projects taken home by the students must have a low cost per student. All designs must be robust and tolerant of error or misalignment in construction and use. While simplicity is important, the technological device should not be so overly simplified as to provide little information about the workings of manufactured versions of the product. For example, it is possible to find highly simplistic designs for electric motors; however, students benefit more from building a motor that includes a readily identified field magnet, armature, and brushes.

To help in activating student curiosity, laboratory projects must have a high degree of novelty and intrinsic interest. That is, a compelling output should result from a modest input of time and resources. A good example of this is a pinhole camera. Nearly everyone is fascinated and intrigued by how a tiny hole in a box can produce not just recognizable images but sharp and clear pictures. This intrinsic interest provides an opportunity to acquaint the student with a scientific principle underlying the observed effect. Projects are most readily explained when it is possible to identify one or two primary scientific principles that are sufficient to understand how the device works.

In activities that aim to bring students into more direct contact with technological devices, it appears to be important to use actual or modified versions of actual commercial products if at all possible. This explains why the effort was made to use real automobile engines rather than small engines or models. This practice of using actual technological devices is comparable to reading from primary sources in history classes. It gives the students a sense of authenticity and promotes a seriousness of purpose.

Summary of Results at Hope College

"Science and Technology of Everyday Life" has been taught at Hope College since spring 1995. As at other institutions, the approach has proven to be very successful in attracting the interest of a diverse group of nonscience majors and increasing the interest of this group in science and technology. Currently, the course continues to attract more students than can be placed in the class. Enrollment has averaged 55 percent women and 30 percent future elementary teachers, and includes students with such majors as art, business, psychology, languages, and music. From these trends, it appears that this type of course appeals to the interests and needs of nonscience and nonengineering students.

Technological Literacy for Engineering Students

Technological literacy courses for nonengineering majors have been found to be both of interest to, and relevant for, engineering majors. Like nonengineering students, many engineering majors are unfamiliar with the inner workings of the technology of everyday life. As technology becomes more complex, and engineering disciplines more specialized, engineering students are increasingly in need of a broad-based exposure to a range of technologies extending beyond the boundaries of a single engineering discipline. Engineering students are also in need of formal instruction to obtain hands-on skills in working with technological devices.

Issues regarding the technological literacy of engineering students have been noticed by those who have begun to offer courses for nonengineers. The "Introduction to Physical Devices and Systems" course at the University of California, Davis, was offered for first-year women engineering students. This was intended to help increase the retention of first-year women in the engineering program. The "Exploring Technology" course at Lake Superior State University included first-year students who intended to major in engineering along with the liberal arts students. At Hope College, engineering majors have expressed interest in having a version of "Science and Technology of Everyday Life" taught for engineers. At the University of Wisconsin, Madison, a survey course of a variety of technological devices is offered for mechanical engineering students.

Institutional Considerations

As encouraging as the results of these examples of initiatives by engineers might be, long-term change requires the proper institutional environment at all colleges and universities. Institutional priorities must include a commitment to science nonmajors and technological education.

The new criteria for accrediting programs in engineering by the Accreditation Board for Engineering and Technology (ABET) provides an opportunity for each engineering department to define for itself the educational objectives of its program. By explicitly including a commitment to reach out to nonscience and nonengineering students, engineering departments can sculpt their own institutional environment into one that includes a role for nonmajors education. This opportunity has the potential to bring all engineering programs to the forefront of efforts to make lasting improvements in the science and technological understanding of all students.

Conclusions

The critical nature of both context and hands-on experience in nonmajors' science and technology education helps to explain why engineering faculty can make a significant contribution in this area of education. The new ABET accreditation criteria provide engineers with a mechanism to make a permanent commitment to the education of all students. In the process of reaching out across campus to liberal arts students, engineers can achieve more equitable relationships with liberal arts departments as well as realize new insights into the educational needs of engineers.

The engineering disciplines include study of both scientific theory and specific technological applications. Engineers as developers of new technology have achieved a balance between theory and application. As educators from other disciplines have discovered, this perspective can be a unique asset in attempting to reach nontechnical students.

Acknowledgments

The author would like to acknowledge the numerous contributions made by the faculty and staff of Hope College. Support for this work was provided by the Hope College Science Division, and the Matthew J. and Anne C. Wilson Foundation. The helpful suggestions and comments made by the reviewers and editors of this volume are also greatly appreciated.

References

1. Goodstein, David. 1993. *Scientific Elites and Scientific Illiterates*, Sigma Xi, Proceedings of a Forum on Ethics, Values, and the Promise of Science, New Haven, CT: Sigma XI, February 25-26.
2. National Science Foundation, 1996. *Shaping The Future: New Expectations for Undergraduate Education in Science, Mathematics, Engineering, and Technology*, NSF 96-139, October 1996.
3. National Research Council, 1996. *From Analysis to Action: Undergraduate Education in Science, Mathematics, Engineering, and Technology*, National Academy Press, Washington D.C.
4. National Research Council, 1999. *Transforming Undergraduate Education in Science, Mathematics, Engineering, and Technology*, National Academy Press, Washington D.C.
5. National Research Council, 1982. *Science for the Non-Specialist: The College Years*, Committee on the Federal Role in College Science Education of Non-Specialists, National Academy Press, Washington D.C.
6. Goldberg, Samuel. 1990. *The New Liberal Arts Program: A 1990 Report*, Alfred P. Sloan Foundation, 630 Fifth Avenue, New York, NY 10011 (December 1990).
7. Trilling, Leon. 1994. *New Liberal Arts Educational Materials*, Council for the Understanding of Technology in Human Affairs, The Weaver: Information and Perspectives on Technological Literacy, 9(1)
8. Ames, Oakes. 1994. A Program for Technological Literacy in the Liberal Arts, *Journal of College Science Teaching*, March/April. 286-288.
9. Dudley, John M., and Bold, Gary E. J. 1996. Top-down teaching in noncalculus-based introductory physics classes, *Am. J. Phys.*, 64(4):418.
10. Bloomfield, Louis A. 1997. *How Things Work: The Physics of Everyday Life*. New York. John Wiley & Sons, Inc.
11. Watson, George. 1997. Science Concepts behind High Technology, University of Delaware, Department of Physics and Astronomy, Available HTTP: http://www.phys.udel.edu/~watson/scen103/. [1997, May 1].
12. Rogers, Eric M. 1960. *Physics for the Inquiring Mind*, Princeton University Press.
13. March, Robert. 1978. *Physics for Poets*, Second Edition, New York. McGraw-Hill.
14. Hacker, M., and Barden, R. 1992. *Living with Technology*, Second Edition. Albany, Delmar Publishers.
15. Maley, Donald. 1987. Integrating Mathematics and Science into Technology Education, *The Technology Teacher*, May/June, pp. 9-12.
16. LaPorte, James, and Sanders, Mark. 1993. Integrating Technology, Science, and Mathematics in the Middle School. *The Technology Teacher*, May/June, pp. 17-21.
17. Spirer, Kathleen G. 1995. Techno Teacher/Techno Kid. *Frontiers in Education 1995*, Budny, D., editor, ETP/Harrison. 4b3.11.
18. Byars, Nan. A., 1998. Technological Literacy Classes: The State of the Art, *Journal of Engineering Education*, 87(1): 53.
19. Henderson, Jerold M., Desrochers Debra., McDonald, Karen M, and Bland, Mary M. 1994. Building the Confidence of Women Engineering Students with a New Course to Increase Understanding of Physical Devices, *Journal of Engineering Education*, 83(4).
20. Bland, Mary M., 1995. Evaluation of "How Things Work:" A Hands-on Experimental Engineering Course, Proceedings of the 1995 WEPAN Conference, Women in Engineering: Is Systemic Change Happening?

21. Mahajan Ajay. and McDonald, David. 1996. Engineering and Technology Experiences for Liberal Arts Students at Lake Superior State University, Proceedings of the 1996 American Society for Engineering Education Conference, Washington DC., June 23-26.
22. Disney, Katy. Vitkovits, Steve, and Pam, Rick. 1995. Designing a Portable Technical Literacy Course for Use in California, *Frontiers in Education* 1995, Budny, D., editor, ETP/Harrison.
23. Krupczak, John. Jr. 1996. Science and Technology of Everyday Life: A Course in Technology for Liberal Arts Students, Proceedings of the 1996 American Society for Engineering Education Conference, Washington DC., June 23-26.
24. Fronczak, Frank, 1997. College of Engineering, University of Wisconsin—Madison, Private Communication.

CHAPTER 15

ANN BROWN

The Museum in the Classroom

Technology in Art

Engineering education has emphasized function and economy over aesthetics for so long that, to many, technology and art seem diametrically opposed. On the contrary, technology and art have shared cultural ties throughout history. Indeed the word *technology*, according to its classical Greek origin, implies art as well as craft.

It is true that the cultural ties between art and technology have frequently been overlooked by contemporary society in its fondness for pursuing separate paths of specialization. This same society has sometimes accused technology of blighting the beauty in its path and has thus weakened the sympathy between artist and engineer. Nevertheless, the ties and sympathy exist as they always have. Therefore, it seems logical, as part of a developing engineering curriculum, to revive this link by including a study of technology in art.

The study of technology in art has many advantages for engineering curricula. An obvious advantage is that combining disciplines helps to develop critical thinking skills as students see the relationships between new analytical experiences and old ones. The new and different visual perceptions of art can enhance and reinforce the visual perceptions that students already encounter in their traditional engineering courses. Furthermore, engineering graduates taking their places in society will benefit from an education that includes the arts, and it can do them no harm to be seen as more human.

The following sections provide a rationale and an approach for bringing art, which includes technology, into the engineering class. The link between art and engineering is discussed. A sample "tour" of selected art representing approximately 300 years is presented (see Appendix). The interpretive comments that are included are appropriate for application to general engineering curricula. Finally, some suggestions are made for more specific applications. A list of the art from the "tour," including available sources, follows the conclusion of the chapter.

The Continuing Link Between Art and Engineering

"Great engineering is great art," said American illustrator Joseph Pennell, whose favorite subjects were the "Wonder of Work" of engineering projects.[1] In his book, which chronicled the construction of the Panama Canal, Pennell praised the "great art" of the project's "magnificent arrangement of line, light, and mass [which] were the last things the engineers thought of."[2]

Pennell's assertion of the natural link between artistic expression and technology has been echoed many times. As recently as 1998, a young award-winning engineer was recognized for a new surgical tool that he invented after spending "most of [his] time . . . sketching and thinking and trying to create." He described his engineering design experience as being "a lot more like art class than . . . math class."[3] Engineer-writer Henry Petroski has noted that "The engineer's dilemma is not unlike that of the artist, who must constantly wrestle with the troublesome horns of substance and form."[4] Another engineer-writer, Richard Meehan, recalled his "pleasure" and "satisfaction" in making an engineering design "that does not differ from any other art."[5] Perhaps Lewis Mumford best expressed the link between art and technology when he said, "[The] moments of balance between art and tecnics . . . represent a high point in any civilization's development."[6]

Bringing the Museum into the Engineering Classroom

Many engineering educators will at first be reluctant to incorporate an artistic study into their curricula. After all, few believe that they have any expertise in the arts per se. What these reluctant educators fail to realize, however, is that as engineers they come equipped with two of the most important characteristics necessary for a study of art: the ability to analyze and keen powers of observation. Furthermore, the lack of artistic training can be an asset to an application of art to engineering study. Unencumbered by formal artistic discipline, teachers and students alike are free to notice details that fit into a technological context. In short, with little practical preparation and with no preparation in the fine arts, any interested engineering educator can introduce art into the curriculum.

Materials for classroom presentation are frequently close at hand and are often unexpected "finds." Any museum is a starting place, and the challenge of locating suitable art can stimulate an interesting museum tour. Once located, many art works are available for purchase on 35-mm slides. Those works not available may usually be photographed, with permission of museum authorities, using extra fast 35-mm film without flash.

Magazine articles featuring museum illustrations are another excellent source of materials. Most copyright restrictions can be upheld if a single illustration is used to produce one color transparency for classroom use only. A third important source of materials is collected books of period or thematic art. One such collection, which provided several of the following discussion topics is Currier and Ives: *Chronicles of America.*[7] An excellent source of ideas and information on nineteenth century art and technology is Julie Wosk's *Breaking Frame.*[8]

Lacking a convenient art historian, those educators wishing to verify their observations and analyses can do so with any of a number of available guides to art. Two such guides are Cole and Gealt's *Art of the Western World*[9] and Marshall Davidson's *History of Art from 25000 B.C. to the Present.*[10] The latter is a particularly helpful first reference because of its handy size, many color illustrations, and annotated index of artists.

Materials in hand, a context needs to be established for the technology-in-art presentations. Two possible contexts are: (1) the history of technology and (2) the development of society's attitudes toward technology as perceived through art. Both contexts can apply to the following "tour," which was prepared from the observations of an English teacher who specializes in engineering writing, but who has never had a course in art appreciation, art history, or the fine arts.

A Sample Classroom "Tour" of Technology in Art

Technology in Pre-Industrial Art

Many paintings demonstrate the significance of technology in preindustrial times. As an example, Rembrandt's *The Mill* depicts a distant windmill as part of its natural, rural environment while portraying the artist's characteristic use of light and shadow. A Rembrandt contemporary, Meyndert Hobbema, painted a country water mill in closer perspective. Hobbema's mill shows the sluice turning the wheel from the top in one of the earliest adaptations of mechanically generated water power. Even this technical detail appears natural and appropriate in the rural landscape. Both paintings clearly indicate that in mid-seventeenth century

Holland, wind and water were natural sources of power just as the technology producing it was accepted as a part of nature and, as such, a subject for art.

Art of the Industrial Revolution

Industrialization in progress was portrayed in later art, such as de Loutherbourg's *Coalbrookdale by Night*, an important turn-of-the-nineteenth century painting of fiery smoke and flames behind industrial buildings and a darkened landscape. This painting of the earliest industrial development in the British Midlands provoked much discussion, even in its own time. Some viewers saw in it the power and progress of the new technology in all its glowing beauty. Still others saw only destructive power and ugliness blasting from the fires of hell. This conflict of opinion can be seen in the partially obscured full moon in the painting's far right sky. Unfortunately, many reproductions of this painting are cropped to eliminate this natural detail. Whether the moon, or nature, is emerging to shine over the explosive scene below or whether the moon will soon be completely obliterated by the furnace's smoke is in the eye of the beholder. The artist's interpretation is unknown. A creator of stage settings in the British theater, de Loutherbourg knew how to achieve a highly dramatic moment in the art of the Industrial Revolution.

Emergence of Technology in Nineteenth-century American Art

Nineteenth-century art often mirrored its society by showing its preoccupation with emerging technology. Particularly in American art, the joint fascination with technology and the ever-popular landscapes is well defined. However, as technology became more celebrated, it grew in significance and scale in the art that represented it. At first the landscape was presented in magnificent, romantic panoramas, in which humans were a small, harmonious part, but where technology did not exist. Asher Durand's *Kindred Spirits*, representing the Hudson River School, is a fine example. The two small figures in the center of the painting are dominated by the majestic nature surrounding them. That the men are amicably talking as they view the scene represents the harmony that characterizes these paintings. Sixteen years later, Jasper Cropsey's *Starrucca Viaduct* at first represents a similar romantic landscape. The small figures in the foreground, one reclining, suggest the Hudson River School. However, a train crossing the viaduct in the background bisects the autumnal mountain scene. Only when the eye catches the movement in the center of the painting does the technology become apparent. The train's smoke, which at first resembles another low cloud, indicates that the train is moving away, leaving man and mountain valley in peace. The distant viaduct, however, remains.

This subtle intrusion of technology is found in other works of the period, particularly in many of Currier and Ives' romantic, commercially produced scenes of

American life and legend. Typical is their lithograph of the *West Point Foundary, Cold Spring*. The small figures in the foreground appear just as amicable as Durand's as they view the lake nestled at the bottom of a peaceful wooded valley. The factory roofs and chimneys rising above the tree tops indicate industrialization below, but do not violate the otherwise harmonious landscape. In the distance a train crosses a trestle on the lake, and still farther in the background a steamboat travels toward, as if to overcome, sailboats.

Developing technology is more intrusive in George Inness's *Lakawanna Valley*. The reclining figure in the foreground surveying the panoramic landscape suggests the Hudson River School. However, in this case the valley is dominated by technology, not nature. The distant factories and even a railroad roundhouse fit naturally into their surroundings, but there the conventional romanticism of the painting ends. The foreground contains only one large growing tree amid a mass of tree stumps. The locomotive in the center of the painting is moving toward, not away from, the foreground containing the large tree and human figure. That the man remains reclining as the train approaches to pass him shows his trust in this technology. His lack of alarm and the soft, romantic light of the painting indicate an acceptance of the burgeoning railroad industry. It is interesting to note that Inness's painting was, in fact, commissioned by the Lackawanna Railroad in an apparently successful attempt to reassure the public that trains were not as dangerous or "disruptive" as many believed.

Technology in Nineteenth-century Allegorical Art

Allegorical paintings, extremely popular in nineteenth-century America, vividly proclaimed increasing technological achievement. John Gast's *American Progress* portrays technology as a benign force overpowering darkness and evil. Gast presents progress as a personified angelic figure wafting her way over the continent from east coast to west. Bright morning light accompanies her as she travels, dispensing technology in the form of telegraph wire that she unrolls from a coil held over her arm. She also carries a large book, representing the wisdom that inevitably comes with progress and technology. Ahead native Americans and wild animals, representing evil, flee from her brilliance into the dark clouds and shadows that hang over the western mountains. Behind her follow three trains, representing the three major east-west railroads. In the foreground, early pioneers precede her on foot and horseback while farmers plow open, fenced land in her wake. An interesting technical detail in the extreme right, eastern portion of the continent is the Brooklyn Bridge, represented in its entirety eleven years before its completion.

A popular Currier and Ives print, *Across the Continent*, presents a similar allegory, using a train to represent progress as "Westward the Course of Empire Takes its Way." The train, moving from lower right to upper left, approaches the center of the lithograph. The train's black smoke, considered a positive image of power at that time (pollution was not yet a concern), rolls to the right, or uncivilized side of the train, where it engulfs two native Americans, who shrink away in awe and fear. To the left of the train, civilization has already begun, as seen in the town of neat log buildings, including church and school. Ahead of the train is vast open prairie, where a few hardy souls venture forth in covered wagons, on the civilized side of the print only. The winding river and distant mountains give the work the romantic panorama that popular taste demanded.

Late Nineteenth-century Art: Technology Triumphant

By the end of the nineteenth century, technology had emerged as a positive dominant force. The art of this period frequently portrayed technological achievement in much the same way as it had celebrated nature in the past. An object of much celebration was the Brooklyn Bridge, the so-called "eighth wonder of the world." The great East River Bridge, as it was then called, was so important that a Currier and Ives print circulated two years before the bridge was completed appears to be crowded from shore to shore with tiny human figures. A later print of the opening fireworks display records what is still regarded as one of the grandest celebrations in American history, rivaling even the V-J Day celebration in Times Square at the end of World War II.

European art at this time records a similar appreciation of technology. Two paintings by Claude Monet show the extent of technological advance into European culture. The dominant features of *Arrival of Normandy Train, Gare St.-Lazare* are technological: the train and the station. The locomotive in the center of the painting is its most vividly defined object. The bright, natural light beyond the station suggests a time of high activity; however, most of the busy human figures are represented by masses of contrasting, dark colors. Beyond the station, the background is an indistinct cityscape, much of which is obscured by locomotive smoke. This departure from earlier works of art indicates the arrival not only of the train, but of technology as a dominant force in late nineteenth-century life. Humans were no longer observing technology as an interesting adjunct to their lives; they were interacting with it.

A later turn-of-the-century Monet painting, *Waterloo Bridge, London, Grey Weather,* is an even more impressionistic presentation of a technological creation as an object of art. The bridge and its reflections in the river suggest that they are one, unified technological-natural structure. In addition, all activity on top of the bridge is treated with the same characteristic blues and greens as the bridge itself, making activity and bridge appear to merge into one dynamic medium. Thus, human interaction with technology seems further intensified. Beyond the bridge, the muted colors of the distant shoreline suggest an industrial landscape, thereby enhancing this celebration of technology.

Technology in Twentieth-century Art

The twentieth century brought to art new media and new perspectives, both in scale and distance and in the ambivalence developing toward technology. Some works still celebrated technology, but others began to shift from the optimistic enthusiasm of the previous century.

As early as the first quarter of the century, some art had changed dramatically, as demonstrated by abstract paintings by French cubist Robert Delaunay and American Georgia O'Keeffe. In its time, Delaunay's *Champs de Mars, The Red Tower* was decried as a sacrilege. The fact remains, however, that the artist chose to portay his return to fundamental, primitive shapes and colors with the well-known, pyramidal structure of the Eiffel Tower, an engineering marvel of the 1880s. Ironically, the Eiffel Tower itself was decried in its own time for lacking those aesthetic qualities thought necessary in the Paris skyline. In the painting, the beautiful ruby-colored tower appears to be breaking loose from conventional art and architecture, as it rises above their duller, conventional shapes and colors.

Georgia O'Keeffe's *Shelton with Sunspots* presents another familiar technological structure, the New York skyscraper. There can be no doubt that O'Keeffe was paying tribute to the building where she lived with her photographer husband Alfred Stieglitz. The magnificent building acquires almost magical power as it radiates heat and light within the confines of other skyscrapers. The influence of her husband's artistic medium can be seen in the painting's unusual portrayal of sunlight, which suggests the photographic distortion that occurs when an unshaded camera is pointed at an object in the sun's path. Thus, the Shelton appears both glorified and distorted at the same time. Just as the "halo" celebrates the Shelton, the bright image of the sun seems to be cutting away at its upper floors. These dual images suggest an underlying ambivalence toward twentieth-century technology and a developing conflict between technology and nature. Indeed, O'Keeffe's later preference for nature and natural subjects is perhaps foreshadowed by the painting's point of view, which looks up toward the sky and not down at the street.

The Power of Technology as Art

A celebration of technological power is the theme of works by Charles Sheeler and Lewis Hine, although the latter work also suggests a conflict between man and machine. Sheeler's famous locomotive is the sole focus of its painting. Here technology has become art itself without the need of any justifying background or framework. Nature has become irrelevant. Painted with photographic realism, the close-up view does not present the entire locomotive, but only its most important mechanical components. Thus, by graphically representing the *Rolling Power* of its title, Sheeler's painting pays tribute to one of America's cultural icons.

On the other hand, Lewis Hine's photograph, *Powerhouse Mechanic*, combines a human element with the machine, suggesting a questionable interdependence of man and technology. The harmony between the arcs and angles of the machine and the man's body is essential to the photograph as a work of art. Without the mechanic, it seems that the engine would be lifeless. Similarly, without the machine, the mechanic would have nothing to do and would appear to have little meaning. However, the title's emphasis on the word "Mechanic" suggests that for the moment the man is vital to power production. A conflict arises because, to maintain control, the mechanic cannot stand upright in a natural human position. Ironically, if the mechanic stands, he gains humanity, but loses control and significance within the context of the painting and its technology.

Technology in Art During the Depression

Two paintings produced during the same year early in the American Depression demonstrate opposing attitudes toward technology. Charles Sheeler's *Classic Landscape* suggests stability and endurance with its presentation of the Ford plant at River Rouge. The view of clean, orderly factory buildings, silos, and a railroad yard suggests a "classic" or timeless presence in American culture. No human presence is expected in this panorama of industrial production. The strong, almost white light of mid-day and the white smoke billowing from the stack in the right foreground indicate that mechanical and human activity abounds within the factory. Even nature is in harmony with this industrial environment as indicated by the distant, grey cloud line, which is about to merge into the same vanishing point as the railroad tracks.

In contrast to Sheeler's panorama is Charles Demuth's close-up painting of an industrial building with twin water towers. In spite of a similar bright light, blue sky, and a stream of white smoke issuing from a single stack, the scene does not suggest the reassuring stability of Sheeler's classic factory-scape. Instead, the contrasting dark colors and near vantage point, which is slightly looking up, make the factory seem slightly top heavy and unbalanced. The almost colorless traffic light in the lower right suggests an absence of activity. Panes in the upper-story windows appear to be missing. If these negative details are not enough, Demuth's title, *And the Home of the Brave*, which ironically parodies the national anthem, defines the painting as a bleak picture of the technological environment.

Anti-technology Art

The negative effects of technology upon humans can be seen in a number of twentieth-century paintings, including Diego Rivera's *Detroit Industry* and Edward Hopper's *Nighthawks*. The cool, impersonal colors and colossal machinery of Rivera's large mural suggest that workers are being exploited as technical drones. Workers at separate levels and departments pursue different tasks, suggesting that little, if any, human interaction takes place. All of the faceless workers lack identity, with the exception of the two figures in the lower right who can be recognized by their coats and ties as managers or engineers. Most condemning is the lower-right border, in which male and female workers wearing street clothes walk in step, in line, robot-like, without speaking and with heads slightly bowed.

By contrast, Hopper's painting presents, with the exception of two coffee urns, no technology at all. However, the night scene outside the all-night diner is obviously in an urbanized, industrial setting. The three customers are not communicating, even though two are seated together. The fact that the diner is situated on a corner allows the viewer to look through two walls of glass to the dark, empty storefronts across the street. Obviously, this restaurant is not an oasis in spite of the refreshments it serves. Even its artificial light offers little relief from the dark, inhospitable environment outside, suggesting that our technologically driven society produces alienation. The title *Nighthawks* also reinforces the painting's human isolation since hawks are lonely predators.

Coda: Two Abstractions of Engineering Achievement

Two concluding works quietly record significant examples of man's technological skill. Alexander Calder's abstract sculpture *Calderberry Bush* is a hand-forged mobile that rests on a solid, but transparent, triangular foot. Calder's characteristic complex design is presented with deceiving simplicity and harmony, even in this early work. Within this harmony, the bush's branches and fruit create a tension that is resolved by the balance of the spherical weighted root within the triangle. This tension and its resolution represent the interaction of important physical laws. Therefore, the sculpture is not only art that incorporates technology. It is a dynamic demonstration of technological principles of weight, distance, and balance in action.

A demonstration of American artist Joseph Stella's love affair with New York City was his series of abstract paintings of the Brooklyn Bridge, each one depicting a different mood of the famous structure. One such painting represents the great bridge at night; the colors are correspondingly dark. The partially lighted cables ascend toward and thus emphasize the twin gothic arches. The city is represented within the arches as small, heavily shadowed New York landmarks that blend with the remaining structure of the bridge. The predominating blues and reds and the contrasting light and dark details within the gothic design give the composition the overall impression of a stained glass window, with all it implies. Stella chose to celebrate his beloved city with one of her best known monuments to technology.

Some Specific Applications of Art to Engineering Education

Visual art can serve as a natural transition between technology and the humanities. This application is already achieving good results at North Carolina State University, where art is being used to link a freshman composition course with a "hands-on" engineering lab. Class journals indicate that students enjoy identifying the technology in art and its significance. Art viewed in class has generated new topics for discussion and composition. The students usually remember what they have seen and look for new examples to extend their appreciation of both art and engineering.

In addition to the general social historical approach suggested earlier, art can suggest important implications to a study of engineering ethics. Important perspective can be gained when students of ethics consider, through art, how society has viewed the engineer and technology, not just in the present, but also in the past. In addition, engineering design and construction courses can be enhanced by graphics from the fine arts as well as the textbook. Many existing engineering drawings from the past are themselves important art and are well worth considering by today's design students.

Finally, it must be remembered that many artistic media, such as photography and flexible alloys, are themselves developing technologies that merit study in artistic, as well as engineering, contexts. Students can more fully appreciate the physical characteristics of artistic materials when they witness how these media have developed. A study of present-day art not only reveals how modern technology has changed artistic media and their expression, but also suggests how these may be influenced in the future.

An interesting example of an early market-driven technology appears in the lithography that was sold in the late-nineteenth century as home decoration. The popular Currier and Ives prints that recorded the history and culture of their day were black and white reproductions that were hand colorized. To keep these prints affordable, a limited number of tints was used, resulting in some unusual images such as trains with bright red cow-catchers. Yet in 1873, while Currier and Ives were busy producing art for America's living rooms, a New York publisher advertised the previously mentioned allegory "American Progress" in ten unbelievably realistic colors. This technologically advanced "chromo," pronounced "one of the grandest conceptions of the eye," was promoted as "richly worth" the then-hefty price of $10. However, as an advertising gimmick, the reproduction was offered free to each new subscriber of a western journal.[11]

Conclusion

A study of technology in art belongs in engineering curricula. Art, like engineering education, is predominantly visual. In addition to this similarity, art offers the benefits of a different perspective and a shift from the analytical emphasis of traditional engineering courses. Because art itself was founded in craft and technology, it is more closely connected to traditional engineering study than many believe. This connection must be recognized and accepted.

Since the days of the earliest cave paintings, art and technology have together represented the consciousness of the culture. Therefore, engineers, as the developers of technology, have traditionally shared a relationship with art. Unfortunately, as they have developed their skills and applied modern technology, many engineers have lost sight of this relationship. Such a loss need not continue. The cultural heritage that was originally implicit in engineering can and should once more be established by expanding traditional engineering curricula to include art.

References

1. Fredrich, Augustine J. 1993. *Great Engineering Is Great Art: Joseph Pennell's Images of American Civil Engineering*. Evansville: University of Southern Indiana Foundation.
2. Pennell, Joseph. 1912. *Pictures of the Panama Canal*. London: Heinemann.
3. Young inventor garners award for surgical device, *Bangor* [ME] *Daily News*, February 7, 1998.
4. Petroski. Henry. 1977. *Beyond Engineering*. New York: St. Martin's Press.
5. Meehan, Richard. 1981. *Getting Sued and Other Tales of the Engineering Life*. Cambridge, MA: MIT Press.
6. Mumford, Lewis. 1952. *Art and Technics*. New York: Columbia University Press.
7. Pratt, John Lowell, ed. 1968. *Currier and Ives: Chronicles of America*. Maplewood, NJ: Hammond Inc.
8. Wosk, Julie. 1992. *Breaking Frame: Technology and the Visual Arts in the Nineteenth Century*. New Brunswick, NJ: Rutgers University Press.
9. Cole, Bruce, and Adelheid Gealt. 1989. *Art of the Western World*. New York: Simon and Schuster.
10. Davidson, Marshall B. 1984. *History of Art from 25,000 BC to the Present*. New York: Random House.
11. Nottage, James, Chief Curator, Autry Museum of Western Heritage, Los Angeles CA. Private Correspondence, November 16, 1995.

Appendix

Works of Art Cited in Order of Presentation
Artist, Name, and Date, Location Available

1. Rembrandt, *The Mill*, c. 1650: National Gallery, D.C.
2. Hobbema, *Wooded Landscape with Water Mill*, c. 1662–64: Art Institute of Chicago
3. de Loutherbourg, *Coalbrookdale by Night*, 1801: Museum of Science, London
4. Durand, *Kindred Spirits*, 1849
5. Currier and Ives, *West Point Foundry, Cold Spring*, n.d.
6. Cropsey, *Starruca Viaduct*, Toledo (Ohio),1865: Museum of Art
7. Innes, *Lakawanna Valley*, 1855: National Gallery, D.C.
8. Gast, *American Progress*, 1872: Autry Museum Autry Museum of Western Heritage, Los Angeles
9. Currier and Ives, *Across the Continent*, 1868.
10. The Great East River Suspension Bridge, 1881
11. The Grand Display of Fireworks and Illuminations, 1883
12. Monet, *Arrival of Normandy Train, Gare St-Lazarre*, 1877: Art Institute of Chicago
13. *Waterloo Bridge, London, Grey Weather*, 1900: Art Institute of Chicago
14. Delaunay, *Champs de Mars, The Red Tower*, 1911: Art Institute of Chicago
15. O'Keefe, *Shelton with Sunspots*, 1926: Art Institute of Chicago
16. Sheeler, *Rolling Power*, 1939
17. Hine, *Powerhouse, Mechanic*, 1920: George Eastman House, Rochester, N.Y.
18. Sheeler, *Classic Landscape*,1931: Prentice-Hall*
19. Demuth, *And the Home of the Brave*, 1931: Art Institute of Chicago
20. Rivera, *Detroit Industry*, 1932
21. Hopper, *Nighthawks*, 1942: Art Institute of Chicago
22. Calder, *Calderberry Bush*, 1932: Whitney Museum of American Art, New York
23. Stella, *Brooklyn Bridge, Variation*, 1939: Whitney Museum of American Art

*Prentice Hall: 1-800-922-0579 (formerly available from Sandak)

CHAPTER 16

KATHRYN A. NEELEY

The Aesthetics of Engineering: Toward an Integrated View of Engineering Design

Why Consider Aesthetics?

The ABET 2000 Criteria do not list an awareness of the aesthetic dimensions of engineering in the list of key abilities that graduates are required to demonstrate, so one might well ask why it is important for aesthetics to be considered within an engineering curriculum. The most compelling reason is that an aesthetic response is fundamentally an integrated response that brings together a range of inputs and modes of perception. An aesthetic response is holistic, it blends empirical observations and value judgments, and it grows out of a multidimensional experience, combining emotional, intellectual, and material elements. The integrative nature of aesthetic perceptions is consistent with the integrative nature of engineering and of engineering education as these are envisioned in the ABET 2000 criteria.

On a more explicit and specific level, an understanding of the aesthetic dimensions of engineering should enhance student ability in design and problem-solving, which are at the core of disciplinary competence. Because a consideration of aesthetics requires students to consider viewpoints that are often radically different from their own, it should enhance their abilities to communicate and to function on multidisciplinary teams. An exploration of the aesthetic dimensions of engineering also helps students develop the perspectives they need to understand professional and ethical responsibility and the social context and impact of engineering. Focusing on aesthetics makes them more aware of the ways in which technology both reflects and affects the individuals and cultures who create it. Aesthetic responses capture what people feel and believe as well as what they can quantify or prove objectively.

In many ways, the voices who tell us that "technology is just a tool," that "engineering is simply the application of science," and that "art is removed from life," are attempting the same thing: to provide us with highly abstracted and one-dimensional views of art, science, and technology. A consideration of aesthetics shows us how multifaceted and culturally grounded all of these entities are. Perhaps most significantly, it provides a useful, rich, and potentially engaging view of engineering.

One need not spend much time in an engineering school to realize that many—if not most—engineering students are unaware of the artistic and aesthetic dimensions of engineering and believe that sensation, emotion, intuition, and tacit knowledge have no place in engineering. As one such student put it, "There is no aesthetic dimension to engineering. This is an engineering school, not an architecture school." In this view, engineering is concerned only with functionality, economics, and quantitative modes of evaluation. This view of engineering also tends to be widely shared by those outside of the profession.

Yet there is ample evidence that sensory perceptions, intuition, emotion, a "sense of rightness," pleasure or revulsion, and other forms of holistic response play a crucial role in evaluating designs and models in engineering and science. Numerous scholars from a variety of backgrounds have identified the important role that creativity and aesthetics play in engineering design.[1-5] Qualities variously referred to as elegance, sweetness, and simplicity are used as criteria for accepting or rejecting models and hypotheses, and in predicting the acceptability and performance of possible designs.[6-8]

Where aesthetic responses to artistic artifacts are usually discussed in terms of beauty, engineering designs are often discussed under the topic of "judgment," a capacity born of experience that cannot be directly taught but is crucially important in engineering design. Eugene Ferguson, author of *Engineering and the Mind's Eye*,[9] describes design using terms strongly associated with aesthetics—as an integrative, creative, and contingent process. He emphasizes the role of visual knowledge, "an intuitive sense of fitness," and "a nonverbal, tacit, and intuitive understanding essential to engineering design." "A successful design," he tells us, combines formal knowledge and experience and always contains more judgment than certainty. . . . no matter how vigorously a 'science' of design may be pushed, the successful design of real things in a contingent world will always be based more on art than on science. Unquantifiable judgments and choices are the elements that determine the way a design comes together.

In *Getting Sued and Other Tales of the Engineering Life*, Richard Meehan offers his own account of the nature and origin of engineering judgment.[10] Like Ferguson, he emphasizes the limitations of systematic knowledge. He also provides some insight into why many approaches to engineering minimize the importance of aesthetic responses.

From years of trial and error, judgment takes rude shape in some inner chamber of the mind. The earth and its material is alive; it shows many faces, cooperates or rebels. The artisan recognizes the personality of earth and stone, just as men come to recognize friendly dogs or troublesome superiors. It is a perishable thing, this judgment; it dies with a person, for it cannot be tamed in books, frozen into codes and regulations. To some technologists its purest form may even be obscene, as in the water witch, the faith healer, the fortune teller. But take note technologists: it was *techne*, artifice, that brought forth the ribbed vaults at Durham Cathedral; *logos*, systematic knowledge, spawned the freeways of Los Angeles.

Meehan is a particularly able guide to thinking about the aesthetic dimensions of engineering because, although he is keenly aware of and highly articulate about the artistic dimensions of engineering, he never loses identity as an engineer and never comes across an artist at heart who was thwarted in his career choice. Equally important, Meehan's accounts make it clear that one need not be a heroic figure—a Leonardo da Vinci or John Roebling—to perceive and benefit from the aesthetic dimensions of engineering. He shows that there can be an aesthetic dimension to any engineering work, provided it is approached in the right frame of mind.

Meehan offers this description of the process of design and the satisfactions of being a designer:

I learned the pleasure in it, in this design, the satisfaction in making a clay bowl or a painting or writing a sentence or a symphony. First the concept, the trial efforts, the crude shape of a good solution, the refinements, balance, and polish until the final arrangement sings with deceiving simplicity and stuns with accuracy of effect . . . I was able to experience technology not as the stepchild of science (which is, after all, impotent) but as an art.

The experience he describes is clearly aesthetic, characterized by pleasure and the perception of elegance. Holistic in nature, this experience cannot be reduced to intellectual or analytical appreciation. From an engineering educator's perspective, it is both noteworthy and regrettable that Meehan first experiences the pleasure of design not during his four years of undergraduate education in civil engineering at M.I.T., or even on his first job, but after many years as a practicing engineer. This chapter describes a strategy for helping undergraduate engineers at least begin to experience technology as an art while they are still students. The heart of the strategy is an innovative approach to using art museums as a context for exploring the aesthetic dimensions of engineering.

Defining the Problem

As Ferguson, Meehan, and numerous others have demonstrated, the problem is not one of establishing that there is a common ground of creativity shared by art and engineering, or to identify the role that aesthetic responses and criteria play in the design, evaluation, and appreciation of the products of engineering. Rather, the problem is one of overcoming the barriers that lead students to overlook the aesthetic dimension of engineering; the need is to make the aesthetic dimension real and to make it relevant to engineering students.

The nature and origin of these barriers are significant subjects in themselves and certainly cannot be treated in depth here. Still, it is worthwhile to consider some of the most important ones. Many of the most difficult to overcome grow out of deeply held and often emotionally charged beliefs and stereotypes, which exert a powerful but often unrecognized influence on the way people think. The belief that engineers and artists reflect opposite ends of a spectrum of cultural values and personality types is so deeply embedded in our culture that most people feel no need to question it, especially once they have chosen a camp. Helping engineers experience technology as an art requires calling attention to these stereotypes, many of which are only partly—or perhaps never—articulated, and opening them up to the critique that exposes both their basis in truth and their inadequacy.

Questioning whether engineers and artists are really as different as they are supposed to be can be threatening to students who hear such questions as suggesting that engineers and artists ought to be exactly alike or that artists are somehow superior. The "artist" identity and personality type tend to emphasize freedom, subjectivity, individuality, and the absence of constraints of all kinds, whereas the "engineer" identity tends to emphasize constraints of many kinds—constraints imposed by physical reality, economics, employers, customers, and the need to come up with a workable final product. Perhaps the most significant of these constraints arise from the fact that the safety and welfare of others often depend on the adequacy of engineering designs. As one student expressed it, "When a work of art fails, no one dies."

Another very important barrier was invisible to me until my students pointed it out: most people do not know enough about both art and engineering to see the similarities between the two. My experience in dealing with engineering students suggests that most of them only need to learn a relatively small amount about the techniques and design process of artists to begin to see a number of useful parallels with engineering.

Overview: Design as the Common Ground Between Engineering and Art

This chapter describes a set of experiences designed to help engineering students cultivate their aesthetic sensibilities and use them in engineering. The students are encouraged to draw on their personal, firsthand experience to increase their capacity to think comprehensively and critically about the products of engineering design and to give them greater awareness of the nature of design as a process. The basic strategy is to get students to think about technology and engineering design in the physical setting provided by an art museum, thereby immersing them in an environment dedicated to aesthetic and artistic considerations. Two interactive tours of a small art museum at the University of Virginia are combined with preparatory and follow-up activities to help students experience and appreciate the artistic and aesthetic dimensions of engineering.

The tours and related activities form the core of a study entitled "The Engineer as Designer," which ordinarily takes up about 20–25 percent of a fourth-year required engineering course entitled "The Engineer in Society." The following sections outline the key concepts and assignments associated with the tours.

Guiding Strategy

The guiding strategy involves blurring but not completely obscuring the distinction between engineering and the fine arts by increasing the students' awareness of design as common ground shared by the two. This common ground is established using four different strategies.

Define all products of design as "artifacts" and the process of design as a series of constrained choices. The word "artifact" literally means "thing made by artifice or skill" and distinguishes human-produced objects from objects created by nature. It helps establish common ground in the sense that it applies equally well to products of art and to products of engineering. The difficulty with the term is that neither artists nor engineers are particularly pleased to have their products defined as artifacts. To many artists, the term suggests an undesirable utilitarian quality. To engineers, the term artifice often carries the negative connotations associated with deceit. These objections notwithstanding, the term "artifact" at least gets the discussion beyond the dichotomy implicit in the vocabulary of "works of art" vs. "machines" and makes it possible to focus on design as a process common to art and engineering. In this view, both the *Mona Lisa* and an advanced microprocessor can be viewed as products of a series of design decisions in which nature, culture, and the designer furnished resources, imperatives, and constraints. The students are also encouraged to think of design as the process of turning ideas into reality and to see all design decisions as important choices that reflect values and beliefs.

Think of ENGINEERING in terms of art. This strategy brings the aesthetic and creative elements of engineering into relief by highlighting aspects of engineering work that are widely acknowledged but not easily taught. It emphasizes the creativity required to develop good solutions to complex problems; the absence of formulaic approaches to design; and the role of intuition, sensation, and emotion in the design, evaluation, and appreciation of engineering designs. The role of intuition and feeling is expressed through the use of phrases such as "that looks about right" or "I have a feeling that this will work." Cultivating intuition is an important but often unarticulated goal of design education. The most important aesthetic value is simplicity—searching for the simplest means to accomplish the objective. To view an engineering artifact aesthetically is to view it holistically and to recognize that intuitive or emotional responses play an important role in the evaluation of engineering designs.

Think of ART in terms of engineering. This move emphasizes the constraints that materials impose on artists, the knowledge of technique and physical processes that artists must possess, and the role of practical considerations such as institutional and financial support. At its foundation, thinking of art in terms of engineering means recognizing that artists make design decisions within a set of practical constraints and with reference to a desired outcome or effect. This also entails recognizing that artistic artifacts are calculated to achieve an effect and do have a purpose.

Distinguish levels of creativity and freedom. We use a hierarchy of invention, design, and routine application to distinguish various levels of creativity and freedom within the two general categories of technology and art.*

> At the highest level—*invention*—none of the available solutions will solve the problem in question. New inventions are usually linked to what already exists, but they require a creative leap. In order to be patented, a new invention must meet

the criterion of not having been obvious to someone skilled in the art. Another significant feature of inventions is that they tend to shape future practice in significant ways.

Design involves following precedents, choosing from the available options, and deciding what to use and how to put it together. Typically, there are many workable solutions but only a few usually work well. Creativity comes into play in envisioning new options; intuitive evaluation is important for moving the process along efficiently since analysis of all the available options is ordinarily prohibitively time consuming.

Routine application involves the use of straightforward, standard solutions and is the level at which technicians are usually presumed to operate. In routine application, procedures are clearly and precisely defined and reliable step-by-step methods exist. Success typically depends on following instructions, and creativity is sometimes undesirable.

Analysis using this hierarchy reveals the varying levels of creativity, skill, and social status within each category since all three levels can be observed in many fields of endeavor. In the area of writing, for example, new genres and classics occupy the invention level; textbooks, best sellers, and most academic publications occupy the design level; and form letters or templates provided by a desktop publishing program occupy the routine application level. In the visual arts, we might conjecture that masterpieces and museum pieces such as those selected for the Museum of Modern Art correspond to the invention level. Like new literary genres or "great books," these works decisively shape future practice. At the design level, we might have portraits, book jackets, and original prints designed for mass production. Significant skill is required to create these kinds of artwork, and they serve important commercial and social functions. If they exemplify outstanding realization of an established genre, they might move toward the invention level. Still, most do not appear to be the product of exceptional insight or creativity, and they tend not to change the way we live or the way we view the world. At the routine application level, we might identify the products that can be created by using standard computer graphics programs.

The point of this kind of analysis is not to assign an artifact definitively to a particular level but to establish the variations that exist within both engineering and art and to reveal the aspects of both engineering and artistic design that tend to be eclipsed by cultural stereotypes. For example, in technology and in fine arts, some people enjoy much greater freedom and prestige than others, and some people are much more creative than others. The students see that many artists deal with significant constraints imposed by the laws of the natural world and by the organizations in which they work. These insights help the students get beyond the stereotyped notion that artists are totally unconstrained while engineers are so constrained that creativity becomes unnecessary or irrelevant to their work.

The analysis also raises some interesting questions. For example, what percentage of practitioners operate at each of the three levels in the respective arenas? How are engineers and artists rewarded for creativity? How do they earn freedom to work on designs of their own choosing? How do economic considerations enter into the various levels? What about skill and fundamental theoretical or scientific knowledge? Would it be fair to say that engineering places the greatest emphasis on those who operate at the design level while art gives most recognition to those who operate at the invention level? These questions are important because engineering students see the roles that expertise, economics, and various constraints play in the design work of artists.

The Visits to the Museum

In the scheme outlined here, both visits to the museum require students to participate actively and to recognize the design decisions made by artists. In addition, the first visit emphasizes aesthetic responses to, and criticism of, designed objects. The second visit asks the students to focus on the ways that design decisions reflect the designer and the culture from which the artifact originated.

Interactivity is a crucial feature of the tours, which primarily involve directed practice in inquisitive looking and dealing with the material reality of the artifact. The people who conduct visitors through the museum are called "docents," from the Latin *docere* "to teach," because their job is not to interpret the work for visitors, but rather to teach them how to construct meaning from observations of the work. The students are told that the analysis of a work of art "is not a research project; it is a visual exercise."[11] This approach emphasizes what the students can learn through visual examination.

In addition to being exposed through readings, lectures, and discussions to the ideas outlined above, the students are given a specific assignment associated with each tour. The students enter the museum with the kernel of a response to the artistic and aesthetic dimensions of a technological artifact. Their immersion in the museum environment and directed practice in examining artistic artifacts provide the stimulation to develop that kernel into a fully developed aesthetic response. These responses are articulated in discussions and written assignments that follow the tours.

First Visit: Analysis of Aesthetic Response to Designed Objects

As preparation, students are given this assignment: Before you visit the museum, you should identify some technological artifact to which you have a strong response and try to determine why you respond to the artifact as you do. During your first visit to the museum, identify an artistic artifact to which you have a strong response. Analyze the basis for that response and compare it to your response to the technological artifact.

In the museum, the students are immersed in an environment that forces them to rely on their eyes, emotions, and intuitions as sources of information. They are taught to read the language of art, which means that they learn how to decode the information conveyed through the artists' use of color, shape, line, texture, light, space, and other sensory aspects of objects. They learn about the decisions that artists must make and the techniques they can use to focus the viewer's attention, create emphasis, or convey a sense of motion. They gain insight into how artifacts are put together.

In many ways, the process they go through as they learn to read the language of art resembles the gathering and interpretation of empirical evidence. One particularly useful piece of art is a *togata,* a piece of public sculpture that originated in classical Rome. The subject of the sculpture is a young man standing with one arm at his side and the other extended at a 90° angle to hold part of his toga. Part of the extended arm has been broken off so that the means of its construction are visible. The head is also missing. At first, most students assume that it, like the arm, has been broken off. When they look more carefully, they see that it was designed to be removable and has been lost, not broken. After speculating for a while on their own, they usually learn from their guide that frequent turnovers in public officials combined with the high cost of marble motivated Roman sculptors to create statues whose identities could be quickly and cheaply changed by exchanging one head for another.

In a similar vein, they see that the extended arm of the statue was inserted into the body, not carved from the same piece. They learn that such a practice meant that the sculptor could save time as well as money by starting with a smaller piece of marble. The docents draw attention to the way that the sculptor has created the impression of soft, folded cloth in a material that is anything but soft or supple, and the students begin to understand the skill in shaping materials that is involved in sculpture. By the time they finish their visit, they are able to see the *togata* not just as an example of classical beauty or a symbol of imperial Rome but as a practical response to a series of design questions in which economics, the physical world, the political environment, and the designer's experience played a significant role.

Their judgments about what makes an artifact appealing necessarily involve holistic perception and evaluation. The docents help the students discern the design decisions that contribute to the viewer's response. They use analogies drawn from engineering and mathematics and provide a supportive atmosphere where students learn to support their own observations and trust their own opinions.

One of the premises underlying this set of experiences is that the students react to visual images and the evocative power of artistic artifacts, but rarely recognize the processes involved in their responses. Both the cultural meanings and the physical setting of the museum can help heighten the students' awareness of visual stimuli and evocative power in artifacts. These sensibilities can then be transferred to the perception and appreciation of technical designs and artifacts.

In the discussions following the first visit, students mention a wide range of artistic artifacts along with a number of recurrent technological artifacts, such as the compact disk, which they admire for its combination of a shiny metallic finish and superior sound quality, and the automobile, whose aerodynamic lines reflect both scientific principle and high performance. They also may identify a wide range of factors that contribute to aesthetic appreciation, including form/function congruence, elegance and economy, order and simplicity, speed and power, size and scale, recognition/comprehension, capacity to operate on different levels, display of skill, intentional aesthetic appeal, and ability to revolutionize. It is interesting to consider which of these factors are brought into the discussion by the inclusion of technological as well as artistic artifacts. It also becomes clear that there is a distinctive set of aesthetic standards that are applied to technical artifacts and that those tend to be different for machines than they are for structures.

Second Visit: Designed Objects as Expressions of Cultures and Designers

For the second visit, the students complete the following assignment: Identify a technological artifact that seems particularly expressive of either the person who created it or the culture in which it was created. While you are in the museum, identify a piece of art that seems to express its creator or culture particularly clearly. In each case, identify the expressive elements. Compare and contrast the two kinds of artifacts with respect to their expressiveness.

Prior to the second visit, students are introduced to the aesthetic and expressive dimensions of technological artifacts designed for everyday use (i.e., those not destined for museums at the outset). Their responses to the first assignment should provide a number of possibilities. The telephone offers an interesting example of a technological artifact with an interesting aesthetic history. The evolution of the telephone from a single black model through a limited range of colors and shapes, to the wide variety of colors, designs, shapes, and personal styles associated with it today offers a visually interesting and conceptually significant example of the ways that aesthetic considerations can become prominent in the design of technological artifacts.

During the second tour, the guides ask the students to gather information about the artifacts they view, but this time the purpose is different: they are looking for ways in which the artifact expresses the culture, natural environment, and the designer who originated it. Perceiving the expressiveness of artifacts means, to a great extent, understanding the information that artifacts carry. African art provides good material for this kind of analysis because it comes out of a significantly different cultural and natural context and because many African ritual objects combine functionality with aesthetic appeal in a way that belies the rigid distinctions that Europeans tend to make between functional objects and art objects. Once students recognize the expressiveness of artifacts from unfamiliar cultures, they are in a better position to see the expressiveness of the technological artifacts produced in their own cultures.

Follow-Up Assignments

A number of written assignments can be used to follow up on the tours, to assist the students in articulating what they gain from the experiences in the museum, and to use the capacities for aesthetic and intuitive perception that they may have developed. They can be asked to write short essays recording their responses, to compose short stories or conversations with characters in paintings, or to write poems about artwork and compare the design process they used to write the poem with the process they use for engineering design. They might also be asked to design a book jacket that expresses the subject, theme, or tone of a book they have read for the course.

Keys to Success

Museum Personnel

The first prerequisite for success in providing these experiences for engineering students is museum personnel (education directors and docents) who rely on inquiry as a mode of teaching and who see themselves as "working to *enable* the students to participate in learning for themselves by beginning to understand visual information found in art."[12] Because they do not entail giving a standard lecture on and interpretation of works of art, inquiry-based tours require extensive preparation and must be specifically developed to achieve the aims described here. The aim of the tours is *not* to enhance the students' capacity for art appreciation *per se* or to give them a particular view of the relationship of art and engineering, but rather to enlarge their capacities for perception and criticism, especially in the visual and intuitive domains. The method requires asking a lot of open-ended questions and the ability to act as a catalyst between the visitor and the art.

An Understanding of How the Museum Is Being Used

These tours are of an art museum, not a museum of science and technology, both because art museums are more common and accessible and because they are more helpful in achieving the aims of the study. Although museums of science and technology could certainly help illuminate the role of aesthetics in the historical development of technologies, they would not provide the opportunities for defamiliarization that help alter the students' perceptions. Both the physical setting and the cultural meanings of art and the museum assist in this process, though they do not automatically serve it. In other words, it would probably be counterproductive either to treat the art museum as a museum of science and technology or to carry on tours as usual.

I believe, however, that it is possible to increase students' awareness of the aesthetic dimensions of engineering without having them visit an art museum. At least a few artistic artifacts would be required because the students need physical objects to which they can respond and analyze, but at least a few of these should be readily available in any university setting. Using the museum as a context for thinking about technology is, at its root, a way of re-framing technological artifacts—of locating them conceptually in a different place than the one where they are usually viewed. It is largely a matter of looking at the technological artifacts differently then we ordinarily do. It is important to remember that the overall objective is to have students learn a little more about the design processes of artists and a lot more about the way engineering design and our responses to technological artifacts work.

Collaboration

The instructor organizing these experiences needs a reasonably complete grasp of both the technical and artistic aspects of a few artifacts that can be used as illustrative examples and points of departure. It is necessary and helpful to draw on the expertise of museum personnel and engineering students. Students might draw on their own technical backgrounds or do research to determine the techniques, materials, and aesthetic considerations associated with various artifacts. In any case, building a full understanding of any artifact, whether technological or artistic, is necessarily a collaborative enterprise. And, like most productive collaborative enterprises, the one described here draws people into that uncomfortable zone where real growth and learning are possible.

Recognition of Important Differences Between Engineering and Art

As mentioned earlier, students typically come into the study with definite ideas about significant differences between engineering and art; these ideas are shared by many members of the culture at large and have a significant impact on the value and social status assigned to activities in each realm. Many proponents of the common culture of creativity drastically minimize these differences. The historian

Harry Eisenman, for example, in speaking of a sculpture, asserts that *"Only* [my italics] the final purpose of [the] work defines it as art rather than an industrial product created from a theoretical design. . . . The process was more important [than] the object." To many—if not most—engineers, the final product and the reason it was created to begin with are not trivial or peripheral matters. Differences in purpose, motivation, and outcome *do* matter a great deal, and they probably have to be dealt with directly in order for students to be able to validate the similarities that exist in the two fields.

The Payoff: What Happens When Engineering Students Experience Technology as an Art?

The tours and related assignments are designed to provide students with a perception, altering experience, not a body of information. A number of factors must come together for the experience to occur as intended, and not all of those can be controlled. Moreover, the effects of enlarged perceptions are not easy to measure and may not be readily apparent. It seems reasonable to suggest, however, that students who experience technology as an art should enjoy a number of advantages in the practice of engineering and the appreciation of technology.

To begin with, they should find it easier to tap their capacity for visual and holistic perception and evaluation. These capacities are directly applicable in engineering design and are useful in a number of other contexts as well, especially when they are understood as complements to, not substitutes for, analytical and quantitative understanding. One advantage of the approach described here is that it requires the students to think of both artistic and technological artifacts to deal with the material reality of the objects they analyze. The process the students go through is much like the one Eugene Ferguson describes when he talks about the ways that engineering students develop an intuitive "feel" for the way the material world works and sometimes doesn't work."[9]

Experiencing technology as an art should lead to a richer sense of the role of creativity and the potential for pleasure in engineering design, provide much needed motivation for engineering students, and set them on the path to greater lifelong satisfaction in their careers. In the right circumstances, realizing the role of aesthetics in engineering can be truly liberating. Richard Meehan had such an experience while on vacation in Haiti when he was asked by a Mr. Hechtman to do an on-the-fly analysis of a retaining wall. Unable to do the calculations and tests he would ordinarily undertake to judge the fitness of the wall, he is initially uncomfortable but finally comes to a judgment of its fitness in which he feels confident. He says afterward:

Not until much later did I realize that Hechtman had unwittingly done me a great favor. I realized then that I had been living too long in the sterile world of analysis, in which theoretical consistency became an end in itself, divorced from the grit and sweat of reality. With Hechtman's wall, there was no opportunity for further studies; I had considered the available information, determined that it was sufficient for the case at hand, and in the end delivered a judgment, which I found good. . . . I had, like Daedalus, lived for many years in a stone tower of logic and analysis. Now at last . . . I had spread my wings and learned to fly.[10]

Appreciating technical work as an expression of the individual who created it also encourages personal identification with work. Martin and Schinzinger, authors of *Ethics in Engineering*,[13] assert that there is a positive correlation between personal identification and pleasure derived from one's work and the capacity to act ethically (i.e., to be socially responsible in the performance of that work). One aim of the study of "The Engineer as Designer" is to help engineering students experience technical work in a way that connects that work both to their personal identity and to their social responsibilities. The premise is that people who understand the aesthetic and symbolic dimensions of their work are more likely to perceive its human and social significance and less likely to fall into the "technology as tool" mentality, which can lead them to overlook important ethical issues.

Understanding the evocative power of artifacts also provides insight into the ways that emotional appeals enter into the appreciation and evaluation of technological artifacts. Aesthetics can be an important part of marketing consumer technologies and also be essential in gaining public acceptance of many nonconsumer technologies. We have a long cultural tradition in which positive aesthetic responses are often associated with truth, beauty, and "the good." Still, where technology is concerned, aesthetic responses can be misleading. As Arnold Pacey explains in *The Culture of Technology*,[14] the aesthetic appeal of technology is often linked to its noneconomic, nonutilitarian significance.

In Pacey's view, technology can be aesthetically appealing and still be either socially destructive or irrelevant. Factors such as technical sweetness; the capacity to enlarge personal capabilities or to master elemental forces; the exhilaration provided by speed and power; the display of skill, scale, and size; or association with adventure can make a technology appealing and encourage investment in a particular technology. Yet such technologies may do little or nothing to meet important human needs and may have very negative consequences in human terms. Students who understand the evocative power of artifacts are in a better position both to use that power to advantage and to recognize when it can be misleading or misused. If they can acquire a larger sense of the choices and alternatives presented by a particular design, they are more capable of thinking critically about technology. The experiences described here are not adequate in themselves to develop the ability to think critically about technology, but they do make students aware of dimensions of technology that are important to its evaluation and often overlooked.

Conclusion: From the Stone Tower to the Runway

Thus, engineers who experience technology as an art come away with a richer view of technology, a broader appreciation of technology, and a greater awareness of the many factors involved in successful engineering design. They should also be better able to appreciate the pleasures that engineering design offers and to integrate any artistic abilities they have into their professional identity.

Such experiences are not easily arranged or guaranteed to have the desired effect, but they do have the potential to be truly illuminating. Because the perceived gulf between artists and engineers, and between art and engineering, is such an important part of the worldview of most engineering students, changing their view of those relationships is a potentially life-changing experience, with ramifications far beyond the student's engineering work. One student who participated in these tours described his experience this way:

> I think that going to the Bayly Art Museum was the most valuable learning experience of the entire course for myself and most of the other engineers. . . . I have developed my creative skills throughout all of my life, and only recently I have used them in an engineering environment. A sense of artistic appreciation, however strong or weak it may be in my person, has always been something that I have been able to turn on or off like a switch. When I work on engineering, the switch is off. . . . The final segment of our lectures was very valuable in helping me reevaluate that switch. I now see more clearly that I can leave the switch on during engineering work, in fact, I should leave the switch on all of my life."[15]

By helping our students appreciate the aesthetic dimensions of engineering, we may be able to help them see their undergraduate education not as Meehan's stone tower of logic and analysis, but as the runway on which they first recognized their ability to fly.

Note

* I developed the concept of the hierarchy based on my conversations with George Bland, former Professor of Electrical Engineering at the University of Virginia.

References

1. Billington, David. 1983. *The Tower and the Bridge: The New Art of Structural Engineering.* New York: Basic Books.
2. Eisenman, Harry. 1990. "One Culture: Creativity in Art, Science, and Technology." In *Beyond History of Science*, ed. Elizabeth Garber. Bethlehem, PA: Lehigh University Press, 284–295.
3. Garber, Elizabeth. 1990. Introduction, In *Beyond History of Science*, ed. Elizabeth Garber. Bethlehem, PA: Lehigh University Press, 7–20.
4. Kranzberg, Melvin. 1984. Confrontation or Complementarity?: Perspectives on Technology and the Arts, In *Bridge to the Future: A Centennial Celebration of the Brooklyn Bridge*, ed. Margaret Latimer, Brooke Hindle, and Melvin Kranzberg. New York: New York Academy of Sciences, 333–334.
5. Topper, David. 1990. Natural Science and Visual Art: Reflections on the Interface. In *Beyond History of Science*, ed. Elizabeth Garber. Bethlehem, PA: Lehigh University Press, 296–310.
6. Kamm, Lawrence. 1989. *Successful Engineering.* New York: McGraw-Hill.
7. Miller, Arthur I. 1988. Visualization Lost and Regained: The Genesis of the Quantum Theory in the Period 1913–27, In *On Aesthetics in Science*, ed. Judith Wechsler. Boston: Birkhäuser, 73–104.
8. Wechsler, Judith. 1988. Introduction, in *On Aesthetics in Science*, ed. Judith Wechsler. Boston: Birkhäuser, 1–8.
9. Ferguson, Eugene. 1993. How Engineers Lose Touch. *American Heritage of Invention and Technology.* 8 (3): 16–24.
10. Meehan, Richard. 1981. *Getting Sued and Other Tales of the Engineering Life.* Cambridge, MA.: MIT Press.
11. Young, Jane Anne. 1996. Private communication. March 11, 1996. The author gratefully acknowledges the assistance of Young and her staff at the Bayly Art Museum in the development and execution of the tours described in this chapter.
12. Young, Jane Anne. 1987. "Short Guide to a Work of Art." Bayly Museum, University of Virginia, Charlottesville.
13. Martin, Mike, and Roland Schinzinger. *Ethics in Engineering.* 1988. New York: McGraw-Hill.
14. Pacey, Arnold. 1983. *The Culture of Technology.* Cambridge, MA: MIT Press.
15. Simon, Xavier. 1993. Private communication.

CHAPTER 17

BARBARA M. OLDS

RONALD L. MILLER

Integrating Humanities and Engineering

Two Models for Achieving ABET Criteria 2000 Goals

Introduction

There is little doubt that the philosophy underlying the accreditation of undergraduate education is rapidly shifting from a "bean counting" or "seat time" model to an approach that emphasizes student learning. In an earlier chapter, Lance Schachterle of Worcester Polytechnic Institute summarizes the three-part approach that regional accrediting agencies are currently requiring for campus assessments. According to Schachterle, the accreditation agencies are looking for (1) a clear statement of mission and goals, (2) a review of curriculum to certify a close connection between stated goals and the curriculum, and (3) evidence of the institution's continuous self-study and use of feedback for improvement in the curriculum.[1]

Professional accreditation agencies, such as the Accreditation Board for Engineering and Technology (ABET), are also shifting their emphasis, as reflected in the *Engineering Criteria 2000*.[2] Specifically, the ABET criteria list eleven attributes that "engineering programs must demonstrate that their graduates have," including "an understanding of professional and ethical responsibility; an ability to communicate effectively; the broad education necessary to understand the impact of engineering solutions in a global/societal context; a recognition of the need for and an ability to engage in life-long learning; a knowledge of contemporary issues." Those of us who teach humanities and social sciences in an engineering context are struck by the number of attributes in this list directly related to our subject areas.

In this chapter we describe two programs that have been developed at the Colorado School of Mines (CSM) that model the assessment process outlined by Schachterle. These programs, both institutionalized for at least eight years, also integrate engineering with humanities and social sciences and thus explicitly address several of the *Engineering Criteria 2000* attributes. We first discuss the Colorado School of Mines' mission and goals. Then we describe how our integrated programs were developed in response to the institution's goals. The first, HumEn (Humanities/ Engineering Integration), is a small-scale, single-course program that has been in place since 1988; the second, the McBride Honors Program in Public Affairs for Engineers, is a larger program involving about 10 percent of the CSM undergraduate student body and approximately forty faculty each year; the program has been in place since 1979. We believe that both of these programs provide viable models for integrating the humanities with engineering. Finally, we will discuss the ways in which these programs assess student knowledge and ability and use the feedback gained from this assessment to improve the curriculum.

The CSM Mission and Goals Statement

The Colorado School of Mines has revised its undergraduate curriculum. In the early stages of this revision, faculty, alumni, students, and other CSM stakeholders worked together to assure that a shared vision of the School's mission was developed. Table 17.1, "Profile of the Colorado School of Mines Graduate," is the document resulting from that process. The Profile, which was approved in 1994, maps well with the attributes outlined in *Engineering Criteria 2000*. In particular, the humanities and social sciences are emphasized in the final four bullets of our Profile, which call for communication skills; cross-disciplinary critical thinking, leadership, and lifelong learning; global awareness; and ethics, professionalism, and good citizenship. Both programs that we discuss in this chapter address these key issues in the context of our students' engineering curriculum.

TABLE 17.1. Profile of the Colorado School of Mines Graduate

The Colorado School of Mines is dedicated to serving the people of Colorado, the nation, and the global community by providing the highest quality education, research, and outreach in all areas of science and engineering and associated fields related to the discovery, production, and utilization of resources needed to become good stewards of the Earth and its resources. To do this, CSM must provide students with perspectives informed by the humanities and social sciences, perspectives which also enhance students' understanding of themselves and contemporary society. CSM is committed to the development of processes and approaches to mitigate environmental damage caused in the past by the production and utilization of minerals, energy, and materials. It is also committed to minimizing such damage in the future, thus helping to sustain the Earth system upon which all life and development depend.

- All CSM graduates must have depth in an area of specialization, enhanced by hands-on experiential learning and breadth in allied fields. They must have the knowledge and skills to be able to recognize, define, and solve problems by applying sound scientific and engineering principles. These attributes uniquely distinguish our graduates to better function in increasingly competitive and diverse technical professional environments.
- Graduates must have the skills to communicate information, concepts, and ideas verbally, in writing, and graphically. They must be skilled in the retrieval, interpretation and development of technical information by various means, including the use of computer-aided techniques.
- Graduates should be capable of working effectively in an international environment, and be able to succeed in an increasingly interdependent world where borders between cultures and economies are becoming less distinct. They should appreciate the traditions and languages of other cultures, and value diversity in their own society.
- Graduates should have the flexibility to adjust to the ever-changing professional environment and appreciate diverse approaches to understanding and solving society's problems. They should have the creativity, resourcefulness, receptivity, and breadth of interests to think critically about a wide range of cross-disciplinary issues. They should be prepared to assume leadership roles and possess the skills and attitudes which promote teamwork and cooperation and to continue their own growth through lifelong learning.
- Graduates should exhibit ethical behavior and integrity. They should also demonstrate perseverance and have pride in accomplishment. They should assume a responsibility to enhance their professions through service and leadership and should be responsible citizens who serve society, particularly through stewardship of the environment.

The HumEn Program

The HumEn (Humanities/Engineering Integration) Program was started in 1988 with help from a grant from the National Endowment for the Humanities. Its original purpose was to explore innovative methods of integrating humanities directly into existing, required undergraduate engineering courses. In HumEn courses we help engineering students make appropriate connections between the humanities and their technical work, connections that will carry over into their professional lives.[3,4] Courses in the program are team-taught by a member of the engineering faculty and a member of the Liberal Arts and International Studies faculty, each an expert in his or her own field, each knowledgeable in the other's.

As the two faculty who received the funding from NEH, we have been involved in the HumEn Program since it was first taught in spring 1989. The course has evolved over the years from a single course with a chemical engineering designation that carried four credits rather than the usual three, to two separate three-credit courses (one in chemical engineering and one in humanities, which were co-requisites for each other), to a completely stand-alone three-credit humanities course that accepts students from any engineering major on campus but is not specifically aligned with any. These modifications have been the result of both changing politics and changing curriculum at CSM, but we feel that HumEn has maintained its integrity and is as strong a course now as it ever was. A list of the features that allowed us to institutionalize the course follows:

1. The two of us are dedicated to the HumEn concept and to keeping the program alive. Although other faculty have occasionally taught HumEn courses, we are the champions who sustain it. Though many people view this kind of reliance on individual "champions" as a drawback, we think our persistent efforts have made the course better over the years, and our clear vision for the program has kept it focused on its original goals.
2. Our institution strongly supports educational innovation and, because CSM is small and has a history of interdisciplinary cooperation, the school did not put barriers in the way of first experimenting with, and then institutionalizing, HumEn.
3. Students strongly support the program. Through the years the students who have participated in HumEn have indicated their satisfaction by recommending the course to their colleagues. This word of mouth advertising has assured us of a continual stream of interested, qualified HumEn participants.

The primary goal of the HumEn course has always been to help students begin exploring engineering in a larger context as a human enterprise. Most of the students who enroll in HumEn are second-semester sophomores, just beginning their engineering studies after three semesters of introductory math, science, and humanities and social science coursework. We find that all too often they immediately become immersed in the world of engineering computations and technical analysis, e.g., mass and energy balances, and never have the opportunity to step back and look at what engineers really do and the implications of that work. The HumEn class encourages them to explore engineering more broadly by emphasizing such attributes as these articulated by ABET:

- An understanding of professional and ethical responsibility,
- An ability to communicate effectively,

- The broad education necessary to understand the impact of engineering solutions in a global/societal context, and
- A knowledge of contemporary issues

We will discuss how the HumEn course encourages growth in all of these areas.

An Understanding of Professional and Ethical Responsibility

One four-week module in the HumEn course deals with "Personal and Professional Ethics." Our purpose in the module is to introduce students to the kinds of ethical issues they will face as engineers, largely through the use of case studies. The three cases we use are Jon Else's film, *The Day after Trinity* (plus handouts); the Challenger incident with a focus on Roger Boisjoly's involvement; and Bertoldt Brecht's play, *Galileo*. The *Day after Trinity* focuses on Robert Oppenheimer and the other people responsible for building the first atomic bomb. The first half of the film emphasizes the spirit of patriotism and the excitement of discovery in the elite group isolated at Los Alamos. Students catch the infectious spirit of real research by viewing interviews with the participants. The first half of the film ends with a fascinating series of slow-motion shots of atomic explosions. The second half of the film is much different, focusing on the decision to *use* the bomb at Hiroshima and on the anguish and guilt felt by the developers even half a century later. We read a variety of articles from a spectrum of experts about the decision to drop the bomb, and students discuss their own views based on careful consideration of the arguments both pro and con. Many are forced to conclude as a *Newsweek* article they read does, that "in a real sense there was no decision, no careful weighing of the pros and cons. Like most acts of embattled governments in time of war, this one was driven by the interplay of temperament and personality and the sheer momentum of events."[5]

Next we explore the Challenger disaster and particularly the role of Roger Boisjoly, Senior Scientist at Morton Thiokol, who argued strongly against the launch and who was labeled a whistleblower after the disaster. Students are particularly intrigued by the reported discussion among the Morton Thiokol managers before recommending the launch when one turned to another and asked him to "take off his engineering hat and put on his management hat."

Finally, we read and discuss Bertolt Brecht's *Galileo*, in which the great scientist is depicted as a true "paradigm shifter" but also as a less-than-admirable human being.[6] Taken together the case studies in this section not only push students to "an understanding of professional and ethical responsibility" but they also force them to admit that not all ethical decisions are clear cut. We want them to recognize this fact, not because we are trying to produce cultural relativists but because we want to start them thinking about how difficult and how complex true ethical decisions often are.

An Ability to Communicate Effectively

Students in the HumEn class have the opportunity to practice their communication skills, particularly writing. Each student is required to keep a journal that counts for 15 percent of the final grade, a term paper that counts for 30 percent, and essay midterm and final exams that count for 40 percent. The journal is essentially the students' "annotation" of the course. They are expected to write frequently (three times a week minimum) about anything they like that is related to the course. The journal is graded holistically and grammar, mechanics, and spelling errors are not marked.[7] Students may discuss the readings, comment on class lectures or discussions, or reflect on their own opinions about class topics. The journals are an excellent way for us to carry on a dialogue with the students. In fact, we find that they often say things in a journal that they would not say to us in person. The journals reflect the students' ability to think deeply and clearly about complex subjects and to relate those subjects to their own lives.

For the term paper, we ask students to select a topic related to the course themes, explore it in some depth, focus on a thesis, and develop that thesis as an argument. We work through the entire writing process with the students and require them to meet a variety of milestones including selecting the topic, developing a thesis statement, compiling a list of references, and submitting rough draft(s), and final product. Allowing students to select their own topic and thesis (with guidance) has definite advantages: first, we don't have to read many papers on the same topic, and second, students are more interested in the topic when they select it. Students choose a range of interesting positions to argue. A few examples from last year's class include: "Medical Care in the United States: Inequalities in Access and Quality Exist Between the Rich and the Poor"; "The Ethical Issues Behind Cloning Animals"; and "Use of the Atomic Bomb on Japan During World War II Was Justified."

The midterm and final exams allow students to synthesize the material they have read throughout the semester and to organize their thoughts under time constraints.

The Broad Education Necessary to Understand the Impact of Engineering Solutions in a Global/Societal Context

One of the areas we explore in the HumEn course is technology transfer and the implications of exporting First World technology to developing countries. Many students in the class have never seriously considered that technology may have negative consequences. Therefore, they are shocked when they encounter an article like one we read by Khor Kok Peng, who argues that the Third World is "where the majority of the *victims* of modern science live" [italics ours].[8] We continue our discussion by reading Chinua Achebe's *Things Fall Apart*, a novel set in the late nineteenth/ early twentieth centuries that describes from an African perspective the coming of Europeans to the Igbo people of modern Nigeria and the consequent disruption and ultimate disappearance of much of their traditional culture.[9] The students come to appreciate that all cultures have admirable and not-so-admirable qualities; they recognize the irony at the end of the novel when the English commissioner decides that he could devote "a reasonable paragraph" in his book *The Pacification of the Primitive Tribes of the Lower Niger* to the story of Okonkwo, the protagonist of Achebe's entire novel.

A Knowledge of Contemporary Issues

We tie a discussion of contemporary Nigerian politics into the reading of Achebe's novel. Using newspaper and World Wide Web resources in addition to information from the Global Studies issue on Africa, we discuss the relationship between Western petroleum corporations and Nigerian politics.[10] Such a discussion inevitably leads us to reflect again on issues of ethics and the global/societal context.

We also take advantage of current events whenever they tie naturally into our course discussions. One year, for example, a student mentioned cloning shortly after the announcement of cloning the sheep Dolly in Scotland. We were shocked that almost none of the other students in the class had heard of this event and immediately assigned each of them to bring a newspaper article about the issue to the next class. The resulting discussion was one of the liveliest of the semester.

Assessment

Since its inception we have used various methods to assess the students' attitudes and learning in the HumEn program, including student surveys, student course evaluations, and the evaluations of outside experts. HumEn students fill out surveys at the beginning and end of the semester; these surveys are designed to measure students' perceptions about the importance of humanities to their education and profession and are compared to responses from non-HumEn students in our introductory chemical engineering class. Three statements from the survey help to illustrate the impact the HumEn course has on our students:

1. "Humanities won't enhance my engineering career."
2. "Humanities are important to my engineering education."
3. "Engineers are only responsible for technical solutions. Others are responsible for any nontechnical ramifications of their actions."

Statement 1 was designed to determine if the students can evaluate the importance of humanistic studies beyond what they are required to take during college. About 75 percent of our HumEn students at the beginning of the semester disagreed or strongly disagreed with the first statement while nearly 96 percent disagreed or strongly disagreed by the end of the course. Non-HumEn students' opinions remained much more constant, with 68 percent disagreeing or strongly disagreeing at the beginning of the course and 76 percent disagreeing or strongly disagreeing at the end.

We noted similar trends for survey Statement 2. At the beginning of the semester only 28 percent of the HumEn students strongly agreed with Statement 2. By the end of the semester, 56 percent strongly agreed and nearly 96 percent agreed or strongly agreed. Once again, much less change in student perception was noted in the non-HumEn students we surveyed.

Finally, survey Statement 3 probes the students' understanding of their responsibilities for the nontechnical aspects of their work. By the end of the course, nearly 75 percent of the HumEn students strongly disagreed with the statement while 50 percent strongly disagreed at the beginning of the course. About 50 percent of the non-HumEn students strongly agreed with this statement, a percentage which remained nearly constant from the beginning to the end of the semester.

Taken together, the results obtained in this survey suggest that the HumEn course does influence the perceptions of students in the course. But do the attitudinal changes we observe in our sophomore HumEn students stay with them after the course has ended? In a follow-up survey of graduating seniors, we found that HumEn alumni retained their attitudes about the importance of the humanities. In fact, nearly 80 percent of the seniors either disagreed or strongly disagreed with the statement that "engineers are only responsible for technical solutions."

Student satisfaction with the course is consistently high. When evaluated using a campus-wide instrument, the course instructors are ranked among the highest in the Liberal Arts and International Studies Division. We have also collected a great deal of anecdotal evidence that students appreciate the learning they have achieved in the course. The following quote from a student journal is representative:

> There were times when this class made me doubt whether I would be a good engineer or not. I hope that I will never have to deal with crooked company ethics. At this time, I do not know all of the constraints which would affect my decision as to

what I should do given a situation like Boisjoly's [Challenger disaster]. What scares me about engineering is having to make a decision that actually affects people's lives. I don't know if I ever want to have that much responsibility. At least, when it comes time to make a critical decision I will have some background and knowledge of case histories from this class to consider.

As part of the project work funded by the National Endowment for the Humanities, two distinguished consultants provided us with an independent evaluation of HumEn. These consultants interviewed HumEn students and faculty, campus administrators, and members of the HumEn campus advisory committee. Their overall impression of the program was very favorable, as the following comments from their project evaluation indicate:

> The HumEn pilot course has been highly successful in engaging the interest of engineering students and helping them understand the positive contribution that study in the humanities can make to their professional education.
>
> [The course] seems to be creating more favorable attitudes toward humanities and social sciences in general as well as greater awareness of non-technical criteria for making engineering design choices, while it seems to leave technical performance not only unimpeded but even improved.

Based on our experiences, we believe that courses like HumEn can serve as powerful bridges between engineering and the humanities and social sciences and can provide models for meeting many of the goals of the *Engineering Criteria 2000*.

The McBride Honors Program

The Guy T. McBride, Jr. Honors Program in Public Affairs for Engineers is another successfully institutionalized program at CSM that took a different route to its current state.[11, 12] Unlike the single course HumEn program, this program is a twenty-seven semester hour sequence of seminars and off-campus activities with the primary goal of providing a select number of engineering students with the opportunity to cross the boundaries of their technical expertise and to gain the sensitivity to prove, project, and test the moral and social implications of their future professional judgments and activities, not only for the particular organizations with which they will be involved, but also for the nation and, indeed, the world. To achieve this goal, the program brings themes from the humanities and social sciences into the engineering curriculum that will encourage the habits of thought necessary for effective management and enlightened leadership.

After a rigorous application process, approximately 10 percent (fifty to fifty-five students) of the first-year class at CSM is admitted to the McBride Program each year. Special features of the program include small seminars, an interdisciplinary approach (faculty from engineering and science disciplines and faculty from liberal arts are co-moderators of each seminar), the opportunity for one-to-one faculty tutorials, the opportunity to practice oral and written communication skills, a Washington, D.C., public policy seminar, the opportunity for a practicum experience (either an internship or foreign study), and the development of a community within a community. This program's success has stemmed from a number of factors:

1. The faculty as a whole (engineering, science, and liberal arts) strongly support the program. A dedicated cadre of faculty serve as moderators in the program. Approximately 30 percent of the total CSM faculty have served at least one term on the tutorial committee, a four-year commitment to teaching and program governance. Engineering and science faculty believe in the program strongly enough to volunteer their time and teach the seminars as overloads.
2. The school realizes that it has a strong recruiting and retention tool in the McBride Program. It is, therefore, in the best interest of the institution to keep the program healthy.
3. The program was endowed when CSM president emeritus Guy T. McBride retired approximately fifteen years ago. The endowment funds give the program a certain amount of independence and immunity from the fickle winds of state and institutional funding.

McBride Program Mission and Goals

Concurrent with the undergraduate curriculum revision at CSM, the McBride Honors Program began to review and revise its curriculum in 1993. The resulting curriculum can be seen in Table 17.2. Before revising the curriculum, the faculty and students in the program developed a mission and goals statement that can be found in Table 17.3. The skills, knowledge, and values articulated in this table mesh well with the *Engineering Criteria 2000* and the CSM Profile. Over the course of their eight seminars in the program, students have the opportunity to develop and practice each of the listed skills, knowledge, and values. Seminar moderators are expected to clearly articulate the skills, knowledge, and values they wish to emphasize in each seminar. These areas of emphasis become the basis for assessment of student achievement, and what students know and are able to do at each step in the Honors program.

By the time a student completes the program, he or she should possess all of the skills, knowledge, and values listed in the mission and goals statement. These map well onto the *Engineering Criteria 2000* as illustrated below.

TABLE 17.2. Colorado School of Mines McBride Honors Program Seminar Titles—New Curriculum

Freshman Year Spring Semester	LIHN101A Paradoxes of the Human Condition: Reflections in the Humanities *or* LIHN101B Paradoxes of the Human Condition: Expressions in Fine & Performing Arts *or* LIHN101C Paradoxes of the Human Condition: Reflections in American Culture
Sophomore Year Fall Semester	LIHN200A Cultural Anthropology: A Study of Diverse Cultures
Sophomore Year Spring Semester	LIHN201A Comparative Political and Economic Systems
Junior Year Fall Semester	LIHN300A International Political Economy *or* LIHN300B Technology and Socio-Economic Change
Junior Year Spring Semester	LIHN301A U.S. Public Policy: Domestic and Foreign *or* LIHN301B Foreign Area Study
Junior Year Summer	LIHN400A McBride Practicum
Senior Year Fall Semester	LIHN401A Study of Leadership and Power
Senior Year Spring Semester	LIHN402A Senior Honors Seminar: Science, Technology and Ethics

An Understanding of Professional and Ethical Responsibility

Although ethics are an important part of every Honors seminar, they are emphasized especially in the first and last seminars, Paradoxes of the Human Condition and Science, Technology and Ethics, respectively. The first seminar, for example, includes a module on the paradox of freedom and responsibility. The course description from the "Reflections in the Humanities" section states: "Each human being feels a drive to let his thoughts and actions range expansively as if unrestrained by any limits whatever. This is the energy of freedom. But each also feels the counter-pressure of heavy restraints imposed by the external world, the claims of other people, the teachings of the ages and his own physical and mental makeup. This is the resistance of necessity." Course readings such as Solzhenitsen's *One Day in the Life of Ivan Denisovich* demonstrate this paradox. The course description for the senior seminar on "Science, Technology, and Ethics" states, "Professor Robert Coles of Harvard discovered some time ago that he could successfully use literary works to help professional students come to grips with ethical problems that arose in their own learning and work experience. In stories that engage the ethical issues of people who have taken professional responsibilities, each

TABLE 17.3. Colorado School of Mines McBride Honors Program Mission and Goals

The McBride Honors Program provides a select community of CSM students with the enhanced opportunity to explore the interfaces between their areas of technical expertise and the humanities and social sciences; to gain the sensitivity to project and test the moral and social implications of their future professional judgments and activities; and to foster their leadership abilities in preparation for managing change and promoting the general welfare in an evolving technological and global context. In preparing to become leaders, students completing the McBride Honors Program should possess the following skills, knowledge, and values:

Skills	They should be able to communicate effectively, verbally and in writing, to a variety of audiences.
	They should be competent in the art of civil discourse.
	They should be able to work effectively both alone and in teams.
	They should have the ability to analyze and critically evaluate both their own ideas and those of others.
Knowledge	They should possess the knowledge necessary to explore the relationships among economic, political, social, and cultural systems an ability central to the mission of the Program.
	Whenever possible, they should have first-hand experience of the concepts discussed in their seminars through internships, overseas experiences, in-depth research, and community service.
Values	They should be persons of high principle and character.
	They should develop reflective minds.
	They should accept personal responsibility for their actions as leaders, as professionals, and as citizens.
	They should be aware of, and sensitive to, diverse languages, cultures, and beliefs, both in this country and abroad, through direct experience wherever possible.
	They should exhibit a love of learning and the promise of continuing it throughout their lives.
	They should appreciate interconnectedness in the changing world.

story reveals the moral character of our existence, and imaginative reflection by the reader on one's own parallel experiences leads to ethical and moral learning."

An Ability to Communicate Effectively

Each McBride seminar emphasizes communication skills, both oral and written, although the type and amount of communication vary from seminar to seminar. In some seminars (for example, the first) students write weekly two-page reaction papers to questions posed by the moderators regarding the readings for that week. In the second seminar, Cultural Anthropology, students produce an ethnographic study of a subculture in the Denver area. Students in Technology and Socio-Economic Change write a lengthy position paper on a technology issue and present their results in a formal oral presentation. Students who graduate from the program frequently comment on how it enhanced their communication skills. One McBride alumnus, Vivek Chandra, writes:

An engineer who cannot communicate is much like an athlete who excels at only one sport; he may be very good in his particular sport, but he will not be able to compete in other events. Today's corporations demand and require their employees to be able to perform a multitude of tasks. The best technical ideas are useless unless they can be communicated to management. The McBride honors program stresses the importance of the written and spoken language. From a personal point of view, I know that classroom presentations and reports allowed me to be ready for the "real" world.

The Broad Education Necessary to Understand the Impact of Engineering Solutions in a Global/Societal Context.

Each student in the McBride Program must complete a practicum, either an internship with a corporation, government entity, or nonprofit organization, or an overseas trip. The practicum is designed to give students "hands-on" experience with the concepts they have been studying in the classroom. If they will be doing an internship, students take a public policy seminar that includes a week-long visit to Washington, D.C. to meet with policymakers and explore policy issues. Students must keep a journal during their internship and must write a substantive report after it is completed; the report must address how the internship met the following program goals:

- To become familiar with the organizational structure, culture, and complexities of a private company or public agency.
- To become aware of how the organization interacts with other elements of society including businesses, communities, governments, and other countries.
- To work on substantive projects.
- To make presentations and write reports on projects.
- To reflect critically on the internship experience and its relationship to the McBride Honors Program goals.

Students who choose to travel abroad spend a semester learning about the history, culture, economy, technology, and public policy of the country they will be visiting. Students travel to the country for approximately one month with two CSM faculty sponsors. The trip gives McBride students the firsthand experience that is an invaluable part of their international education. The program currently travels to the following countries/regions on a rotating basis: Chile, South Africa, China/Hong Kong, Turkey, Brazil, and Southeast Asia. Students must keep a journal during the trip and write an extensive paper on their return. In the paper they are asked to discuss in depth some aspect of the cultural, economic, technical, or political scene in the country they visited. The report should synthesize the "book learning" they gained in the preparatory seminar with the hands-on experience they gained through travel.

A Recognition of the Need for, and an Ability to Engage in, Lifelong Learning

This recognition is instilled in McBride students through the example of their faculty moderators. Most of the people who serve as McBride moderators were not originally content experts in these fields. In fact, one of the richest components of the program is the fact that engineers and scientists serve as role models by teaching in it. These people demonstrate an eagerness to continue to learn that the students admire and seek to emulate. As McBride graduate Gary Womack states:

> Society is demanding different things from the scientist of today than it did of the scientist of yesterday. I personally feel that the McBride Honors Program has helped develop in me the tools and the insights that will allow me to be more than simply a good scientist. I recognize that first and foremost I am a citizen of society, and that I have a responsibility and the capability to function within the framework of this ever-changing world.

A Knowledge of Contemporary Issues

McBride students gain this type of knowledge in nearly every seminar. In some cases, large research projects are devoted to contemporary issues. In the 1996 Technology and Socio-Economic Change seminar, for example, student teams researched such topics as censorship on the internet, social ramifications of nuclear waste disposal, uses and abuses of technology in K-12 education, and ethical and social issues surrounding the "morning after pill." In other seminars, such as Reflections in American Culture, students are assigned to bring current newspaper clippings or magazine articles related to course topics. The seminar participants spend approximately fifteen minutes at the beginning of each three-hour seminar discussing these current events.

Assessment

As we redesigned the Honors Program curriculum, we also put in place a longitudinal portfolio assessment to provide feedback both to individual students and to the program as a whole. We are in the process of developing a web site for the program that will allow students to store their portfolios electronically. In brief, the assessment program works as follows:

1. The student selects a representative work from each seminar to include in the portfolio. He or she also writes a brief reflection about progress toward meeting the program goals during that semester.
2. The student meets with a moderator near the end of the semester for a tutorial in which they discuss the student's progress, and the moderator completes an assessment of the entire portfolio to date. This meeting provides formative feedback to the student about the student's individual growth. Feedback from the meetings is also used to revise seminars to better meet student expectations and programmatic goals.
3. The portfolio must be current and the student must be in good standing before he or she can register for the next McBride seminar.
4. At the end of the students' three- and- one-half years in the McBride Program, they write a more lengthy reflection about their accomplishments, growth, and areas that are in need of additional work. This reflection, in addition to the entire portfolio, is reviewed by a committee of faculty members from the Tutorial Committee. They meet with the student to review the entire Honors Program experience and to provide summative feedback.

We are currently working to encourage alumni of the program to continue to update their portfolios after graduation and to continue to reflect on their McBride Program mission and goals as they mature and contribute in the workplace. Although it is still relatively new, this assessment program is already paying dividends in terms of explicit formative feedback for the students and excellent input to the faculty for improving the program.

Conclusion

From our experience over the past two decades, we conclude that programs designed to integrate the humanities and engineering can be successful. In our view, these integrative programs work best when they have both faculty champions and strong institutional support. In addition, a clear set of educational objectives, a curriculum or syllabus that addresses these objectives explicitly, and a strong assessment component help to ensure success.

In addition, this may be a particularly propitious time to propose a course or sequence that addresses such topics as ethics, communication skills, and knowledge of contemporary issues. Now that all engineering programs are concerned with demonstrating that their graduates have achieved the ABET *Criteria 2000* attributes, deans and department heads may be more willing than in the past to support experiments in the integration of humanities and social sciences into the engineering curriculum. Educators interested in implementing a modest experiment such as HumEn, or even a more ambitious one such as the McBride Honors Program, may find that the time is ripe. Their case will be even stronger if they propose a program that has clearly articulated goals in harmony with the ABET criteria, as well as a thoughtful and clearly presented assessment plan.

References

1. Schachterle, Lance. Some Consequences of the 'Engineering 2000 Criteria' on Liberal Education. *Proceedings of the 1996 ASEE Conference*, Washington, DC, Session 3661.
2. Accreditation Board for Engineering and Technology. Engineering Criteria 2000. February 11, 1997. Http://www.abet.ba.md.us/EAC/eac2000.html.
3. Olds, Barbara M., and Ronald L. Miller. Meaningful Humanities Studies for Engineering Students: A New Approach. *Proceedings of the 1990 ASEE Conference*, Session 2230, pp. 1040–1043.
4. Olds, Barbara M., and Ronald L. Miller. A Model for Professional Education in the Twenty-First Century: Integrating Humanities and Engineering Through Writing. *Studies in Technical Communication*, ed. Brenda R. Sims, 1990, pp. 105–121.
5. Thomas, Evan. Why We Did It. *Newsweek*, July 24, 1995, pp. 22–30.
6. Brecht, Bertolt. 1996. *Galileo*. New York: Grove Press, 1996.
7. Fulwiler, Toby, ed. 1987. *The Journal Book*. Portsmouth, NH: Heinemann.
8. Peng, Khor Kok. Science and Development: Underdeveloping the Third World. *In The Revenge of Athena: Science, Exploitation and the Third World*, ed. Ziauddin Sardar. London and New York: Mansell Publishing Limited, 1988, pp. 207–215.
9. Achebe, Chinua. 1959. *Things Fall Apart*. New York: Anchor Books.
10. Ramsay, F. Jeffress. 1995. *Global Studies: Africa*, 6th Ed. Guilford, CT: Dushkin Publishing Group.
11. Olds, Barbara M., and Ronald L. Miller. A Liberal Education Model of Leadership Preparation: The McBride Honors Program in Public Affairs for Engineers. *Proceedings of the Frontiers in Education Conference*, Salt Lake City, Utah, November 6–9, 1996.
12. Olds, Barbara M. Educating the Engineer of the Future: The Colorado School of Mines' McBride Honors Program. *College Teaching*, Winter 1988: 16–19.

CHAPTER 18

HEINZ C. LUEGENBIEHL

Responding to ABET 2000

A Process Model for the Humanities and Social Sciences

The humanities and social sciences (HSS) component of the engineering curriculum constitutes a significant portion of the total educational program. It can even legitimately be argued that the HSS component is becoming more central to engineering education than it has ever been. In the following I will provide an overview of the current state of, and future prospects for, the mission of the humanities and social sciences in engineering education within the context of the general aims of liberal learning and within the more specific task of educating future professionals. The discussion will review the current Accreditation Board for Engineering and Technology (ABET) requirements for the HSS component, my home institution's interpretation of these requirements, and some alternative models for meeting the requirements. I will then consider the new requirements found in *ABET 2000*, the new accreditation standards, and my institution's method of integrating HSS offerings with those requirements. The chapter concludes with some reflections on the role of humanities and social sciences in engineering education, with emphasis on the opportunities and challenges to be faced in the process of curriculum revision.

Liberal Education in Engineering Education

Rose-Hulman Institute of Technology, my home institution, prides itself on providing a liberal education in the sciences and engineering. While its primary aim is certainly to ensure that its graduates are technically prepared for their careers, it also recognizes that professionals must be problem solvers within the context of a larger societal whole and that their adult lives are not completely circumscribed by their professional roles. Future professionals, as do all students, need a broad background of knowledge and the capacity for independent thought as part of their course of study. With this recognition, the Rose-Hulman educational philosophy fits squarely within the Western educational tradition.

The role of liberal education in the West is an ancient one. The concept originated with the Greeks and signified for them that education necessary to become a free person, free in the sense of being able to participate fully and in an informed fashion in the business of the polity. In later centuries, this was seen to be the core function of a university education. Even as universities moved to specialized curricula on the German model, they therefore retained the ideal of a broad, general education as the basis of further study. Independent, reflective thought, on the basis of a strong foundation of inherited wisdom, was thought to enable both a successful career and a fulfilling professional life. An ideal course of study would include not only the disciplines of the humanities and social sciences, but also courses in the arts, physical sciences, and mathematics.[1]

During the last few decades a gradual shift away from this traditional perspective has occurred, with fewer and fewer schools requiring a broad range of subjects as a liberal education core, a major exception to the trend being some of the small, elite liberal arts colleges. Other colleges have moved more and more toward forms of specialized education, where increasing emphasis is placed on courses in a major and perhaps a minor. Some of the broad, general education has, therefore, been left behind, to the extent that when Stanford University reestablished a quite unregimented core it made national news. Even when the semblance of a core remains, most often the emphasis is on providing a multitude of choices. As a result, a variety of American commentators have called for the reestablishment of a unifying core of liberal learning.[2] Interestingly, most of these come from the political right and thus put the call for liberal education in the context of the learning of traditional values or basics, thereby neglecting to some extent the freeing function of liberal education.

Lost, to a large extent, in the debate over the appropriate nature of higher education, has been a recognition that engineering education has actually been a remaining bastion of the ideals of liberal education. Due to the existing accreditation standards, all engineering students are required to take a number of courses in the humanities and social sciences. They, of course, also take many courses in mathematics and the physical sciences. When combined with their study of technology, which, given its societal importance, ought probably to be a part of all curricula, engineering education can be seen a providing a truly liberating education. Some engineering educators have started referring to engineering education as "THE liberal education for the 21st Century."[3] Perhaps this is not as far off the mark as some academics in colleges of arts and sciences might believe.

Current ABET HSS Requirements

The Accreditation Board for Engineering and Technology has played a central role in the discussion about the aims of engineering education. The educators responsible for setting the minimum requirements for accreditation have recognized the need for instantiating the philosophy described above. As a result, very specific requirements were established to ensure that engineering departments would achieve the goals implied by the philosophy. The following briefly reviews the existing requirements and their justification.

The overall curricular objective of the ABET criteria clearly enunciates the relevance of broad perspectives for the future professional life of engineering students. Included in the goal of applying their knowledge in a "professional manner" are developing "(1) a capability to delineate and solve in a practical way the problems of society that are susceptible to engineering treatment (2) a sensitivity to the socially-related technical problems which confront the profession (3) an understanding of the ethical characteristics of the engineering profession and practice (4) an understanding of the engineer's responsibility to protect both occupational and public health and safety, and (5) an ability to maintain professional competence through life-long learning" (Criterion IV.C.2).

To achieve these aims the basic ABET requirement is "one half-year of humanities and social sciences" (IV.C.3.a[2]). A half-year is defined as sixteen semester or twenty-four quarter hours. This requirement is then refined by the following stipulations:

- HSS coursework must be designed to fulfill objectives "appropriate to the engineering profession and the institution's educational objectives." To fulfill this requirement the HSS curriculum "must provide both breadth and depth and not be limited to a selection of unrelated introductory courses" (IV,C.3.D[2][a]).
- Courses that emphasize "routine exercises of personal craft" are not acceptable as HSS courses. Any performance course must include history or theory (IV.C.3.d[2][b]).

- Skills courses such as accounting, engineering economy, or industrial management are not to be counted for HSS graduation requirements (IV.C.3.d[2]c).
- Competence in written and oral communication skills must be demonstrated throughout the total curriculum (IV.C.3.i).
- As a minimum, engineering faculty must infuse professional concepts regarding ethics, society, economics, and safety throughout the curriculum (IV.C.3.j).

The ABET requirements, within a limited scope, attempt to assure a minimum adherence to the traditional ideal of liberal education. They recognize that all students need exposure to the humanities and social sciences. Of course, engineering students are required to take more science and mathematics courses than would be the case within a traditional liberal arts core. The "breadth and depth" requirement ensures both that students are exposed to several HSS disciplines and that they study one with some thoroughness. Since the overall requirement is for only six quarter-length HSS courses, depth generally means only a sequence of two courses. Yet it establishes the proposition that learning in HSS disciplines is continuous just as it is in technical fields. The nonskills requirement indicates a commitment to the traditional liberal arts, rather than simply an emphasis on job training. The HSS requirement is intended to aid the development of engineering students as whole persons. The need for such a perspective has been verified by the fact that the real value of the HSS component becomes apparent to many students only years after graduation, when they regularly report that they wish they had received more humanities and social science content in their education.

Rose-Hulman's Interpretation of the ABET Requirements

Rose-Hulman's HSS original graduation requirements, while more substantial than is the case at most engineering colleges, reflect the traditional ABET position. They were designed to meet the ABET accreditation standards, while also emphasizing the tradition of disciplinary study in the humanities and social sciences. Much of the innovation that has occurred in recent years can be traced back to outstanding disciplinary specialists developing new courses and modifying the content of existing courses.

The HSS department consists of eighteen full-time faculty members in eleven disciplines, including four foreign languages. The largest group of faculty members is in English language and literature, with five. The normal teaching load is three courses per quarter, with some faculty having release time for other Institute duties. In addition, about ten adjunct courses are offered per quarter, most in the arts and foreign languages. All the courses count as four credit hours, which means that they normally meet four times per week, for a total of forty hours per quarter. On average, courses require three to four examinations, one major or several shorter writing assignments, and participation in class discussions. The nominal size limit for the classes is thirty students, with beginning English classes being limited to twenty.

The humanities and social sciences graduation requirements can be summarized as follows. Students are required to take nine HSS courses, for a total of thirty-six hours. This was approximately 18 percent of the typical major's total required hours, as opposed to the 12 percent ABET mandated minimum. On average, students actually take between ten and eleven HSS courses before they graduate. Of the nine courses, one must be the beginning English course, "Literature and Writing." In addition, each student must take three courses designated as a humanities courses (HU) and three social science courses (SO). Of these at least three must be upper-level courses (junior or senior), with at least one each from the humanities and social sciences. Disciplinary choices are left to the student. However, some majors at Rose-Hulman require a specific course such as "Principles of Economics" or "Technical Communication." The nine courses must also include one course designated as providing an insight into non-Western cultures.

Some courses offered by the department are designated as "limited credit," which means that a student may take only one of these for HSS graduation credit. These are courses seen to have a primarily skills component and were originally so designated in light of existing ABET requirements. The ABET standards were also partially responsible for the emphasis on literature in the beginning English course, so that this would not be considered a skills course. Some flexibility in the total requirements exists for students choosing to take a large number of foreign language courses.

Generally students thus have a great deal of latitude in the courses they choose, while at the same time ABET's "breadth and depth" requirement is preserved. Students may enhance the depth of their studies by focusing more deeply in one area through the attainment of an area minor. These minors are available in all disciplines as well as in Asian Studies, European Studies, Latin American Studies, and Science, Technology, and Society. The department also offers a major in Economics and students who study German for four years can earn a certificate in technical translation. Given the size of the department and the Institute, beginning students are not always able to register for their first-choice courses. However, in total a large number of selections are available for students during any given term.

Alternative Models for Meeting the Requirements

Given that ABET has also stressed the idea of creativity in meeting its requirements, Rose-Hulman's model is by far not the only one currently used in engineering education. A number of innovative approaches exist on other campuses. This section will briefly review several atypical approaches to meeting the ABET standards.[4] In this way a broad overview of humanities and social science education in the context of engineering education will be provided.

University of Virginia

The University of Virginia is one of the few remaining with its own humanities and social sciences department in the College of Engineering. Engineering students take four to six HSS courses in the College of Arts and Sciences and at least four within the College of Engineering itself. The intention is that some of the HSS requirements be met in courses where engineering students have the opportunity to interact with other students at the university, while other courses have a more immediate connection to the students' career plans. The courses internal to the College of Engineering are designed to be an integral part of engineering education, concerned especially with the relation of "technology and the world that produces it." These required course are one freshman writing course, one STS course, one history course in Western technology and culture, and one course concerned with values called "The Engineer in Society." The latter two are taken at the senior level and are integrated with an engineering thesis project that students are required to write.

Colorado School of Mines

The School of Mines offers an interdisciplinary honors program in public affairs, resulting in a minor in public affairs. It is seen as "active liberal learning," which emphasizes the role of technology. It includes five interdisciplinary seminar courses, an internship or travel during the summer following the junior year, and a capstone senior course. Each seminar has three faculty members, an HSS leader and two from science or engineering. For those students not following the honors track, two freshmen seminar courses focusing on the interdisciplinary nature of knowledge are required.

Harvey Mudd College

Harvey Mudd has probably the most extensive, but the most traditional, HSS requirements in engineering education. Students are required to take a minimum of twelve HSS courses. At the freshman level they take Rhetoric (a writing course) and Humanities II (introduction to college level research and discussion skills in HSS). Humanities II is offered in eight different versions and has an interdisciplinary emphasis. After the freshman year students take nine electives. Courses, as a rule, are not tailored to engineers, with most being discipline specific. The program has distribution and depth requirements. The final course is a senior seminar in a particular HSS discipline.

Worcester Polytechnic Institute

Probably the most innovative HSS program is at Worcester. It requires the completion of a humanities sufficiency. This consists of five or more thematically related courses, followed by an individual research activity, known as a mini-thesis. The program can be thought of as a thematic minor, which synthesizes knowledge, as opposed to many survey courses in different fields. Students also take two required social science courses. Finally, as part of their engineering requirements, students must complete an interactive qualifying project that explores the two-way relationship between technology and society. The project is undertaken at the junior level and equals at least three conventional courses. The project must contain both a technology and a humanities/societal dimension.

While the above illustrate a variety of approaches to fulfilling the ABET accreditation requirements, it should be noted that many engineering schools require only the minimum number of courses mandated by ABET and leave much of the choice up to students. Most do require a freshman level English course and many fulfill the depth requirement through the use of a two course introductory humanities sequence. The rather haphazard fashion in which many engineering schools treat their HSS requirements is, however, likely to no longer be a viable approach in the near future.

ABET 2000 and HSS requirements

Engineering education in the U.S. is about to undergo a significant, and perhaps profound, period of change as a result of a new vision for the future. This vision is outlined in a new set of accreditation requirements found in the document *ABET 2000*. Included in the new requirements are proposed changes for the HSS component of the curriculum. The most fundamental changes affecting the HSS curriculum are the elimination of the half-year minimum requirement and the establishment of a mission and assessment based curriculum.

ABET 2000 establishes that each institution must have:

- "Published educational objectives consistent with institutional mission.
- A process of determining and evaluating the objectives.
- A curriculum to achieve the objectives.
- A system of evaluation and improvement in relation to the objectives." (From ABET Criterion 2)

Included in the objectives must be a demonstration of the following outcomes relevant to the HSS component of the curriculum:

- "an understanding of professional and ethical responsibility
- an ability to communicate effectively
- the broad education necessary to understand the impact of engineering solutions in a global/societal context
- a knowledge of contemporary issues." (From ABET Criterion 3)

Instead of requiring a minimum number of HSS courses or distribution requirements, the new criteria establish a convergence of aims between the technical and liberal education components of the curriculum. Criterion 4, called the professional component, only states that included must be "a general education component that complements the technical content of the curriculum and is consistent with the program and institution objectives." Emphasis is given to the fact that the total curriculum must prepare students for a major culminating design experience that includes recognition of the following types of considerations: "economic, environmental, sustainability, manufactur-ability, ethical, health and safety, social and political."

The new criteria appear to imply less emphasis on the ideal of a traditional liberal arts core in favor of a more applied conception of the humanities and social sciences. This is verified by the fact that the nonskills requirement is being eliminated. The basis behind this might be that some believe that, given the pace of knowledge accumulation in the sciences and technology, there is insufficient room in the curriculum for meeting all traditional educational objectives. Thus as much practical material as possible, even at the expense of traditional liberal learning, needs to be integrated into all the components of the curriculum. Another interpretation might be that the technical content of the curriculum has become too narrow and should, therefore, be expanded in two directions. One, within the technical courses itself, and two, by relating the technical component to the humanities and social sciences in HSS courses. Whatever the intention, it is clear that humanities and social science departments related to engineering colleges will have to respond to the change in requirements.

Rose-Hulman and the New Requirements

At Rose-Hulman, the humanities and social sciences department is committed to the ideals of the traditional liberal arts. However, we also recognize that we function within the context of engineering education. In order to preserve our understanding of our mission in a time of changing requirements, we have, therefore, taken a proactive approach to the new accreditation standards by engaging in a fundamental reexamination of our role within the total educational process. Although this is an ongoing task, it is already possible to report on some significant results and on innovative new requirements we plan to propose.

In accord with ABET guidelines, the first step in the process was to develop a mission statement for the department, based on which a rationale for the curriculum could be established. The mission statement reads: "The Department [HSS] seeks to encourage the emotional and intellectual growth of Rose-Hulman students, enabling them to become sophisticated thinkers, active citizens, and effective leaders and to lead rewarding lives." Eight goals for student outcomes were then established: growth, values, critical reasoning, communication skills, breadth of knowledge, systemic thinking, open-mindedness, and flexibility. Each goal was then given more specific form. For example, growth was interpreted as meaning "curiosity, creativity, an awareness of self, and a commitment to life-long learning and achievement." In addition, goals for the department as a whole were established: "Maintain faculty of high quality, continue program and curricular development, and support excellent teaching through professional development," and to "support and encourage service to the department, the Institute, appropriate professions, and the community."

Once the mission framework was approved by the HSS department, some rapid initial curricular modifications were undertaken. Since nonskills courses were no longer a barrier to counting courses for accreditation purposes, the prior "Literature and Writing" course was changed to "Freshman Composition" to reflect a greater emphasis on the development of student writing skills. The same justification resulted in the elimination of the limited credit designation. The non-Western requirement was also eliminated as a result of a more general emphasis on global studies. Finally, the distinction between the humanities and social science was eliminated in recognition of the increasingly integrated nature of knowledge.

In place of these requirements the department proposed a new curricular structure that reflects the emphases in the new ABET criteria and also maintains the minimum number of HSS courses required for graduation. Each student is now required to take "Freshman Composition." In addition to this course, each student must take two courses from each of four categories: global studies, rhetoric and expression, self and society, values and contemporary issues. These categories are specifically intended to reinforce the goals of the HSS department and of ABET 2000. Special accommodations will be made in the requirements of students desiring to emphasize foreign languages in their studies. As a result of these changes, the department is redistributing existing courses into the new categories. For example, both "Principles of Economics," a social science course, and "Shakespeare," a humanities course, become society courses. Along with these changes, the requirement for distribution among lower and upper-level courses has also been eliminated.

The most innovative aspect of the proposed curriculum is the establishment of a new "Sophomore Seminar." The seminar will be a new graduation requirement for all students. It will be given on one evening and the next full school day. The aim of the seminar will be to help students see that all knowledge is linked in an integrated web. This will be accomplished through small group discussion of a contemporary topic such as the global environment or the economy in the twenty-first century. All Institute faculty members will be invited to participate as discussion facilitators. A variety of disciplinary approaches to the same topic will thus be demonstrated and interaction between groups should result in a sharing and wider examination of these perspectives. Since we hope to make this an ongoing concern of students, all HSS courses during the following year, and other Institute courses as appropriate, will devote some attention to the seminar theme as well. In addition, several outside speakers will be invited to discuss the topic. In this way we hope to help students recognize not only that learning is cumulative, but also that their study of the humanities and social sciences has the potential to contribute to engineering and science solutions to contemporary problems.

We thus hope that our curricular revisions will have a fundamental effect on our students' perceptions of the role of the humanities and social sciences in their professional lives. The proposed changes will perhaps also have some long-term effects on the curricular offerings in the department. Potential changes include changes of focus within existing courses, elimination and restructuring of existing courses as a result of the establishment of the new categories, addition of new courses emphasizing the aims of a particular category, changes in disciplinary hiring patterns, the involvement of engineering and science faculty in HSS courses, the integration of HSS material into technical courses, and the opportunity to have students apply their liberal learning in a direct way. However, as we engage in the curricular revision process we also need to be aware of some potential problems: the possible loss of the core of liberal learning, the absorption of too great a portion of the HSS curriculum into engineering courses, too many courses with a practical or skills focus, and the possibility of a decrease in the number of required courses. By taking a proactive approach we hope to demonstrate the centrality of the humanities and social sciences to engineering education as reflected in the *ABET 2000* standards.

A Vision for HSS In Engineering Education

In conclusion I would like to offer just a few reflections on the role of the humanities and social sciences in engineering education as a way of summarizing the previous discussion. In order for us to do our job well we need the cooperation and support of our colleagues from other departments, both in terms of professional activities and in terms of the intellectual nurturing of students. However, it seems to be the case that not all engineering faculty are convinced of the importance of the liberal education mission. Some have convinced themselves that the HSS component is not critical, that students should instead focus on their technical courses. Perhaps their own liberal education was too limited, perhaps they had a bad experience in some HSS course, perhaps someone convinced them they were lacking in the appropriate talents. Whatever the reason, such people might well see the change in accreditation standards as an opportunity to reduce the HSS content in the curriculum. I believe this would be potentially dangerous course on which to embark.

Today, more than ever, life is a complex enterprise. Our students will be asked to make more and more decisions, both as professionals and in the personal lives. These decisions will sometimes entail far-ranging consequences. In order to be aware of these, engineers will need to be able to reflect on the societal and global forces that surround their decisions. They will need to recognize their own motivations, to analyze their decisions, and to find meaning and justification in their activities. They will need to take into account issues of justice, benefits, and rights. In short, they will have need of an educational background that only the humanities and social sciences can provide.

It is important to integrate these considerations into the overall education of engineering students and to offer HSS courses that students will see as being relevant to their career paths. Courses such as "The Engineer in Society," "The History of Technology," or "Engineering Ethics" have a valuable role in the curriculum. Equally important, however, are traditional humanities and social science courses that prepare students to live freer and fuller lives, whatever might happen in, and to, their careers. The strength of such courses lies in the fact that they can force students to think independently, to react critically to inherited wisdom, and to formulate their own solutions. These are qualities we ought to value not only in the general citizenry, but in our future engineers as well.

References

1. Leugenbiehl, Heinz, Applied Liberal Education for Engineers, *The International Journal of Applied Philosophy*, Vol. 4, No. 4, Fall 1989, pp. 7–11.
2. Cheney, Lynn, 50 Hours—*A Core Curriculum for College Students*, Washington, D.C.: National Endowment for the Humanities, 1989.
3. The report by the Engineering Deans Council on *Engineering Education for a Changing World* states that "engineering is an ideal undergraduate education for living and working

in the technologically-dependent society of the twenty-first century." Washington, D.C.: American Society for Engineering Education, 1994, p. 15.
4. Johnson, Joseph Susan Shaman, and Robert Zemsky, *Unfinished Design—The Humanities and Social Sciences in Undergraduate Engineering Education*, Washington, D.C.: Association of American Colleges, 1988.

CHAPTER 19

JOSEPH R. HERKERT

STS for Engineers

Integrating Engineering, Humanities, and Social Sciences through STS Courses and Programs

This chapter discusses the use of science, technology, and society (STS) courses as a means of integrating perspectives drawn from engineering, humanities, and social sciences, with particular focus on the STS general education requirement (GER), the STS Minor Program, and the Benjamin Franklin Scholars dual-degree program in engineering and humanities and social sciences at North Carolina State University. Discussion of these programs follows a brief introductory section on the role of STS in engineering education.

Why STS for Engineers?

MIT's first Dean for Undergraduate Education, the late Margaret MacVicar, once noted that the challenge for educators with respect to integration of engineering, humanities, and social sciences is to bring about:

> a true educational partnership among the technical, arts, social and humanistic disciplines so that on some level students see the interrelationships between science and technology on the one hand, and societal, political, and ethical forces on the other.[1]

One approach to exploring such interrelationships is through the STS courses and programs that have sprung up over the past quarter century.

In recent years it has become increasingly fashionable to denigrate STS courses and programs with a humanities/social science perspective as "anti-science" and/or "anti-technology." In the so-called "science wars," scholars in the field of science and technology studies have been attacked by scientists in a manner that Lowenstein has characterized as dangerously oversimplified:

> The oversimplified stereotype of science-and-technology studies as some sort of anti-science, postmodern, literary quagmire is analysis by sound bite that is dangerous and polarizing.[2]

Within the engineering community, this mistrust of science and technology studies predates the recent science wars by nearly two decades, having found its strongest voice with the 1981 publication of Florman's *Blaming Technology*:

> The blaming of technology, as we have seen, starts with the making of myths—most importantly the myth of the technological imperative and the myth of the technocratic elite. In spite of the injunctions of common sense, and contrary to the evidence at hand, the myths flourish.
>
> False premises are followed by a maligning of the scientific view; the assertion that small is beautiful; the mistake about job enrichment; an excessive zeal for government regulation; the hostility of feminists toward engineering; and the wishful thinking of the Club of Rome.
>
> These in turn are followed by distracted rejoinders from the technological community, culminating in the bizarre exaltation of engineering ethics.[3]

Cutcliffe[4] identified three stages of course development in the STS field. The first phase was indeed perceived to have an antitechnology flavor, in that it sought to educate scientists and engineers about the negative social impacts of their work. Ironically, this phase included the participation of many science and engineering faculty. A second generation of courses, however, that was directed at liberal arts majors as well as scientists and engineers, attracted the participation of larger numbers of faculty in the humanities and social sciences, and focused on the mutual interaction between technology and society. The third phase, which began in the mid-eighties, aimed at increasing the technological literacy of liberal arts students, and attracted renewed interest among science and engineering faculty.

In looking back on two decades of STS curriculum development, Cutcliffe[5] argued that an integrated approach, incorporating both the methodology and context of science and technology, is essential if students in STS courses and programs are to gain a thorough understanding of science and technology:

we should strive for an increasingly sophisticated understanding of the internal workings of science and engineering, yet one that places that understanding of science and technology in their societal context.

Contrary, then, to Florman's blanket indictment of any critical analysis of engineering and technology as having an anti-technology bent, proponents of STS education have for years been advocating a balanced view of science, technology, and society based on the sort of educational partnership promoted by MacVicar.

In my view we are embarking on a fourth stage of STS curriculum development, in which STS courses will come to be seen as an integral component of engineering education. This development has been encouraged in part by the ABET Engineering 2000 Criteria, which in recognizing the importance of such factors as multidisciplinary teams, ethics and social responsibility, effective communication, societal impacts of engineering solutions, and contemporary issues,[6] has set Florman's diatribe against technology studies on end. Indeed, the critical reflection on the interactions among engineering, humanities, and social sciences typically found in successful STS courses and curricula have caused a number of educators to call for major changes in the engineering curriculum in order to accommodate STS concepts and issues.[7]

At North Carolina State we have found it possible to expose engineering students to STS courses and programs even in the absence of major changes in the engineering curriculum, both through the institution of an STS general education requirement, and the development of a minor and a dual-degree major program that provide engineering students and others the opportunity for extended STS studies.

An STS General Education Requirement

All students entering North Carolina State since fall 1994, including engineering students, are required to take at least one STS course. The rationale and goals of the STS requirement are as follows:

Rationale

North Carolina State University, as a land grant university, has a mission that stresses the application of science and technology for the betterment of humankind. It is essential, therefore, that students be exposed to the vital interactions among science, technology, society, and the quality of life.

Goals

1. Developing an understanding of the influence of science and technology on civilizations,
2. Developing the ability to respond critically to technological issues in civic affairs, and
3. Understanding the interactions among science, technology and values.

Engineering students, and other students with majors in science and technology, are required to fulfill this requirement with a course developed from "a humanities and social science perspective," or with an interdisciplinary course designed to incorporate perspectives drawn from science, technology, humanities, and social sciences. Conversely, students with majors in the humanities, and social sciences are required to take an interdisciplinary STS course, or one that draws heavily on science and technology perspectives. The STS requirement is thus an explicit statement that all students, including engineering students, should partake in study that in some fashion integrates perspectives drawn from science, technology, humanities, and social sciences.

The STS requirement for engineering students is incorporated in the required humanities and social sciences courses. Students select a course from an approved list of approximately thirty, about half of which are multidisciplinary studies courses (MDS) on such topics as Environmental Ethics; Science & Civilization; Contemporary Science, Technology, and Human Values; Humans & The Environment; Ethics in Engineering; Technological Catastrophes;[8] Alternative Futures; and Bio-Medical Ethics. Offerings on the list that are based in the traditional departments include a number of courses on the history of science and technology; Technology in Society and Culture (sociology and anthropology); Environmental Economics; Philosophy of Science; and Science, Technology and Public Policy (political science). A complete list of the courses fulfilling the STS requirement for engineering students is included in Table 19.1.

TABLE 19.1. College of Engineering STS List

ANT (SOC) 261:	Technology in Society & Culture
ARE (EC) 336:	Introduction to Resource & Environmental Economics
ARE (EC) 436:	Environmental Economics
HI 321:	Ancient and Medieval Science
HI 322:	Rise of Modern Science
HI 341:	Technology in History
HI 480:	Scientific Revolution: 1300–1700
HI 481:	History of the Life Sciences

TABLE 19.1. *(continued)*

HI 482:	Darwinism in Science & Society
HI (MDS) 485:	History of American Technology
MDS 201:	Environmental Ethics
MDS 214:	Technology & Values
MDS 220:	Coastal and Ocean Frontiers
MDS 301:	Science & Civilization
MDS 302:	Contemporary Science, Technology, and Human Values
MDS 303:	Humans & the Environment
MDS 304:	Ethical Dimensions of Progress
MDS 320:	Ethics In Engineering
MDS 322:	Technological Catastrophes
MDS 323:	World Population and Food Prospects
MDS 324:	Alternative Futures
MDS 325:	Bio-Medical Ethics: An Interdisciplinary Inquiry
MDS 326:	Technology Assessment
MDS 402:	Peace and War in the Nuclear Age
MDS 405:	Technology & American Culture
MDS 410:	Toxic Substances & Society
MDS 412:	Entering the 21st Century: Agricultural, Technological & Environmental Perspectives
PHI 311:	Philosophical Issues in Medical Ethics
PHI 322:	Philosophical Issues in Environmental Ethics
PHI 340:	Philosophy of Science
PS 314:	Science, Technology & Public Policy

An STS Minor

A minor in STS serves to meet the need of students who desire more comprehensive study of the relationships between science, technology, and society. In addition, minor programs provide "institutional legitimacy" to STS efforts, as well as a means of drawing together faculty with STS interests and of advertising to students the availability of STS courses.[9]

The STS minor at North Carolina State consists of five courses, including Issues in Science, Technology, and Society, an independent study technology assessment project. Of the remaining four courses, at least one must be taken in each of three content areas. The Group I courses provide a historical perspective on science and technology while the Group II courses focus on science, technology, and values. The third group consists of courses with social science perspectives and courses focusing on specific problems and issues posed by science and technology. As in the case of the STS GER requirement, the courses in the minor program can be drawn from multidisciplinary studies and/or the traditional disciplines. In addition, as a pre- or co-requisite, students must take at least one advanced course in science or technology and either a statistics course or the first semester of calculus. The minor requirements thus ensure that each student has some exposure to the methodology of science and technology in addition to the five required minor courses that emphasize the historical, ethical, and sociopolitical contexts of science and technology. The program is overseen by a committee of faculty from throughout the university with STS background and interests, including such fields as chemistry, engineering, english, history, psychology, and multidisciplinary studies. The requirements for the minor in STS are detailed in Table 19.2.

TABLE 19.2. Requirements for the Science, Technology, and Society Minor

The minor in Science, Technology, and Society consists of fifteen hours of coursework, including one course each from Group I, Group II, and Group III, plus an additional course from any of the three groups listed below. Also required is MDS 490, an independent study course especially for the Science, Technology, and Society minor.

A grade of C or better is required in all courses satisfying the minor requirements. A pre- or co-requisite for the Science, Technology & Society minor is one course in a science or technology above the introductory level, and mathematics through a first calculus or statistics course.

Students should declare their interest to the STS minor adviser by their junior year or earlier if possible. It is recommended that MDS 490 be taken in the junior year, or as soon as possible after declaring the minor.

Each student will be assigned an adviser with whom to discuss coursework for the minor. Upon petition by a student and adviser, the Science, Technology and Society Minor Advisory Committee may allow a course not on the lists below to count for the minor, provided that its subject matter is pertinent and relevant.

Group I	HI 321:	Ancient & Medieval Science
	HI 322:	Rise of Modern Science*
	HI 341:	Technology in History
	HI 480:	Scientific Revolution: 1300–1700,
	HI 481:	History of the Life Sciences
	HI 482:	Darwinism in Science and Society
	MDS 301:	Science & Civilization*
	MDS 405:	Technology and American Culture
	MA 433:	History of Mathematics
Group II	ENG 376:	Science Fiction
	MDS 201:	Environmental Ethics*
	MDS 214:	Technology & Values
	MDS 302:	Contemporary Science, Technology & Human Values
	MDS 320:	Ethics in Engineering
	MDS 324:	Alternative Futures
	MDS 325:	Biomedical Ethics*
	PHI 311:	Philosophical Issues in Medical Ethics*
	PHI 322:	Philosophical Issues in Environmental Ethics*
	PHI 323:	Nuclear Arms: Philosophical Issues*
	PHI 325:	Theories of Human Nature
	PHI 340:	Philosophy of Science

TABLE 19.2. *(continued)*

Group III	GN 301:	Genetics in Human Affairs
	MDS 220:	Coastal and Ocean Frontiers
	MDS 303:	Humans and the Environment
	MDS 322:	Technological Catastrophes
	MDS 323:	World Populations and Food Prospects
	MDS 323:	Technology Assessment
	MDS 402:	Peace & War in the Nuclear Age*
	MDS 410:	Toxic Substances & Society
	MDS 412:	Entering the 21st Century: Agricultural, Technological & Environmental Perspectives
	PS 314:	Science, Technology & Public Policy
	SOC (ANT) 261:	Technology in Society & Culture
	SOC 281:	Sociology of Medicine

*Because of some duplication in subject matter, students may take either one or the other but not both of the following courses: HI 322 or MDS 301, MDS 325 or PHI 311, MDS 201 or PHI 322, and MDS 402 or PHI 323

The Benjamin Franklin Scholars Dual-Degree Program

Some of the strongest students at North Carolina State enroll in the Franklin Scholars Program, a dual-degree program specifically designed to integrate engineering, humanities, and social sciences throughout the five-year course of study.[10] Franklin Scholars earn a BS degree in an engineering field or computer science, and a BA or BS degree from the College of Humanities and Social Sciences (CHASS). Also, since the GER requirements in humanities and social sciences are more extensive in CHASS than in engineering, the scholars must also take additional basic courses in the humanities and social sciences beyond those required of other engineering students. The CHASS degree can be in a traditional discipline, such as English or philosophy, or a multidisciplinary studies degree that is designed by the student with the advice of a faculty sponsor.

While not an STS program per se, the Franklin Scholars Program incorporates STS concepts in two ways. All scholars, regardless of their engineering and CHASS majors, are required to take a core sequence consisting of three STS courses: Technology and Values (first year), Ethical Dimensions of Progress (second or third year), and the Franklin Capstone (fourth or fifth year), a team-oriented project course in technology assessment and public policy.

The descriptions of the first two Franklin Scholars courses are as follows:

> MDS 214H-*Technology and Values*. Introduction to the relations of technology and society. Emphasis placed upon the nature of technology, contrasting attitudes toward technology, technology's relation to the individual and to values, and to the future relations of technology and society.
>
> MDS 304H-*Ethical Dimensions of Progress*. Multidisciplinary examination of

traditional western notions of progress, focusing on ethical issues raised by the concept of progress, and connections between science, technology, and society. Places relationships such as engineering and social responsibility within the context of present day redefinitions of the notion of progress.

The principal focus of the capstone is semester-long "technology assessment" projects conducted by teams of four to five students, in which the students must integrate their technical skills and knowledge with the background provided by their studies in the humanities and social sciences. The capstone topics considered to date have been

> Health and Safety Impacts of Electromagnetic Fields
> Alternatives to Tobacco Production in North Carolina
> Low-Level Radioactive Waste Disposal in North Carolina
> The Internet and the First Amendment
> Alternative Fuel Vehicles
> Digital Libraries

More information on the Capstone course, including a link to the complete text of the digital library report, can be found on the course World Wide Web page at http://www4.ncsu.edu/unity/users/j/jherkert/bfscapst.html.

With the exception of a member of the engineering faculty who participates in the capstone course, all of the required courses are taught by members of the Division of Multidisciplinary Studies, a component of CHASS, who specialize in teaching STS courses (two philosophers, a political scientist and an engineer).

In addition to the required courses, many of the scholars who choose to pursue their second degree in multidisciplinary studies design major concentrations in STS areas. The concentration consists of nine to ten courses, at least half of which must be in the humanities and social sciences, focusing on a coherent, multidisciplinary theme that the student is required to justify in an essay that is evaluated by a committee composed of faculty from across the university. Concentrations have been approved for Franklin Scholars in such areas as: Ethics and Technical Management; Environmental Ethics, Policy and Science; Technology, the Environment and Public Policy; Science, Technology and Society; Issues in Medicine and Bio-technology; and Technology and Communication in a Cross-cultural Context. The courses constituting some typical MDS degree programs are shown in Table 19.3.

Plans are currently under way to develop an STS track for the degree in multidisciplinary studies. It is anticipated that this new track will be utilized by a number of the Franklin Scholars as well as by scholars in North Carolina State's other dual degree programs involving CHASS and the colleges of Textiles, Agriculture & Life Sciences, Forestry, and the School of Design.

TABLE 19.3. Illustrative Multidisciplinary Studies Degree Programs of Benjamin Franklin Scholars

Bachelor of Science in Multidisciplinary Studies with a Concentration in Science, Technology, and Society

E 497S:	Franklin Scholars Capstone
HI 322:	Rise of Modern Science
HI 341:	History of Technology
MDS 214H:	Technology and Values
MDS 301:	Science and Civilization
MDS 304:	Ethical Dimensions of Progress
MDS 322:	Technological Catastrophes
MDS 324:	Alternative Futures
PS 314:	Science, Technology and Public Policy

Bachelor of Science in Multidisciplinary Studies with a Concentration in Issues in Medicine and Biotechnology

E 497S:	Franklin Scholars Capstone
EC 301:	Intermediate Microeconomics
EC 437:	Health Economics
GN 411:	Principles of Genetics
MB 351:	General Microbiology
MDS 214H:	Technology and Values
MDS 304:	Ethical Dimensions of Progress
MDS 324:	Alternative Futures
MDS 325:	Biomedical Ethics

Bachelor of Arts in Multidisciplinary Studies with a Concentration in Technology and Communication in a Cross-cultural Context

ANT 261:	Technology in Society and Culture
ANT 392:	International and Cross-cultural Communication
COM 462:	Cross-cultural Communication
E 497S:	Franklin Scholars Capstone
FLF 308:	Advanced Conversation in Contemporary French
FLG 307:	Technical and Commercial German
HSS 294S:	East Central Europe: Commerce & Politics After 1989
HSS 298:	Great Britain and WWII
MDS 214H:	Technology and Values
MDS 304:	Ethical Dimensions of Progress

Conclusions

Contrary to the perception of many scientists and engineers, STS courses and programs have matured from a singular focus on the negative impacts of science and technology to a balanced perspective that takes seriously both science and technology and their societal context. The mandates of the ABET 2000 criteria, which call for more serious attention to perspectives on engineering drawn from the humanities and social sciences, suggest that there is an integral role for STS courses in engineering education.

As the experience at North Carolina State has shown, existing and new STS courses and programs are an excellent vehicle for integrating perspectives drawn from engineering, humanities, and social sciences for the benefit of engineering students as well as students majoring in the natural sciences, social sciences, or humanities. Such integration can be effectively achieved at the level of a single required course in STS or through more ambitious programs such as an STS minor or a second, multidisciplinary major in STS.

References

1. MacVicar, Margaret L. A. 1987. General Education for Scientists and Engineers: Current Issues and Challenges. *Bulletin of Science, Technology & Society* 7 (5 & 6): 592–597.
2. Lowenstein, Bruce V. 1996. Science and Society: The Continuing Value of Reasoned Debate. *The Chronicle of Higher Education*, June 21, p. B1-B2.
3. Florman, Samuel C. 1981. *Blaming Technology*. St. Martin's Press, New York.
4. Cutcliffe, Stephen H. 1987. Technology Studies and the Liberal Arts at Lehigh University. *Bulletin of Science, Technology & Society* 7 (1 & 2):42–48.
5. Cutcliffe, Stephen H. 1990. The STS Curriculum: What Have we Learned in Twenty years? *Science, Technology, & Human Values* 15 (3):360–372.
6. Schachterle, Lance. 1996. Some Consequences of the "Engineering 2000 Criteria" on Liberal Education. 1996 ASEE Annual Conference Proceedings.
7. Herkert, Joseph R. 1990. Science, Technology and Society Education for Engineers. *IEEE Technology and Society* 9 (3):22–26.
8. Herkert, Joseph R. 2002. A Multidisciplinary Course on Technological Catastrophes. Elsewhere in this volume.
9. Mack, Pamela E. 1987. Building a Program Gradually: Science, Technology and Society at Clemson University. Proceedings of Conference on the State of S.T.S. in North America, Western Europe, and Australia, Worcester, Massachusetts.
10. Porter, Richard L. and Herkert, Joseph R. 1996. Engineering and Humanities: Bridging the Gap. Proceedings of the 1996 Frontiers in Education Conference. Institute of Electrical and Electronics Engineers and American Society for Engineering Education (available on the World Wide Web at http://www.caeme.elen.utah.edu/fie/procdngs/se8c4/paper5/96407.htm).

CHAPTER 20

SCOT DOUGLASS

Teaching Students, Not Texts

The Utility of the Humanities in Fulfilling ABET 2000 Criteria

ABET 2000 and the Focus on "Ability"

The quickest of glances at the ABET 2000 criteria of "program outcomes and assessment" reveals a clear emphasis on the word "ability."[1] Seven of the eleven criteria begin with the literary formula "an ability to." Such an emphasis only underscores the fundamental shift in the accreditation process toward what a graduate should be *able* to do. "To do" well, that is, to have "an ability," requires "having done" in a context that provides feedback and evaluation. Constructing curricula to produce demonstrable ability in the technical aspects of the criteria—designing and conducting experiments; identifying, formulating and solving engineering problems; etc.—seems relatively straightforward.[2] But how does one create learning spaces that produce abilities in the nontechnical aspects of the criteria: functioning on multidisciplinary teams, communicating effectively, taking ethical responsibility, understanding the impact of engineering solutions in a global and societal context, and being abreast of contemporary issues?[3]

This chapter will wrestle with this last question in two ways: (1) it will explore the value of a particular model of interactive liberal education in the formation of "abilities" in these nontechnical criteria and, (2) it will extract pedagogical principles from this model that could be incorporated across the engineering curriculum. As the subtitle suggests, I will argue for the utility of humanities in fulfilling the ABET 2000 criteria.

The Context

The pedagogical principles to be examined in this chapter find their expression in the Herbst Humanities Program for Engineers at the University of Colorado-Boulder. The Program, in its ninth year of existence, was made possible by the generous gift of an alumnus, Clarence Herbst, Jr. (Ch. E. '50). Pioneered by Athanasios Moulakis and Leland Giovannelli, we currently have a faculty of four. Offered as an upper-division elective, our main, two-semester seminar interacts with approximately a quarter of the juniors and seniors among the college's 2,400 undergraduate students.

Within the framework of the question, "What does it mean to be human," our program encompasses more than the integration of humanities into the profession of engineering; it addresses the current student and future engineer as citizen, wife, husband, parent, child, friend, employee, employer—a self-reflective subject in a complex set of relationships and responsibilities. Like engineering, such learning is very practical. It involves skills—the very problematic skills of living well. These skills, or "abilities," necessarily overlap with the ABET 2000 criteria, and are better cultivated in a humanities model and then applied to an engineering context, than directly cultivated within an engineering context.

Teaching Students, Not Texts: The Classroom as Wrestling Mat

Which brings us to the title of this chapter, "Teaching Students, not Texts." In one sense, this phrase reveals a method. In another sense, it speaks more of a goal. In the Herbst Program of Humanities for Engineers, it is both. As a method, it informs how we conduct seminars, organize the learning environment, and expand our requirements outside of the classroom. As a goal, it enlivens our passion, sharpens our focus, and pitches our expectations.

Ironically, we have concluded that the best method to "teach students, not texts," is almost completely text-based. The text, though, does not embody what they learn, but rather is the arena in which they learn. The text both circumscribes and colors the space within which students interact. Our seminars are like a wrestling match. One student gets on the mat to grapple with Dostoevsky, only to find himself wrestling with another student who has jumped into the fray because he disagrees with how his fellow student is wrestling. Then perhaps another student, in tag-team fashion, adds her strength to the first student's. In the end, an entire class can be productively spent, students wrestling with one another—even though Dostoevsky's full wrestling weight is never unleashed by the superior reading of the teacher. This teacher restraint is initially frustrating to students who desire the "right" answer. The reason for this restraint will be discussed below. As a result, we regularly, especially at the beginning of a semester, "sacrifice" texts on the altar of a student's intellectual formation.

How can we say, though, students have learned if we know they walk away only partially understanding or even misunderstanding Dostoevsky? Because the students have learned to wrestle better. They have engaged a complex question and have discovered the limitations of inadequate preparation. *They have been pinned.* They have learned the power of using the text in the formulation of their understanding. *They understand the importance of training.* They have grown in their ability to discern fruitful directions of inquiry from rabbit trails. *They have exhausted themselves by running on and off the mat.* They have been surprised by the limitations of their own self-referentiality. *They have been overpowered by new and innovative maneuvers employed by other wrestlers.* They have worked together with peers, their future peers, to refine their understanding of what it means to be human. They have gotten upon the mat of humanism with the expectation of gaining from the experience.

The cultivation of this positive expectation is perhaps the most significant goal of our program. The gap between taking refuge in the universal truthfulness (and, therefore, untouchable nature) of personal opinion and participating in a rigorous, sometimes murky, inquiry into philosophical or literary meaning is frequently maintained by a deep sense of insecurity and inadequacy to find such "answers." For many engineering students, their attraction to the field is a combination of its practicality and quantifiable precision-as well, of course, that they think they are good at it. In contrast, they, generally, do not think they are very good at doing humanities, and, like the typical person who actively avoids playing sports at which he or she is not good, they avoid the humanities.

The manner in which they live out this avoidance was made clear in a recent technical writing class I taught for the college of engineering-a non-Herbst class. I took an informal poll on how the students selected their required eighteen hours of humanities electives. There were twenty engineering juniors in this class. Here are the results, with number one being the most important reason and number five being the least important:

1. Proximity of class to engineering building.
2. The class that looks like it will use the most mathematics (e.g., an economics class).

3. The class that will have the least amount of writing in it.
4. The class that will have the least amount of reading in it.
5. A class that looks interesting to me.

Reasons 2, 3, and 4 all have something to do with limiting the demand of doing something they believe is beyond them.[4] In further discussion, they spoke of a fear of failure, a sense they would be exposed as inadequate—especially in comparison with English and humanities majors taking the same classes. They brought the same concerns and sense of dread to this required writing class. The cost of such an orientation to the potential value of their taking eighteen humanities and social science hours is incalculable.

In the Herbst program, even though our students choose to take this humanities elective, they share many of these sentiments. Therefore, we find it productive to spend more time encouraging their pursuit than focusing on correcting their conclusions. By spending more time on *how* they came to their answers than the answers themselves, we attempt to build a set of humanities skills and an expectation that the more wrestling they do, the more fruitful will be their efforts. Getting engineering students on the mat of humanism is not easy. We attempt to follow the pedagogical admonition of Michel de Montaigne who said the good tutor will make his pupil trot before him and then attempt to match his approach to the student's gait and pace.[5] Like good coaches, our task is to create training conditions that stretch the abilities of our students but do not exasperate them. The success of our efforts in this area is perhaps best seen by the formation of continuing, private "Herbst seminars" that students have organized on their own after completing the two-semester sequence.

It is our experience that students engage each other and the texts much more productively, enthusiastically, and efficiently by the end of two semesters of wrestling. Texts they productively misread the first week of class, they later integrate back into the discussion with refinement. The initial "sacrifices" are necessary to get to this point. In the imagery of Montaigne, at the beginning of the semester we match our pace with the awkwardness and inefficiency of their running—glad they are even on the racecourse. To demand perfect, swift, and efficient running at the outset—even showing off our own speed—usually produces exasperation in the students and results in their stopping. We work hard to construct an environment that entices their engagement, does not stifle their initial clumsiness, subverts their expectations of the role of teacher and the role of student, and honors the honest pursuit of knowledge.

The literary and philosophical text is essential to our method. It is the one story that is common to all the stories represented in the classroom. It provides the quality of the content and keeps all the wrestlers on the same mat. In many ways, the text is the wrestling mat. We strictly enforce the general rule: you cannot bring in an outside source (another text, a scene from a movie or a television show, a lengthy personal story, etc.) if the majority of the class is unfamiliar with it. The texts are granted a privileged status as the central reference point for all class discussions. Students, following the example of their teachers, learn to wrestle by asking each other, "That sounds good to me, but can you show me how you got that from the text?" or "Where is that in the text?" or "How does that fit with what Socrates says here?," in response to a statement like "Well, I think Socrates is just saying that everyone needs to stand up for what they believe."

We desire to avoid the two extremes of "this is what you should think" and "whatever you think is wonderful"—the sterility of memorizing and regurgitating the "right answers" and the chaotic, self-referential blindness of speaking from a belief in the inherent truthfulness and equality of personal opinion. The balance we seek is having students actively and productively listen to well-developed voices, first through their own reading and then via classroom dialectic; and, then, we have them clarify/justify their understandings of the relevance/nonrelevance of this voice to their own lives. We respect their abilities to engage an issue enough to challenge their initial thinking, invite them to think again in light of this or that and, generally, ask for more from them. In being so "text-centered," we ensure the rigor and content of the discussion. I reiterate this point to avoid the potential appearance that our seminars are about the robust exchange of ignorance and wild theories of interpretation. Class time is a wrestling mat, not a dance floor—neither the type of dance floor that invites self-indulgent, self-referential movements nor demands previously prescribed choreography.

In creating a learning environment in which students are expected to struggle with one another in the pursuit of understanding, we are giving students much more than humanistic knowledge. Students are living out the ABET 2000 nontechnical criteria in an environment where they receive feedback in "how they are doing" from both their peers and their teachers. As a result they are learning skills. In a supervised context, they grow in their "abilities" to articulate complex problems and solutions; to function as a team member by productively disagreeing, asking questions, and working through personality conflicts; to evaluate the identifications, formulations, and solutions of problems endemic to the human condition, as articulated by some of the greatest thinkers of all time. They are allowed to wrestle in the presence of a wrestling coach.

Guidelines for Creating and Using a Learning Space: Getting Students to Wrestle

In this section, I will attempt to be more specific, more concrete, in how these general principles are enacted. This is necessary for the wrestling metaphor to move beyond being an interesting idea to something that can have utility in a wide context. Much of the following discussion is designed to spur your own pedagogic imagination. We have failed too many times to believe that anything we do is formulaic and prescriptive.

We indicate in the paragraphs below how we create an environment where students are willing to wrestle with each other over substantive questions.

Text Orientation

As stated above, we are very "text-centered." We emphasize close reading and use the "text" (including operas, movies, art, and music as types of text) to create a learning space. In this space, how students learn and interact with the text is as important as how well they understand the text. In this sense, we are very methods-oriented. The text is granted authority, not in respect to arbitrating truth and reality, but in presenting its own meaning. The goal of each class is to understand the text in its own terms and then question to what degree it accurately describes or informs life.

By such a clear, textual circumscription of interactive space, there is a heightened valuation of the text's currency. This serves to both focus the conversation and storm the fortress of personal opinion. It also addresses the following common problems: a lack of vocabulary in approaching complex problems, inexperience in different modes of thinking and understanding, and a reticence to challenge the life philosophy of people to their face (but the willingness to challenge a nonpresent author).

Short Term vs. Long Term

We understand that our program may constitute the last formal humanities learning environment in which these future engineers participate. Because of this, we attempt to train them in both reading skills, which will allow them to engage texts in the future with a realistic expectation of gaining something from them, and dialectical skills, which will allow them to engage others. To promote this, we exercise teacher restraint. They will not have access to "authoritative readers/teachers" in the future. We hope and suggest, though, that they will be able to duplicate a Herbst style reading group with future colleagues and friends. These dialectical skills, of course, have great additional currency both in terms of ABET 2000 criteria and their engineering careers.

Class Size

We have experimented both intentionally and unintentionally, as a result of variable enrollment, with different class sizes. We have had as few as seven students in a particular seminar and as many as eighteen. In any case, a wrestling orientation requires relatively small classes. Small numbers, in themselves, though, are not enough. Success in enticing engagement seems to be contingent on the relationship between the number of people in a class and the physical dimensions of the classroom. Given the physical dimensions of our specific seminar room, if there are more than fourteen students, it is very easy for an individual to feel justified in not participating. When a class feels too crowded, students privately argue that participation would mean forcing themselves into the conversation in such a way that would require rudeness and heroic boldness. In classes smaller than nine, they feel their participation is too highlighted. They relate that they feel too isolated from the rest of the class. In speaking, a student draws too much attention to himself. There is a balance in setting up a situation in which students feel enough anonymity to risk participation and not enough anonymity to think they can hide. When we have had smaller numbers of students in a section, there has been a noted improvement in the quality of conversation when we have moved to a smaller classroom with a smaller table. In one such case, a class of only seven students moved to a conference room whose table seated ten. The positive difference in participation with the same students who had occupied eight of sixteen chairs and who now occupied eight of ten chairs was very instructive. In the larger class, they felt isolated and distant. Participation was perceived as less of a discussion and more of a series of individual statements.

Number of Moderators

In the first semester of our two-semester seminar, we have two moderators (instructors) in each class. The term "moderator" is used to subvert the passive expectations associated with the term "teacher." We use two moderators for a number of reasons. First, it undermines the students' expectations of the student-teacher relationship. They cannot focus their attention, even their physical gaze, on one source of information—one locus of "correct answers." Seated at a round table with a moderator on either side of the room, a student cannot engage both moderators without at least looking at the majority of their classmates. Second, the dual moderator models the idea of community learning. Third, the two moderators can demonstrate participation skills by asking clarifying questions of each other, by referring a student back to a question or comment made by the other moderator, or by taking a comment by the other moderator and pointing out a supporting text or adding an additional thought. The moderators are able to embody with each other how students should interact with other students.

Establishing Participation Expectations

Because what we offer is different enough from their normal, engineering classroom experience, we have the freedom to establish our own set of expectations. In the first class, we explain that the goal of our seminars is not merely the gaining of humanistic content, but learning to actively engage themselves and their peers about what it means to be human. We structurally support this by basing a full third of their grade on the quality of their participation and having no content quizzes or exams. We attempt to clarify from the beginning that good participation is not a function of the quantity of comments made in class, but is based on a number of fairly well-defined criteria. In respect to participation, each student is evaluated according to the following six criteria:

1. *Quality of listening* as demonstrated in the coherency of his comments compared with those of others.
2. *Commitment to group learning* as seen in her respect for the ideas of others, a certain civility in the treatment of others, an active commitment to including the entire class, etc.
3. *Preclass preparation* as demonstrated by an ability to interact with the main substance of that day's text, to find relevant passages, and to be able to place those passages within their overall context.
4. *Self-challenge* as seen in the willingness to suspend personal judgment until a more accurate understanding of the text is obtained.
5. *Personal presentation* as observed in how they sit, how low they wear a baseball cap, whether or not they talk or make noises while others are speaking, whether or not they distract the class via fidgeting, twirling a pen, etc.
6. *The amount of their participation* as measured by the amount or volume of quality contributions.

These categories are somewhat subjective, but we have found that when moderators coteaching a class independently have calculated grades for individual students, there is agreement in the vast majority of cases. This is another advantage of having two teachers in each class-the ability to compare notes and minimize the impact of a teacher's particular disliking or liking of a student. Additionally, as will be discussed below, the requirement of individual conferences to discuss participation heightens the students' awareness of these categories and opens up a specific dialogue regarding how they are doing.

Enticing Engagement

As moderators, we attempt to structure the classroom experience in such a way that it helps students "find their voice." At the beginning of the semester, this can take the form of requiring each student to speak briefly about something that is relatively safe. "What specific passage from Frederick Douglass did you find especially striking? Read it out loud and tell us why." "What is one question you have about a particular part that you found confusing or unclear?" As can be seen in these examples, we are asking something of the student that doesn't involve a single, correct answer. The value of a question like the one regarding the impact of Frederick Douglass' text is manifold. Although it doesn't seek the right answer, it does ask something personal. It focuses upon literary impact and asks the student to share something of herself. It shows students that others are influenced in the same way they were as well as models the diversity of the class. By having to say both *what* and *why*, it requires students to go beyond personal opinion or preference to an informed opinion or preference. Requiring students to all say something, anything, seems to make it easier for them to volunteer their own comments later.

To underline the idea of the ongoing, larger discussion, we might open the second seminar on Douglass by asking the question, "What did we talk about last time?" This connects the two seminars and implicitly values the importance of what they said in the last class.

As the semester progresses, we attempt to structure participation less and ask for more initiative from the students. As moderators, we are committed to outlasting their reticence, to not rescuing them out of their silence and waiting for them to get frustrated enough with the boredom of the moment to do something. We let the students live with the weight of the class experience they have helped to create. At the end of a boring or frustrating class, one of the moderators might simply state, "I hope you were bored enough today to never want to spend another seventy-five minutes of your life like this again." We desire them to take responsibility for the quality of their classroom experience.

Attention to Dynamics

The moderators occasionally stop class to make observations about the dynamics of the discussion—"How many of you feel left out of this conversation?"—when the class discussion has turned into a repeated back-and-forth between two students. "How many of you are bored with this topic and want to move onto something else?" and "Why are you bored?" "How many of you think we're repeating ourselves and are stuck?" and "Can anyone identify where we're stuck and ask a question that will help get us unstuck?" "How do any of the last four comments relate to each other?" or, more specifically, "Wait a second, how did what you just said relate to the question that John just asked?"

We also require two individual conferences per semester whose main focus is how each student is doing in respect to class participation. During these conferences, we ask the students to evaluate themselves in light of the six criteria of participation. We ask them which classes "worked" and *why* they think they worked. We ask them which classes "didn't work" and *why not*.

These conferences have proven to be invaluable supplements to the classroom experience and the development of the student. Privately, we are able to give each student individual coaching about how he or she is wrestling. We will frequently help to develop personal strategies tailored to the individual student's needs. For example, we have suggested the following: write down a question from the reading before class and bring it up in class (for the student who has trouble thinking on his feet and gets lost in the flow of discussion); attempt to ask one follow-up question of a fellow student (for the student who feels that she can't follow the conversations); use your confusion to ask a clarifying question of another student or simply proclaim that you are lost in the discussion (for the student who has a hard time connecting their reading to the discussion); pretend that you are one of the moderators in today's class and your main goal is to get everyone involved (for the student who has a tendency to dominate discussion); etc. The development of personal strategies forms a basis for discussion and future accountability.

Moderator Restraint

The moderators *do not* end class by saying something like "This was a good discussion, but let me tell you what the text really means." Early in the first semester, students will frequently turn to the teachers and ask for a definitive reading. They will proclaim their frustration at the end of the class because we have not come to any real conclusion and haven't answered the main questions we were asking. At such times, the moderators ask the class to remember such feelings for the next class and focus on being more efficient. We also make ourselves available in office hours for more interaction about the text as well as encouraging the students to continue the discussion with each other outside of the classroom. The goal is not to be evasive, but rather to allow the class to bear the responsibility of the quality of the discussion. By providing "less" we seek to ask "more."

When a discussion of a secondary point is struggling, we will occasionally ask the students, "Would you rather keep searching for an answer to this question or get an answer from us (if we have one) and use that to move onto other more central questions?" About two-thirds of the time, the students desire us to provide a solution to the immediate problem so they can move onto other problems.

As moderators we ask more questions than we answer. We focus more on the process of seeking answers and pay special attention to the relationships between their comments and the text.

Relevant Observations of the Engineering Student

As stated above, we have designed our program with the engineering student in mind. The following is a list of what we consider to be key observations regarding the students we serve. It goes without saying that this list is very general and does not apply to every engineering student or collectively to any particular student.

- Engineering students are very diligent, responsible, and accustomed to working very hard. They even pride themselves on the amount of work they do relative to students in other colleges in the university.
  Application: We give them, on average, one homework assignment per week that focuses upon some reading skill as well as prepares them for better class participation. Typical literary "homework sets" this past semester were finding ten examples of irony in Oedipus Rex and briefly explaining why/how each is ironic; rewriting Book 1.1 of Aristotle's Nicomachean Ethics in their own words and using their own examples; as well as exploring interpretive problems such as what does the Grand Inquisitor mean when he says, "But I awakened and would not serve madness." These assignments are done in advance of discussion on that text to increase the quality of their interaction in class.
- On the whole, they are very intelligent.
  Application: We can have high expectations of them.
- They are very practical in the way they think and the way they learn. Unlike students in the "hard sciences," they seek pragmatic answers that work when given a variety of competing variables/limitations.
  Applications: Similar to the principles behind the above comments regarding homework, we attempt to utilize literary modes of apprehension that they can apply to any text they might read. They are less interested in theories of how to live and more in how theories to live can be applied to their own lives. They frequently ask the question, "So what?"
- They believe science and engineering are about facts. Philosophy, literature, and art are about opinion. They are surprised to learn there are better ways and methods to approach complex issues of life, that their opinions can be exposed to be based upon faulty thinking and are, thus, subject to criticism and evaluation.
  Application: This perspective is very difficult to overcome. We attempt to address this by underscoring that our goal is not just to know what we think/believe, but why we think/believe what we think/believe. In class discussion, homework, and papers, the emphasis is upon demonstrating the sources and reasons behind their conclusions and the necessity of allowing opposing voices to challenge these conclusions.

As alluded to above, part of this perspective is based on a fear of engaging a world with nonquantifiable questions and answers. The real breakthrough for most students in respect to this issue is when something from a particular class touches on personal problems—problems with roommates, family, a significant other, etc. A student boldly proclaimed at the end of one semester, "Epictetus, My Man!" Such an exclamation resulted from his life making more sense and somehow working better because of this ancient wisdom.

- They understand the intentional complexity of engineering design but have no category for the intentional sophistication and complexity of literary structure, word choice, etc.

    Application: This is also very difficult to overcome. The only thing that seems to have had any level of success is the continual attention to detail and the significance of these details in arriving at any conclusion. Once again, it seems that there are psychological barriers to embracing this. When they see a complex equation or engineering solution that they do not comprehend, they are glad someone figured this out and it does not offend them. When they look at a Picasso and don't get it, or read an Eliot poem and are confused, or read Aristotle and get lost in the first sentence, they get angry. They defensively ridicule the work. I am continually surprised, even though it happens frequently, when a student stands in utter condemnation of the writing skills of Flannery O'Connor, the logic of Aristotle, the simple-mindedness of Epictetus. As one of the first and only classes that require the student to speak and interact from somewhere inside, the experience is sometimes akin to turning on a faucet in an old Victorian home and initially getting two minutes of rusty water. Not being patient enough to wait for the pipes to clear is frequently a reason for failure in this model of education.

- They generally do not possess much of a cultural or historical matrix in which texts can be contextualized. "Didn't people think this back then?" is a common statement. "Well, you've got to remember that Sophocles' view of guilt would have been forced on him by the Catholic Church." "He's one of those BC guys."

    Application: In the first semester, we focus almost exclusively on the immediacy of the text and the development of reading skills. The focus is on ideas and values, not history and schools of interpretation. We desire to disabuse them of the excuse that they must be experts to read productively. At the same time, we provide, in the form of brief handouts, relevant historical information. In the second semester, we organize our curriculum around landscapes of the mind that focus upon a particular theme, period, genre, etc., which explores the significance of coherency, intertextuality, and context. Attempts at massive remediation via concurrent lectures have not been successful.

- They are insecure about their ability to read and comprehend. Many have not written an essay since their junior year in high school.

    Application: We are patient and gear our expectations toward participation for its own sake at the beginning of the first semester and slowly engage the question of the quality of the participation as the semester progresses. Having the text be the final arbiter of its own meaning as opposed to the teacher's reading of the text seems to help students dig in. They cannot use the excuse that they are not experts because the "experts" don't put them in their place. By constantly returning the attention to particular moments in the text, the student learns, at least in this class, that he must utilize his own resources and those of his classmates.

These are just some of the factors we take into consideration when designing our courses.

Application of This Model across the Engineering Curriculum

The purpose of these guidelines is not to reproduce the Herbst model in other institutions. Rather, I have included these details to stimulate thinking about the creation of learning environments that consciously attempt to promote the types of abilities required by ABET 2000. Most of these principles could be applied in different degrees across the engineering curriculum.

Perhaps, the most direct application would be an increase in the supervision of how teams function. Conferences could be held with individuals and with teams to discuss the frustrations and successes of their team experiences. Classtime could be devoted to discussions about group dynamics. A student recently replied to my question, what have you learned from group work, with, I hope atypically, "I learned to do all the work myself and tell my group if they don't like it, 'Tough!'" She went on to recount her first group experience in which one irresponsible group member cost the whole group a good grade. In her case, a supervised group debriefing might have been very beneficial.

More could be asked from students during classtime, even in somewhat large lectures. The typical rhythm of an engineering student seems to be something of the following: attend class and obtain a somewhat shaky idea of what's going on, don't ask any questions because you won't understand it until you do the problem sets, and then, if you still don't understand it, read the textbook. Asking the student to be engaged in a predominately lecture-oriented class can take many forms. Individuals can be asked to summarize key points, formulas, principles, or processes either from a reading of the text or from a previous lecture. Students can be required to occasionally walk the professor through a particular problem. The teacher can split the class into groups of three to four and have them all attempt to find the solution to a particular problem in a given amount of time and then collectively work through the problem. The value of these types of exercises is the asking of a student to be present, to be engaged. Consistent with the ABET 2000

268 | Integrating Engineering and Humanities

shift in focus, the teacher would be using classtime to move beyond the passive transfer of content to an underlining of the importance of obtained skill.

The Two-Semester Sequence

The two semesters are organized differently. In the first semester, all the sections have a common curriculum. The texts we are currently using have been chosen with the following considerations in mind: accessibility; genre, geographical, and historical variety; the specific themes of freedom, the human condition, and happiness; length; and "teachability/discussability." Table 20.1 lists texts used this past semester. The second semester is organized around what we call "landscapes of the mind." These landscapes are organized around a coherent theme, a historical period, a particular genre, etc. The second semester utilizes only one moderator per class. Each class is composed of two landscapes and teachers switch sections at the halfway point. This past semester, these were the landscapes offered: *Love and Friendship*, *Imperial Russia*, *The Black Death in Art and Literature*, *The Bible as Literature*, *American Literature*, *The Drama of Desire*, *Julius Caesar*, *Futurism*, and *Epic*. One of the reasons we moved to a variable curriculum in the second semester was for the moderators. We all desired (needed) the opportunity to teach different texts, be able to experiment, and feel a greater sense of ownership.

TABLE 20.1. Summary of "Texts"

First Semester

Douglass: *Narrative of the Life of Frederick Douglass*
Epictetus: *The Enchiridion*
Stockdale: *Courage Under Fire*
Sophocles: *Oedipus Rex*
Orwell: *Politics and the English Language*
Montaigne: *On the Education of Children*
Western Poetry
Trip to the opera: "Spanish Hour" and "Gianni Schicci"
O'Connor: *A Good Man Is Hard to Find*
Trip to the Denver Art Museum in which students were required to copy a painting using pastel crayons
Plato: *The Apology of Socrates*
Aristotle: *The Nicomachean Ethics: Books 1–3*
Dostoevsky: "Rebellion" and "The Grand Inquisitor" from *The Brothers Karamazov*
Shakespeare: *Hamlet* (including having to act out a scene)
Haiku
Conrad: *Heart of Darkness*
Film: *Un coeur en hiver*

References

1. The eleven criteria of "program outcomes and assessment" can be found in the "ABET Engineering Criteria 2000, Third Edition," revised in December, 1997, section II, Criterion 3. Program Outcomes and Assessment.
2. The greatest challenge in the technical aspects of the new criteria is not curricular, but assessment. How does an institute demonstrate abilities in their graduates, as well as give "evidence" that the results of this demonstration/assessment "are applied to the further development and improvement of the program?"
3. It is interesting to note that the three criteria that are most concerned with a liberal education, "an understanding of professional and ethical responsibility, the broad education necessary to understand the impact of engineering solutions in a global and societal context" and "a knowledge of contemporary issues," shy away from using the word "ability." In emphasizing "knowledge" as the expected outcome, each of these criteria are weakened. If a student can "identify, formulate and solve engineering problems," they should be *able* "to identify, formulate and solve ethical problems as they relate to being an engineer." In the same manner, they should have "the broad education necessary to identify, formulate and predict/solve the impact of their engineering solutions in a global and societal context." This choice to emphasize "understanding" in respect to the more liberal concerns of being an engineer and "ability" in respect to the technical concerns of being an engineer, in my opinion, only underscores the perception of the radical distance between these two discourses.
4. The fifth reason, the question of personal interest, is a function, in many cases, of this insecurity. Their projected lack of interest, as well as their skepticism of the relevancy of the humanities, is frequently a product of their seeing the humanities as a hopeless labyrinth of personal opinion and non-provable propositions.
5. Montaigne, "On the Education of Children," *Essays*, tr. By J. M. Cohen (New York: Penguin, 1993).

CHAPTER 21

ANN BROWN

STEVE LUYENDYK

DAVID F. OLLIS

Implementing an English and Engineering Collaboration

Introduction

An intensive, yet varied, introduction to engineering has been achieved through pairing first-year English composition with an Engineering laboratory.[1] Through the former, students read imaginative literature about technology, essays that deal with technological issues and history, and biographies of engineers, then respond verbally and in writing. The laboratory experience requires students to use, take apart, and teach each other about modern devices. Through reviewing engineering in the past, and taking apart present day technology, these students develop an engineering identity in year one, rather than waiting until year four, in which the conventional capstone engineering design course finally establishes this integrative accomplishment. This course pairing represents an interdisciplinary integration in which the collaborating colleges staff and support these individual courses at levels traditional to each college. The gradual institutionalization of these two course formats has occurred through different paths, and each has the potential to achieve permanency, provided that student participants and concerned academic stakeholders are identified and involved positively all the way.

The present chapter describes our motivations leading to this English and engineering collaboration, the experience itself, and our near universal local circumstances that could lead to convenient adoption elsewhere, and especially at larger, public universities. In order to cast our experience in the guise of an implementation, we first consider the characteristic path of successful educational reforms.

A Rationale for Implementation

The present decade is seeing a revolution in the attitudes toward, and the methods of, teaching engineering. The changes to date have been spurred on by industry, professional engineering societies, and various federal agencies through their funding activities, including the NSF Engineering Education Consortia, the Technology Reinvestment Projects, and the Engineering Scholars Program.

The institutions involved in engineering education changes are soon to be increased from the funded few to the mandated many, primarily through the restructured national accreditation standards for engineering curricula. These new criteria, developed by ABET criteria for the next millenium, will shortly require all engineering schools to define individual education missions, devise curricula suited to those particular missions, and develop assessments that test graduate achievement against curricular claims.

The implementation process for reforms in engineering education will have substantial evolutionary or chronological structure. Inasmuch as such structure exists in education reform, its study is important because it may provide common milestones against which individual champions may gauge the progress of a given reform, and against which the varying maturity of differing reforms may be assessed. Accordingly, we outline the phases commonly arising when curricular change evolves into permanency, i.e., when reform implementation occurs.

The Structure of Full-scale Implementation

At a 1994 NSF Project Impact meeting, Susan Millar first surveyed curricular adventurers who discussed the academic trials and tribulations accompanying new technical course development and implementation. From these conversations emerged "a schema that shows how moving to full-scale implementation entails a reform project's 'whole story.'"[2]

TABLE 21.1. Phases for Full-scale Implementation of a Curricular Reform

Phase	Function
1	Articulate the need.
2	Develop promising reform ideas and reform-ready colleagues.
3	Gain administrative support for reform and formulate a plan.
4	Engage in initial development.
5	Implement and evaluate the reform on a pilot basis.
6	Revise and beta test the reform.
7	Locate and work with a publisher.
8	Work for change on the national level.

To guide would-be academic revolutionaries, Millar and Courter[3] reduced the process to eight typical elements (Table 21.1) and illustrated development of a first-year engineering design course using these phases to chart their progress.

To chart our progress, the story of our NCSU English and Engineering courses are framed below using Millar's implementation structure. This framing will display not only how far we have come, but will also signal the remainder of the road to be traveled if a truly full-scale implementation of such reform is to be realized. Additionally, education reforms are undergoing progressively more intense scrutiny, and must increasingly include evaluation and assessment. Thus, eight phases plus evaluation and assessment activity constitute a plausible general structure for an implementation.

The 'Whole Story': Implementing Collaboration

We now consider the eight phases for our own side-by-side stories.

Articulate the Needs

Engineering: After a first-year engineering class tour of an IBM PC assembly plant in North Carolina, one student remarked that she had never been in a factory before. An on-the-spot survey of the students by the instructor revealed her experience to be universal: none of the sixteen incoming engineering students had viewed before the act of construction of a commercial product! This after-the-tour comment suggested that most of our entering engineering students do not understand what engineering is about, in part because their prior experience, in work and school, is devoid of design and manufacturing examples. So academic need number one was identified: connect entering engineering students early with assembled devices.

English: Teachers of literature and composition have apparently taken the "two cultures" so seriously that there are virtually no courses that consider the "literature" of engineering and technology. Engineers apparently think similarly. The result is a contrived divorce within academia that even C.P. Snow would not have believed. If students should study the past to avoid repeating its mistakes, then part of every engineer's education should be the arena of technology through literature, history, and biography. We are led to need number two: find a way to embrace the engineering past, don't avoid it!

Develop Promising Reform Ideas and Reform-ready Colleagues

Engineering: An antidote to the standard lecture class in which students work individually and competitively would be a laboratory in which students work collaboratively, in teams, and without formal faculty lecture. We conceived a first-year "take-apart" or mechanical dissection laboratory involving newer technologies of some familiarity to many of the entering students. The combination of team collaboration and practical "take-apart" experiments we thought would offer an inviting and inclusive introduction to engineering. At the same time, this laboratory appears to provide a clear definition of engineering activities on day one instead of year four, and to offer an atmosphere that would better prepare students for professional and academic work.

English: The spirited opening arguments in two recent texts on "writing about science" provided all the assertion needed to justify why beginning engineers might profit from writing about their field.

In the first, Gladys Leithauser and Marilynn Bell created *World of Science: An Anthology for Writers*[4] to provide new students some firsthand encounters with writings of the scientists such students may wish to become. The varied writings offered in the anthology approach offset the weaknesses of the single-voice textbook. In consequence, the utility of writing from the English literature of science emerges. One motive for creation of their collection was that student writers often do better work when their readings reflect their special interests, yet anthologies for such readings are rare for students in the natural sciences. And rarer still for students of engineering, we add.

Another editorial pair, Elizabeth C. Bowen, a writer, and Beverly Schneller, an English professor, attempted to reverse the stereotype of scientists as bad writers by using good counterexamples, provided in their collection *Writing about Science:*[5]

> We know that science students are ill-served by traditional freshman English anthologies and that writing can be taught equally well when the subjects of composition are scientific ones.

Having established the plausibility for writing about engineering, there remained the search for reform-ready colleagues.

One of the authors, Ann Brown, had previously taught all levels of first-year English for years, then moved to the Writing Assistance Program within the Engineering School, where she advised students on specific writing assignments, provided foreign students with tutorials on the structure of English, and lectured engineering departments on various subjects, from writing lab reports to Ph.D. theses. An avid owner of a 320-acre farm, Ann was as familiar with agricultural technology as with analysis in teaching. She agreed to pilot a literature of technology course for a first-year section of honors engineering students.

Gain Administrative Support for Reform and Formulate Plan

Engineering: One of us (David Ollis) received 1993–1994 funding through NSF/SUCCEED, an engineering education consortium, to offer two years of pilot trials for such a first-year laboratory. In this course structure, each device in the "dissection" laboratory involved four laboratory periods, during which the student teams played the successive roles of device user, assembler, problem solver, and teacher/presenter. Each role required a progressively deeper understanding of the device and its development. Honors lab pilots were offered in the summers of 1993 and 1994. Laboratory OPEN HOUSES held at the end of each summer effort involved a "walk-through" during which student teams pitched explanations of their take-apart devices (bar code scanners, CD players, photocopiers, optical fiber communications, videocameras and videocassette recorders, and water purifiers) to engineering faculty and administration. This approach bypassed any initial attempt at formal evaluation of learning, and relied instead on the more informal judgment of quality learning characterized by the familiar phrase "I can't define it, but I know it when I see (hear) it!" The university administration's judgment of this experiment was an award of $1500 for a permanent freshman engineering laboratory that opened in fall of 1998.

The students who helped develop and demonstrate the lab concept were its best salesforce. This important implementation lesson learned, we used it advantageously for the next education experiment, the English composition pilot.

English: In fall 1994, the honors lab transferred one-third of its time and budget so that Ann Brown could introduce students to readings in the literature and biography of technology and of engineers. In creating a sequence of readings for her course, her criteria were two: All had to concern engineering or technology, and all had to be well written, thus providing positive models for the students' composing. Example topics for reading and discussion included Homer's *Hephaestus*, the mythical metallurgical engineer in *The Iliad*, and the Roeblings (John, Washington, and Emily), the remarkable engineers of the Brooklyn Bridge. Emily's presence also introduced the attitudes and circumstances challenging the woman engineer. The science fiction of Jules Verne and Ursula LeGuinn provided contrasting portraits of engineers. Poetry selection from Emily Dickinson, Walt Whitman, and Carl Sandburg highlighted engineering achievements, while the destructiveness of technology was shown through the eyes of William Wordsworth and Henry Reed. The personal vs. professional conflicts for a civil engineer attempting to build a modern dam in ancient-culture China is revealed in John Hersey's *A Single Pebble*. The anti-technology movement was introduced through Vance Packard's *The Wastemakers*. Several biographical readings from civil engineers were provided, including the "existential pleasures" of writer-engineer Samuel Florman, the bridges displayed in Henry Petroski's essay "Imagine," and Richard Meehan's "Snowbound on the Rio Pangal," a description of engineering's psychological rewards. When we considered how to scale-up the instructional resources for such reading and writing about technology, funding was an immediate roadblock. While funding a "take apart" lab would be challenging enough for an engineering school with no tradition of such hands-on options in year one, the provision of engineering school resources for teaching the reading and writing portion of the course was more problematic.

Our Assistant Dean of Engineering for Academic Affairs, Ric Porter, was quick with the proper rescue: "If you want freshmen to read and write about engineering, let English do it." His motivation was clear: each semester, our pilot engineering lab struggled for funds, while English routinely offered two hundred sections of freshmen composition. Surely among these English instructors we could identify some "reform-ready colleagues" at our technology saturated NCSU campus.

Accordingly, at semester's end, we devised an OPEN HOUSE for the English and engineering faculty to hear from the students who had completed Ann's initial version of the reading and writing component. To prepare the OPEN HOUSE, color posters of covers from books that were read were prepared and mounted. Members of the English Department, especially its chair and freshman council, were invited, as were all engineering faculty and administration. About a dozen members from English and half of that from engineering came to hear ten first-year University Scholars who stood next to posters of their choosing and conversed with the academic passers-by. A following open discussion included student comparisons of the new vs "regular" English writing materials and that led to a consensus supportive of the Bowen and Schneller assertion: "writing can be taught equally well when the subjects of composition are scientific ones."

The OPEN HOUSE experience convinced the English department administration that this "Writing Across Engineering" format could provide a credible vehicle for first-year composition, and was worthy of further investment. The department head agreed to provide appointment resources sufficient to teach as many sections of English as were needed to provide course pairs with the elective freshman engineering laboratory.

Engage in Initial Implementation

With the assignment of responsibilities for the College of Engineering(COE) and English components to their academically proper colleges, the challenge was now to implement individual courses for both COE and English. The engineering laboratory was reduced to a two-unit course, so that it could fit reasonably as an add-on to the first or second freshman semester, providing typically a seventeen-unit load in place of the more usual fifteen. The honors composition course was now offered (Fall '95) for entering university scholars as a three-unit honors section, taught by Ann Brown along the lines presented earlier, and including not only the literature noted above, but also references and discussions from film and art (see Ann's companion chapter in this volume).

Implement and Evaluate the Reform on a Pilot Basis

Having demonstrated the combining of Engineering lab and the English writing with honors students, the next challenge was to democratize the experience by implementing multiple, elective sections of both lab and lecture course for all entering engineering students.

English: In the context of liberal education, the challenge was to broaden the "Writing from Technology" option from the honors group (approximately top 15 percent of class) to all eligible students. This challenge required development of versions of the composition-from-technology offering that were suitable for individual first-year courses.

The academic tasks for our three NCSU levels of first-year English are summarized below:

> English 111: Conventions of edited American English, exposition and argument, idea generation and organization, revision and editing, analyzing and gathering information, library research, and source documentation.
> English 112: Improve English 111 skills, composition for academic writing and interpretation, evaluation and writing about literary works.
> English 112H: Advanced placement, honors version of English 112.

The honors student OPEN HOUSE also succeeded in identifying our next "reform-ready colleague" for this growing experiment. Steve Luyendyk, an English instructor who attended the OPEN HOUSE and was a member of the department's

Freshman Council, agreed to take on the tasks in Fall 1996 and Spring 1997 of teaching pilot versions of English 111 and English 112, recast to use examples from the engineering related literature.

From Fall 1996 through Fall 1997, we have taught four sections of the appropriate English composition course to a regular cohort of entering engineering students, in addition to three more honors sections.

In implementing this broadened range of composition offerings, Steve considered two key goals for the entering students: improve, if not initiate, their critical thinking and reading, and prepare them for writing for an academic audience for the balance of their undergraduate career.

His approach included drafting as a creative exercise, multiple drafts to improve writing, teacher and peer evaluation of drafts, and worksheets to train students for editing their own work. The engineering technology readings helped create a specific academic and professional context that encouraged students to take their writing seriously.

Well-established methods for teaching composition were used and found to work satisfactorily with this literature base. These methods include discussion, lectures, group work, computer communication, writing, and conferencing. Discussion was found to dominate more than in a conventional first-year course, because the content of the readings was of interest to the students, and because the students had a sense of "mastery" of the subject area. In short, the students considered themselves engineers and had a stake in readings about their field.

Some materials used for his first course, English 111, are summarized in Table 21.2. These include early science fiction of Jules Verne, a historical account of the evolution of utensils, the characteristic two-edged nature of technological change, the anti-technology and anti-consumer response, and a film that seriously considers the future of technology. Florman's piece provided grist for a final exam.

TABLE 21.2. Readings (partial) for English 111, Fall 1996

A Manned Moon Shot in 1865, J. Gleick, Invention & Technology, spring 1994 (relation between science and art).
How the Fork Got Its Tines, H. Petroski, *The Evolution of Useful Things* (two-way relation between culture and technology).
Prologue and Journey's End by J. Burke and R. Ornstein, *The Axemaker's Gift* (ideal of technology and its consequences).
Forward, S. Hawking, *The Physics of Star Trek* (elicit science fiction interests).
Technology and the Tragic View, S. Florman, *The Existential Pleasures of Engineering* (rhetorical analysis for final exam).
Progress through Planned Obsolescence, V. Packard, *The Waste Makers* (analyze advertisement that sells technology).
Film: *Blade Runner*, articles and novel *Do Androids Dream of Electric Sheep?* (Philip K. Dick) upon which the film is based.

Two comments from the instructor's view are apropos. First, English departments currently speculate whether courses should be created that are discipline specific. Steve found that the transition from conventional to technology-related literature provided a more specific context for writing. In this setting, the purpose of each writing assignment was clearer to the students, leading to a more advantageous discourse situation in which they could take stronger positions.

Also, the English instructor may experience two anxieties while teaching from an engineering centered literature: "I am not an engineer. How can I teach about engineering?" and "The Engineering School is taking over my department!" These anxieties are relieved through the course-pairing approach developed here. Each instructor of lab or rhetoric is independent and works largely on his or her strengths. For English in particular, the instructional focus remains that of strengthening the students' skills in creating, developing, and presenting written discourse. The present pairing of individual courses, each in a format recognized easily by its college, works by utilizing the strongest skills of the respective instructors, and is capable of being implemented and scaled-up to reach a progressively larger number of students. The English courses now have permanent acceptance as designated sections of English 111, 112, and 112 Honors, and the freshman laboratory moved into new, permanent quarters in spring of 1998.

Revise and Beta Test the Reforms

The last two milestones in Table 21.1 to "full-scale implementation" are "Locate and work with a publisher" and "Work for change on the national level." These challenges remain. While Thomas Kuhn bemoaned the monolithic text approach traditional in the sciences (and we echo his plaint on behalf of engineering), the appearance of new lab and writing texts may be the *sine qua non* for each engineering education reform component to become permanent through movement beyond the campus of origination. So, to the (writing) monasteries, dear crusaders!

Coda on Collaboration

The independent, yet side-by-side, offerings of composition and laboratory provided us, the instructors, with time and proximity for collaboration in various circumstances. The student lab team oral presentations were often heard and graded by both lab and writing instructors. A composition team writing assignment involves either the redrafting of a lab manual chapter, or the creation of a draft for a new lab experiment. Reading materials for the composition course are suggested by both instructors; the particular excerpts and choices utilized for pedagogical reasons remain those of the composition instructor.

The first-year students were also important collaborators of ours. Student enthusiasm for both the lab and the readings and writings has been conveyed firsthand to other English and engineering faculty and administration. These conversations have been key to achieving the third phase of Millar's development: "Gain administrative support."

In this collaboration, everyone got into the act!

References

1. Brown, Ann, and David F. Ollis. Team Teaching: A Freshman Rhetoric and Laboratory Experience, ASEE, Washington D.C., June 1996.
2. Millar, Susan B. Full Scale Implementation: The Interactive "Whole Story," in Conference Proceedings of Project Impact: Disseminating Innovation in Undergraduate Education, National Science foundation, pp. 14–19, 1994.
3. Millar, Susan B. and Sandra Shaw Courter. From Promise to Reality: How to Guide an Educational Reform from Pilot Stage to Full-scale Implementation, ASEE PRISM November, pp. 31–34, 1996.
4. Leithauser, Gladys G. and Bell, Marilyn P., *The World of Science: An Anthology for Writers*, New York. Holt, Rinehart, and Winston, 1987.
5. Bowen Elizabeth C. and Schneller, Beverly E., *Writing About Science* (second edition), Oxford University Press, 1991.
6. Beaudoin, Diane and David F. Ollis. A First Year Engineering Laboratory, *Journal of Engineering Education*, ASEE annual meeting, Edmonton, Canada, July, 1995.

CHAPTER 22

MARSHALL M. LIH

The Parable of Baseball Engineering

Once upon a time, a multimillionaire was granted a new franchise in a baseball league. He had ambitious goals and hired a general manager to achieve them.

As luck would have it, just prior to spring training, the GM heard from one of his nephews, who was an engineering student at a large university. The nephew told his uncle that engineers had one of the most rigorous curricula around. Impressed, the GM asked the nephew to get a copy of the curriculum so he could emulate it.

As it turned out, his nephew's school had a traditional engineering program. Because of the reputation of the school, and the fact that his nephew was a very bright student, the GM had confidence in the curriculum. So he followed it, and, after adjusting the time scale from eight semesters to eight weeks, set up the training schedule (Table 22.1).

The new players came to the team with different levels of competency and experience, and they were from various backgrounds. The GM wanted to make sure everyone started on the same footing in all aspects of the game: knowledge, techniques, rules, physical condition, and so on. And he did not want them to perpetuate their bad habits. So he delayed full-scale games until the last two weeks of training, just like the engineers, who took their capstone design course during the senior year. The last two weeks would be the culmination of everything the players had learned in the first six weeks.

Spring training began, but midway through it, most players found it so boring that half of them resigned. Undaunted, the owner had great confidence in his plan. Together they figured that if those players could not survive the rigorous training program, it was probably better they wash out early.

TABLE 22.1 Baseball Training Schedule, Engineering School Style

Week	Activities
1	Sports Fundamentals, Electronic Sports Lab
	Baseball Rules I
	Running
	Calesthenics
	Physics
2	Sports Physiology
	Baseball Rules II
	Electronic Baseball Lab
	Jumping and Diving
	Economics
	Biomechanics
3	Throwing
	Batting
	Base-running
	Psychology
	Business Practices
	Aerodynamics
4	Catching
	Bunting
	Sliding and Base-Stealing
	Sportsmanship
	Management
	History of Baseball
5	Offense Strategies, Electronic Baseball Simulation
	Pitching
	Signaling
	Baseball Business
	Teamwork
6	Defense Strategies, Advanced Lab Simulation
	Home-Plate Catching(E)
	Tagging(E)
	Classical Games
	Coaching (E)
7	Pitcher-Catcher Coordination(E)
	Business/Sports Ethics
	Substance Abuse
	Infield Strategies(E)
	Umpiring(E)
	Games I
8	Sports Law
	Contracts and Negociations
	Verbal Abuse (E)
	Baseball Management (E)
	Baseball Greats
	Games II

(E) Indicates elective

With more recruits, there were still enough players left at the end of the training period to field a team, just not a good team. The players had a miserable season. Although they were knowledgeable about all aspects of the game, knew the basics and the best strategies, they kept losing. Could it have been a lack of full-scale, team practice? After all, they'd only had two weeks of that.

For example, once the team lost a close game in a ninth-inning, bases-loaded and nobody-out situation. The other team's batter hit a grounder. The shortstop and second baseman were so consumed with making one of their beautifully rehearsed double plays that they forgot the other runners, who scored from second and third. Although the team got better as the season went along and the players gained more real experience, the early-season problems were just too much to overcome. At the end of the season, the team's standing was so poor and the players were so discouraged, the team disbanded.

Ah, but while there wasn't a happy ending, there was a silver lining. Because of their extensive knowledge of baseball and sports, many of these players ended up having very successful and lucrative careers as sports columnists and broadcasters for some of the nation's most prestigious newspapers and networks. No one became a great ball player, but they made pretty good emcees and after-dinner speakers in the sports world.

CHAPTER 23

JOSEPH R. HERKERT

A Multidisciplinary Course on Technological Catastrophes

This chapter describes a multidisciplinary course on technological catastrophes, *Multidisciplinary Studies 32:Technological Catastrophes*, that is offered as a general education elective in Science, Technology and Society (STS) at North Carolina State University.

All students entering the university since fall 1994, including engineering students, are required to take at least one STS course.[1] The goals of the STS requirement are threefold: (1) develop an understanding of the influence of science and technology on civilization; (2) develop the ability to respond critically to technological issues in civic affairs; and (3) contribute to an understanding of the interactions among science, technology, and human values.

While a course on technological catastrophes begins with the study of technology, it inevitably leads to an examination of public and private organizations, individual and social values, and a critical assessment of one of our most important contemporary issues: technological risk. In so doing, the course not only addresses all three goals of North Carolina State's STS requirement, but also the ABET 2000 criteria that engineering students have "the broad education necessary to understand the impact of engineering solutions in a global/societal context" and "a knowledge of contemporary issues."[2]

Course Background

The course was originally developed with this author at Lafayette College as a required senior colloquium for liberal arts majors, including students enrolled in the Bachelor of Arts in Engineering Program. Senior colloquia were designed as multidisciplinary capstone experiences with substantive focus on human values.

The course, initially team-taught by two instructors, was developed by the author (an engineer) and two chemists, with advice from colleagues in economics, sociology, and religion. It was immediately recognized that technological catastrophes lend themselves to multidisciplinary analysis. A familiarity with the technology itself is essential to an understanding of why such accidents occur and what their effects are. Likewise, the behavior of complex organizations is a relevant topic. The economics of risk and risk management also play an important role. Finally, social and behavioral aspects of how risks are perceived have a great bearing on the success or failure of risk assessment and risk communication.

The team chose to structure the course around three major case studies: the Bhopal chemical leak in 1984 and the 1986 explosions of the space shuttle Challenger and the Chernobyl nuclear reactor. A fourth case, the Exxon Valdez oil spill, which occurred in 1989 during an early offering of the course, was later added.

At the heart of this inquiry is a critical examination of risk and methods of risk assessment, topics that involve fundamental value judgments. The impacts of technological catastrophes also have values dimensions, as, for example, in the case of Bhopal where third-world exploitation is an issue, and where the impacts are international, crosscultural, or intergenerational in scope, or all three of these, as was the case in the Chernobyl accident. The social and ethical responsibilities of individuals and of corporations are also brought to light by technological catastrophes, for example, through the whistleblowing engineers in the Challenger case and in Exxon's response to the Valdez spill.

Course Description

The catalogue description of *Technological Catastrophes* is as follows:

> Interdisciplinary examination of the human, organizational and technical factors contributing to the causes and impacts of recent technological accidents, such as the Bhopal chemical leak, the space shuttle Challenger explosion, the Chernobyl nuclear accident, and the Exxon Valdez oil spill. Evaluation of risk assessment, risk perception and risk communication strategies. Consideration of options for living with complex technological systems.

The course objectives and course outline are shown Appendixes I and II, respectively. Further information on the course, including a complete syllabus, can be found on the course World Wide Web page (http://www4.ncsu.edu/unity/users/j/jherkert/mds322.html).

The course is designed to approach the topic of technological catastrophes through integration of perspectives drawn from engineering, the natural and social sciences, and the humanities. The required readings are written by engineers, scientists, sociologists, psychologists, journalists, political scientists, and management analysts. The course seeks to develop an understanding of the underlying causes and impacts of technological catastrophes through synthesis of the four major case studies.

Lectures and discussions focus on the human, organizational, technical, and external environmental factors underlying the causes and impacts of technological catastrophes (see Table 23.1), with significant emphasis on interdisciplinary approaches to risk assessment and risk management. Human factors, often mistakenly regarded as the sole cause of technological accidents, generally entail some form of operator error or lack of proper training for the operators, as in the case of Chernobyl when the operators shut down virtually all of the plant's safety systems during an on-line test, or when the intoxicated captain of the Exxon Valdez left an inexperienced third mate at the helm of the ship. Organizational factors include such items as production pressures and cost-cutting, which resulted in most of the plant safety systems in Bhopal being either inadequately designed or inoperable at the time of the accident, and overly rigid organizational hierarchy, exemplified by NASA's failure to focus on serious safety concerns raised by the engineers closest to the design of the space shuttle. Technological factors are typified by poor designs such as the booster rocket o-rings on the space shuttle and the unstable Chernobyl-style reactor. External environmental factors include regulatory laxness, typified by the Coast Guard's deteriorating vigilance with regard to oil tanker traffic in the port of Valdez, and lack of emergency preparedness, a failure of both Union Carbide and the government of India with respect to Bhopal.

Two models of risk perception are explored, the quantitative approach employed by most engineers and other technical experts ("technical rationality")

TABLE 23.1. Contributing Factors in Technological Catastrophes

Factors	Examples
Human	Misperception, misjudgment, negligence, lack of skills
Organizational	Systems, strategies, structures, culture, resources, skills
Technological	Design, equipment, procedures, operators, supplies
Environmental	Regulations, infrastructure, preparedness, risk management policies

Adapted from Shrivastava.[3]

TABLE 23.2. Technical and Cultural Rationality: Selected Factors

Technical Rationality	Cultural or Social Rationality
Trust in scientific evidence	Trust in political culture
Appeal to expertise	Appeal to peer groups and traditions
Narrow, reductionist boundaries of analysis	Broad boundaries of analysis
Risks depersonalized	Risks personalized
Statistical risk emphasized	Impacts on family and community emphasized
Appeal to consistency and universality	Focus on particularity

Adapted from Plough and Krimsky.[4]

and the "social or cultural rationality" approach favored by many social and behavioral scientists (Table 23.2). With respect to quantitative risk assessment, students are exposed to the rationale and fundamental methodology of risk assessment,[5] as well as some difficulties and criticisms of this approach. The social/cultural rationality approach examined relies heavily on sociologist Charles Perrow's theory of "Normal Accidents."[6] Perrow's theory, originally derived during his participation in investigations of the Three Mile Island accident, posits that certain types of accidents are inevitable, i.e., normal, as a result of the degree of complexity and coupling of the system components, which include such factors as design, equipment, procedures, operators, supplies and materials, and system environment. As opposed to conventional explanations of technological failures that focus on failures of individual components, Perrow's emphasis is on the entire system and how the components interact with one another. Perrow defines a normal accident (or system accident) as one emerging from "unanticipated interaction of multiple failures" in highly complex and tightly coupled systems. One interesting implication of Perrow's theory is that technological fixes that increase system complexity or coupling may actually decrease system safety.

Each of the four case studies is evaluated in light of Perrow's theory, and in each case it is demonstrated that the causes go far beyond the simplistic explanations popularly assumed (e.g., "the Soviets were unsophisticated" or "the captain was drunk"). Indeed, all four cases are shown to have significant human, organizational, technological, and environmental factors. Each case is also evaluated in light of conventional risk assessment, or the lack thereof, as in the case of Challenger, where the late Nobel physicist, Richard Feynman, a prominent member of the presidential commission investigating the accident, compared NASA's risk management strategy to a game of Russian Roulette.[7]

The implications of the different risk models for designing efforts at risk communication between experts and nonexperts are also evaluated.[8] In particular, the traditional model of risk communication that sees only technical experts as possessing valid risk information is critiqued.

Throughout the course, students are challenged to evaluate the nature of "catastrophe" in terms of human safety (as in the case of Bhopal and Chernobyl), environmental impact (Chernobyl and Exxon Valdez), and the role of advanced technology in society (all of the cases, but especially Challenger). Consideration is also given to options for living with complex technological systems (or modifying or abandoning them in some instances), and the relative merits of preventative versus response strategies in dealing with technological catastrophes.

In Perrow's study of normal accidents, for example, he concludes that some technologies are not worth the risk (nuclear power and nuclear weapons) while others should be tolerated but subject to careful regulation and modification where possible. Another sociologist, Lee Clarke,[9] takes a different tact, arguing in the case of the Exxon Valdez that meaningful response to large oil spills is impossible. Consequently, much more attention to preventative strategies is warranted, such as double-hulled tankers, better navigational and communication systems, and improved operating procedures.

Pedagogical Methods

The course makes extensive use of cooperative learning techniques, including student group projects on past and potential catastrophes, and other innovative pedagogical methods, such as use of the World Wide Web and video documentaries discussion of the causes and impacts of the case studies from the perspectives of the analytical models introduced in the course.

In one exercise, for example, groups of three to four students are asked to characterize various technological and social systems using Perrow's concepts of complexity and coupling (Figure 23.1), as well as his notion of catastrophic potential:

> Place the following systems on the Interaction/Coupling Chart and indicate which have catastrophic potential:
> 1. Final exam week at North Carolina State
> 2. Tailgate parties at Wolfpack football games
> 3. 1996 Summer Olympics
> 4. The Internet

Through this exercise, students gain increased insight into Perrow's concepts (including the importance of defining system boundaries), a sense of the similarities and differences in technical and social systems, and an awareness of how individuals and groups perceive situations differently.

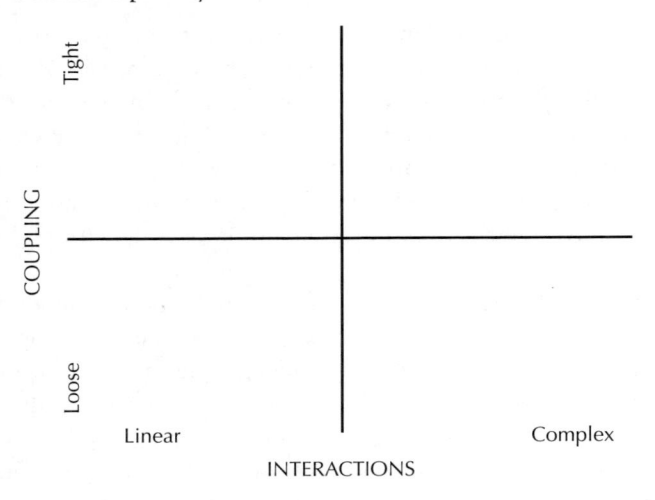

FIGURE 23.1. Interaction and coupling chart (Perrow[6]).

In another exercise, designed to demonstrate difficulties in judging probabilities, students are confronted with the following problem:

> Monty Hall, a thoroughly honest game-show host, has placed a car behind one of three doors. There is a goat behind each of the other doors. "First you point to a door," he says. "Then I'll open one of the other doors to reveal a goat. After I've shown you the goat, you make your final choice, and you win whatever is behind that door."
>
> You want the car very badly. You point to a door. Mr. Hall opens another door to show you a goat. There are two closed doors remaining, and you have to make your decision. Should you stick with the door you chose? Or should you switch to the other door? Or doesn't it matter?

The answer, which is counter-intuitive (you are twice as likely to win if you switch), demonstrates the problems most people have in dealing with probabilities. A news account about mathematicians and other Ph.D. scientists who solved the problem incorrectly underscores the elusiveness of probability, even to experts.[10]

In a third exercise, students are asked to do a back of the envelope risk assessment:

> Assume you are a member of a group at North Carolina State that includes about 500 students (residents of a particular dormitory, senior class of a particular department or college, students enrolled in two or three large lecture sessions of the same course, etc.)
>
> 1. How many members of this group do you expect will die in automobile accidents?

2. How many will be injured in automobile accidents seriously enough to require major hospitalization?
3. What sort of information would you need to more accurately assess the risk of death or injury to members of the group due to automobile accidents?

Here again, the answers are unexpected (10–15 deaths and 100–150 serious injuries), and the ensuing discussion helps students focus on the methodology, assumptions, and limitations of risk assessment.

Group projects consist of research and presentation during an entire seventy-five minute class period of a case study of a past or potential technological catastrophe that was not included elsewhere in the course. Recent topics have included Airline Safety, Dam Failures, Computer Network Crashes, Breach of Biological Containment, and High-Speed Police Pursuits. The presentations, which are required to actively involve the rest of the class, often include simulations, talk show formats, and mock hearings. Presentations in the fall 1996 semester, for example, featured a simulation game designed to highlight risk factors in the airline industry, a simulated train accident (using a model train set) resulting from a failure in automated control systems, a mock town meeting regarding an actual dam failure, and in-car police video from actual high-speed pursuits.

The World Wide Web is an important new teaching tool for courses dealing with engineering ethics and the societal implications of engineering.[11] Web links utilized in *Technological Catastrophes* include primary source archives from the Three Mile Island, Chernobyl, Apollo 13, Challenger, and Exxon Valdez cases, and other resources such as one focusing on Roger Boisjoly, a whistleblower in the Challenger case, that is maintained at the World Wide Web Ethics Center for Engineering and Science. Documentary videos, obtained from PBS and various broadcast and cable channels, include footage of the disasters and/or their aftermaths for all of the major case studies considered in the course, as well as such cases as a ship explosion that leveled downtown Texas City, Texas; Three Mile Island; Apollo 13; and a gas leak that occurred at a Union Carbide plant in Institute, West Virginia. Students are also required to screen the feature film, *Apollo 13*. The videos are extremely effective in bringing the impact of the catastrophes considered home to the students, especially the lesser-known cases such as Texas City and Bhopal. Future plans for the course include requiring a technical media exhibit (e.g., student web page or student-produced video) in conjunction with the group project.

Students are evaluated on the basis of preparation, class attendance, and participation in class discussions; two writing assignments; the group project; and a take-home final examination.

The group project incorporates peer evaluations by group members and the classroom audience. The class members evaluate the presentations by answering the following questions, with each question scored on the basis of a numerical scale ranging from 0 (very poor) through 5 (excellent):

1. Overall, how do you rate this project?
2. Informational content of the presentation was?
3. Relevance of presentation to theme of course:technological catastrophes?
4. Staging of presentation was?
5. Encouragement of class participation was?

Students in the audience are also given the opportunity to provide written comments on their classmates presentations. Contrary to what might be expected, the students are reasonably rigorous in their evaluations of the performance of their peers as compared to the instructor's evaluation of the presentations. In addition, the student presenters are asked to evaluate the contribution of the other group members toward the preparation of the presentation. This process is accomplished using a simple form with which each group member distributes 100 points (or 100% of the project effort) among the other members of the group.

Student Response

At this writing, the course has been offered multiple times at North Carolina State, during which student response has been very positive. As shown in Table 23.3, student ratings of course content have been well above the departmental average. Instructor ratings were also well above the departmental average in the initial offering of the course when the enrollment was only thirteen students. The instructor ratings were much lower, however, for the second offering when the enrollment nearly doubled. This was in part attributable to a poor room assignment, a very small room with auditorium style seating that makes collaborative learning exercises awkward to stage especially when filled nearly to capacity as was the case in fall 1996.

TABLE 23.3. Course Ratings: MDS 322–Technological Catastrophes

Semester	Students	Course and Instructor Evaluations: MDS 322			Departmental Mean Scores		
		Course as a Whole	Course Content	Instructor Overall	Course as a Whole	Course Content	Insructor Overall
Fall 1995	13	4.31	4.46	4.38	3.78	3.84	3.97
Fall 1996	24	3.52	4.10	3.43	3.79	3.87	4.03

Student comments have also been generally favorable. The following is a selection of comments from early course evaluations in answer to the question, "What were the best aspects of the teaching and content of the course?"

Fall 1995
The course was very interesting. It involved some history, philosophy, science, and sociology. I really enjoyed the class. The movies (videos) were interesting as well, rather than just being filler material.

All sides of an issue were presented fairly. The technological aspects were not overwhelming. The ethics brought forth were applauded by this student. The readings were interesting and the videos brought the issues to life by using news & documentaries. [The instructor's] teaching style involves the student in participating & he offers various class activities to involve the student. Overall, I came away with many new issues I had not considered prior to course & knowledge details of catastrophes (& the events surrounding them) I would not have known otherwise.

Relevant to today's society & good historical perspective. Discussion encouraged & videos very helpful for giving pictorial information and opinions of those directly involved in situations.

Fall 1996
The content is real and is here more than ever in today's time.

The accidents were interesting to study because they happened during our lifetime. Because of the complexity of the topics, the videos helped a great deal in understanding the events.

Course content was excellent. This was one of the most enjoyable classes I have taken. The material was very good and I learned much that I did not know.

Course content was very interesting in the fact that it dealt with catastrophes and their impact on people, animals & the environment.

Looking at some of the catastrophes that I remember as a kid happening. Now finding out later on what the causes of the catastrophes were. I like the videos, they give you the firsthand look at what went on before & after. Enjoyed the group project, at first I was upset about doing one but now I have a good time working with it.

The videos really allowed me to see the destruction of accidents that I could not by the reading alone. I was not able to really imagine what happened but the videos allowed me to put everything together and get a different perspective.

The following lengthy comments were submitted by an engineering student in conjunction with a student self-evaluation for the fall 1996 term:

MDS 322 has been an interesting class which has given me a new perspective on technological catastrophes. I have read many articles and learned quite a bit about the classification and methodology of defining accidents in general.

The most interesting thing I believe I will find useful was in the World Wide Web pages that were by Roger Boisjoly. The continual reinforcement of "engineering ethics" and having to maintain a personal standard of quality that you can live with really hit home. I've always heard of keeping a personal "diary" or account of events, but I never really understood why until now. It really seems important to do this. This bit about integrity may not be directly a part of this class, but I feel it is very worthwhile and something I will use as an engineer.

Another interesting section which I thought was useful was the discussion of probability and heuristics. This is the first class in which any mention of the downside of probability has been mentioned.

The most work I have done in the class was in the group project. I investigated the potential for catastrophe with high speed police pursuits. Though I did not mention it in the presentation, I already knew that high speed police chases could be very dangerous due to my personal involvement in a few of them. I have even lost a family member who was involved in a high speed police pursuit. The project really made me step back and think about how dangerous these needless pursuits can be.

The last student comment presented here is extracted from a letter of support written by a student in the fall 1996 semester:

Technological Catastrophes was one of the most interesting and beneficial classes I have taken. [The instructor's] ability to combine the social implications with the technical information into something easily understood is only one example of his professional qualifications. This class provided each student with the knowledge to make responsible and informed decisions regarding the use of technology in our society.

Relevance to Contemporary Issues

A course on technological catastrophes is relevant to both the general educational needs of students and the specific needs of science and engineering majors and students interested in STS. All students will evaluate risks in their daily lives as well as their inputs into the processes of democratically controlling technology (e.g., through voting or purchasing products). Most science and engineering graduates will implicitly, if not explicitly, deal with risk assessment during the course of their careers. In the context of STS education, few topics are more germane to the relationship of contemporary technology and human values than the issues of risk assessment and equitable risk management.

The course addresses all three goals for the STS component of the general education requirement at North Carolina State. The course helps students develop an understanding of the influence of science and technology on civilization by considering how complex technologies are developed and operated, and how dependent individuals and communities have become on their safe and efficient operation. The course assists students in developing the ability to respond critically to technological issues in civic affairs by exposing them to four detailed case studies of complex technologies, their potential catastrophic impacts, and the controversies surrounding their continued development. The course contributes to an understanding of the interactions among science, technology, and human values by continually raising questions concerning the acceptability of risk, the appropriateness of technical versus cultural perceptions of risk, equity issues in the distribution of risk, and the ethical responsibilities of individuals, corporations, and governments concerning the safe operation of complex technological systems. In addition, the course qualifies as an elective in the five-course minor program in science, technology and society.

While enrollment in *Technological Catastrophes* is not limited to engineering students, engineers have composed 40 to 50 percent of the course enrollments. Interaction with students from other academic and social backgrounds, especially those with nontechnical majors, enhances the experience of the engineering students by giving them exposure to the views on risk and catastrophe held by others. Through this course, engineering students are given grounding in technological risk, an important contemporary issue in a framework that places technology in a broad societal context.

APPENDIX I. Technological Catastrophes: Course Objectives

The student will:
1. Learn fundamental concepts relating to the assessment of risk and critiques of conventional risk assessment.
2. Learn and evaluate alternative models of risk perception, including technical rationality, bounded rationality, and social/cultural rationality.
3. Learn the human, organizational, and technical factors contributing to the accidents and their aftermaths in the four case studies.
4. Evaluate the nature of "catastrophe" in terms of human safety (Bhopal and Chernobyl), environmental impact (Chernobyl and Exxon Valdez), and the role of technology in society (Challenger).
5. Consider questions of professional ethics (Challenger) and corporate responsibility (Bhopal and Exxon Valdez).
6. Evaluate the roles of governments as owners (Challenger and Chernobyl) and advocates/regulators (Bhopal and Exxon) of high-risk systems.
7. Consider transnational impacts of technological catastrophes (Chernobyl) and potential exploitation of developing countries through the export of hazardous technologies (Bhopal).
8. Develop an understanding of the underlying causes of technological catastrophes through synthesis of the case study materials.
9. Evaluate options for living with high-risk technological systems, including proposals to abandon, restrict, or tolerate such systems, and the relative merits of prevention and response strategies.
10. Evaluate traditional and emerging models of risk communication between experts and nonexperts.
11. Develop critical thinking skills and collaborative learning techniques.

Multidisciplinary Student Teams

APPENDIX II. Technological Catastrophes: Course Outline

Course Introduction

I. Risk, Accidents and Safety
   Living Dangerously
   Normal Accidents
   Three Mile Island
   Complexity and Coupling
   Risk Assessment
   The Risks of Risk Assessment

II. Challenger
    Causes of the Challenger Accident
    Roger and Out
    Risk Assessment and the Shuttle

III. Chernobyl
     Back to Chernobyl
     Causes of the Chernobyl Accident
     Chernobyl's Fallout

IV. Bhopal
    Who Should We Tell?
    Causes of the Bhopal Accident
    Ten Years After

V. Exxon Valdez
   Anatomy of an Oil Spill
   A Tale of Two Catastrophes
   Sea of Oil

VI. Group Projects

VII. Living with Risky Systems
     Who Should Decide?
     Abandon, Restrict or Tolerate?
     Are Catastrophes Inevitable?

Acknowledgments

I wish to express my thanks and acknowledge the contributions of Dr. Zexia Barnes and Dr. Laylin James who participated in development of the course at Lafayette College.

References

1. Herkert, Joseph R. 2002. STS for Engineers: Integrating Engineering, Humanities and Social Sciences. Chapter 19 in this volume.
2. Schachterle, Lance. 1996. Some consequences of the "Engineering 2000 Criteria" on liberal education. 1996 ASEE Annual Conference Proceedings.
3. Shrivastava, Paul. 1994. Technological and Organizational Roots of Industrial Crises: Lessons from Exxon Valdez and Bhopal. *Technological Forecasting and Social Change* 45:37–253.
4. Plough, Alonzo and Sheldon Krimsky. 1987. The Emergence of Risk Communication Studies: Social and Political Context. *Science, Technology and Human Values* 12 (3&4):4–10.
5. Wilson, Richard, and Earl Crouch. 1987. Risk Assessment and Comparisons: An Introduction. *Science* 236:267–270.
6. Perrow, Charles 1984. *Normal Accidents*. New York. Basic Books.
7. Feynman, Richard P. 1986. Personal observations on the reliability of the shuttle. Report of the Presidential Commission on the Space Shuttle Challenger Accident, Volume I, Appendix F (available on the World Wide Web at http://www.ksc.nasa.gov/shuttle/missions/51-l/docs/rogers-commission/Appendix-F.txt).
8. Slovic, Paul. 1987. Perception of Risk. *Science* 236:280–285.
9. Clarke, Lee. 1990. Oil-spill fantasies. *Atlantic Monthly* (November):65–77.
10. Tierney, John. 1991. Behind Monty Hall's Doors: Puzzle, Debate and Answer? *New York Times* (July 21):1;20.
11. Herkert, Joseph R. 1997. Making Connections: Engineering Ethics on the World Wide Web. *IEEE Transaction on Education* 40 (3).

CHAPTER 24

CHARLES W. N. THOMPSON

Prolegomena for Evaluation of Multidisciplinary Student Teams

Introduction

Measures of group performance are used in research on factors believed to affect group activities, including laboratory experiments (an area sometimes referred to as "small group theory"), in field research on engineering and other groups, and as a basis for various reward structures, such as the Scanlon Plan. While most of these applications do not necessarily concern themselves with measures of the performance of individual members of the group, the need for this additional step is often expressed in the context of groups in the classroom, where the academic requirement for an individual grade exists.

Over a period of nearly thirty years, a one- (later two-) quarter course in systems engineering management has required students to work in five-member groups (and an "all class" group) to choose a real world project, organize itself to develop a solution, and present the results both verbally and in a final report. Except for the intended artificiality of allowing the class to choose the project and organize itself, the students were allowed to carry out the project with a minimum of direct intervention. Individual grades were based primarily upon a combination of a final examination and a project performance grade, the latter being a "multiple" of an individual, small group, and total project grade. The students were strongly encouraged to develop a "lifeboat" mentality, and the multiple sources of information and measures of performance were drawn from processes generally similar to those that a real world project manager would employ.

While it seems clear that the problem of individual measurement has been long recognized, and that little has been offered in the literature toward solution, the increased emphasis in ABET on outcome measurement suggests it is a timely subject.[1]

In general, measures of performance of individuals and groups are provided in an extensive literature covering both industrial and academic organizations. Decisions on whether to hire, or fire, to promote or transfer, or to retrain usually require some measure of individual performance. And these measures bridge both the performance of the individual and, where the individual is part of a group, the performance of the group. The emphasis on team activities as well as the often difficult problem of measuring the performance of the individual within the group has caused a focus on measuring group performance, in some cases providing a substitute for measures of individual performance.

In the engineering classroom, attention has been given to the issue of how to assign individuals to groups,[1] the importance of team projects,[2-4] the importance of design in projects,[5,6] the use of experiential education,[7] teaching groups how to write,[8] and capstone courses.[9]

Evaluating Individuals in Groups

In the work area, particularly, many individuals are involved in group activities, and evaluation of performance is directed at both the group and individual in many cases. While there may be some instructive guidance in the work group experience, the focus here will be on the evaluation of individuals working in groups in a classroom situation.

Groups in the classroom may vary from short-term ad hoc groupings for some experiment or exercise to course long (or longer) groupings on a single project. Where the group performance is a small component of the overall grade, there is likely to be much less concern than where the group performance is a significant part of the individual grade. Where the performance of the individual in the group is easily identified and separated, there is likely to be much less concern than where it is difficult to identify the contribution of the individual. In an industrial setting, the often significantly long lifetime of the group and the close and continuing observation by supervisors provides a variety of ways to measuring the individual. In the classroom, the group lifetime may be small, and the instructor may be limited in both the closeness and number of observations. Finally, the work setting may be able to accept a group evaluation as a basis for measuring individual performance. In the classroom, it would be unusual, at least, for students and teachers to choose the group measure.

A review of "project-oriented capstone courses summarized the evaluation of student performance as follows:[10]

> One of the most difficult assignments for instructors of capstone courses is the evaluation of student performance and the distribution of grades. The very nature of design courses often leads to subjective evaluations. Many design courses contain no formal quizzes or examinations, leaving only required reports and completed design work subject to evaluation. Other courses do have formal examinations over specific material covered in the course, thus allowing the instructor to follow a grading procedure similar to that of a traditional course.
>
> Evaluation of student performance becomes even more difficult when students work in design teams. The individual effort of a student on a project team is often difficult to identify and reward. Many techniques for evaluating individual students within project teams have been employed in capstone design courses.

The authors describe four examples:

1. Peer reviews—fellow team members do the evaluating.
2. Dividing the project into individual and team portions,
3. Using industrial panel members to evaluate (oral and written) reports, and
4. Projects are divided up into individual tasks.

Another interesting reference to freshmen design teams commented as follows:[11]

> The most difficult problem in evaluating an individual's performance on a team project is the amount of effort that the individual has contributed to the team. There was a fear that an indiviidual could do little but would reap the benefit of being associated with a dynamic group and thus receive a grade that he/she might not really have earned.

Anyone who has taught team project courses is likely to have heard, directly or indirectly, that this is also a major concern of the students.

Course History and Practice

History of the Course

This course originated nearly thirty years ago as the introductory course in the systems area of the department. From its beginning, students worked in small groups on real world projects, and there were additional project courses that allowed for individual, small group, and class-wide projects. In 1972, this course, for the first time, focused on a single class-wide project, the subject of which was "The Crisis in Recreational Sports Facilities at Northwestern University." The class report was hugely successful, resulting in a major change in the organizing and financing of intramural sports and other major changes and additions to facilities.

The objective of the original course, and, generally, of its subsequent versions, was to introduce students to systems—systems design, systems engineering, systems analysis, systems management. The intent was to turn the students loose on real problems, problems that were unstructured and multidisciplinary, problems where the problem itself was not well defined, where the models and tools and theories to solve it had to be identified, where the students had to deal with real people who might, at best, be marginally tolerant of their intrusion. In addition to having to identify the problem, and its boundary, they had to negotiate with the potential client(s) for access to information on the project and for acceptance of the final report. They were also responsible for organizing themselves and the work into five-member groups, and they had to do all the planning and scheduling and directing and controlling

After a few years, the course was expanded to two quarters, the first quarter being, as was true of the original course, an introduction to systems, but also providing small group activities preliminary to the single, classwide course in the second quarter. During this first quarter, a major part of the work was for each five-member group to choose a potential problem to work on during the quarter; their focus was to find a project that would, potentially, be a "good" one for the entire class to work on during the second quarter. They were charged with finding information that would help their classmates in deciding whether or not to choose their project. Part of each student's individual grade was the group grade, without any attribution, and the group grade was based on how well they had provided the information, not on whether or not their project was chosen. And this judgment was the instructor's.

The original course was taught by the senior faculty member in the systems area of the department, Gustave Rath, who was joined by the author who has taught the two-course sequence since 1972. Primarily because of geographical convenience and a higher likelihood of access to problems that they, as distinguished from the client, had chosen, most of the projects concerned on-campus organizations and events. Some of the projects were major successes in terms of having a highly visible major impact; others were modest accomplishments, and one or two could only be described as failures. However, both the students and their potential clients were made to understand that the priority in the class, and the grading, was on the skills the students acquired, although it was also emphasized that the students must avoid "harming" the client.

Organization of the Course

Each of the two-quarter courses covers about ten weeks; both courses meet twice a week for one-and-one-half hours. The first week of the second course is devoted to choosing the class project. The second week is devoted to getting organized. And the last week is devoted to the written report and oral presentation. For the weeks in the middle, the first meeting is primarily lecture, and the second meeting is primarily in the form of a progress review. Lectures are prescheduled to cover salient systems management issues, so the students may get information either too early or too late to meet their needs during the course of the project. The instructor suggests that they read ahead, that they use the text (and the outside readings they are required to find), and review their class notes from the previous quarter. In addition to these "content" lectures, the instructor may provide comments on the immediate project—raising issues to be considered, or providing potential answers that may be helpful. The only project specific direction is limited to intervention if there is a high likelihood that someone could be grievously hurt. The design reviews are conducted by the class with little or no interference, directly, by the instructor. Classes range in size from about 25 to 40, although some classes have ranged up to 70 students.

While criteria were suggested during the first course, the class was free to use any criteria it wanted in choosing the project, and were free to choose from among the projects developed during the first quarter, or go outside for a different one. Most of the time they chose from ones that had been worked on to take advantage of the head start provided and because the relationship with the potential client was apt to be clearer.

Choosing the project, the class was invited to set the boundaries of the project and decide how to divide up the project. The only constraint was the pedagogical requirement that they be organized into five-member groups. The instructor would provide the necessary signup sheets, but the class had to decide how to do this.

An important feature of the class was the giving of no more than two midterm examinations each week, and these were unannounced. Some eight to twelve of these were given, using a system that preserved the anonymity of the student; each exam was ten questions, usually true/false or short, written answer. Exams were, in part, based upon lectures, but they also included project related questions, such as identifying the probable stakeholders by name, which had not been covered in lectures but which the instructor felt that the cleass should know. In this sense the exams were part of the instructing. At the end of the course, the student could decide whether to use the midterm exam results or take a final exam.

Overview of Course Materials

In addition to the textbook, a number of handouts were provided either in paper form or electronically. Students were advised to keep their class notes from the previous quarter, and notes prepared by members of the class for the current lectures were reproduced. Previous class reports, both individual group and single class projects, were available, as well as a few videotapes of prior presentations. Finally, the instructor keeps a number of texts and journal articles on reserve in the library.

Course Handouts

A few more than twenty handouts were used in each course, with some overlap. In the second course they included a course schedule, course description, some related handouts, and a number of short didactic papers on how run a project. Some of these that are strongly related to the present topic will be discussed below.

Each week each student was required to write an "application note." This required the student to find a project problem and apply some theory or method from the lectures or from the literature. The immediate purpose was to encourage the student to see the relationship between the two.

Each week, each group was required to present a "progress report." Suggestions were provided, and each was graded with extensive written comments.

Each week, the entire class was required to conduct a "progress review." The instructor attended these but almost always neither commented nor provided any indication of approval or disapproval during the review.

At the end of class, each student was invited but not required to provide a "self-serving statement" which would help the instructor to identify the contributions of the student or other classmates. Generally, almost all students provided these statements.

Measures of Performance

Individual

Each student received an individual grade, which was a function of the student's project grade and the student's grade on the mid-terms or final examination. The methods for calculating the exam grade and of calculating the final grade were provided.

While each student received more than ten individual grades on weekly written assignments, these were used only as tie-breakers, i.e., if the matrix result was ambiguous. These individual grades were for recording a set of lecture notes, abstracting articles from the literature, and submitting application notes.

The project grade and the examination grade were the major determinants of the student's final grade, and the project grade was weighted a little more heavily. The project grade, in turn, consisted of three grades that were added "non-linearly." The first of these three grades was a total class project grade; it was based primarily on the final oral presentation and final report. The second of these three grades was a group grade, and this was based on both group activities and the contribution of the group to the project. It was made clear that a significant part of that "contribution" was whether or not that group helped or hindered others in the class. The "lifeboat model," or Patrick Henry's "hang together" sentiment, was repeated in many lectures, and some of the interim grades for group products, e.g., progress reports, were explicitly presented in terms of this sense of "contribution." The last component was the individual's grade on the project, with the same criteria applied to the student's relation to both the group and the entire class. Note that the individual grades above do not contribute here.

Group Related

It was usually clear to the students that their grade was largely dependent upon the success of the project and how they contributed to their group and to the class as a whole. Individual brilliance in contributing ideas and stellar individual efforts could be easily outweighed if their effect upon others resulted in a negative net effect. It was necessary, however, for the instructor to be able to measure each student's and each group's contribution.

Students were told that they would run the project, and it would be run as closely as possible to a real world project with the only differences being the insistence on five-member groups and their freedom to choose the project and to organize themselves. Paralleling this, the instructor tried to use methods analogous to those that would be used by supervisors on a real world project. Some of the methods were directly comparable. Some were substitutes for the observational opportunities that a real world project would have.

The "Application Notes" required a student to find a project problem and apply lecture materials, and this had a central purpose of requiring the student to see the relation (and, ideally, learn to use the prior art in solving problems). However, the student had to find a real problem, and a potential solution, and, in this process, the students provided feedback on serious problems and problems which worried this student. This kind of information, a real world supervisor normally would find by "walking around" or in other day-to-day observation, something an instructor would be less able to do.

"Progress Reports" provided summaries of who did what to whom on a weekly basis, usually identifying the individual who contributed (or helped others to contribute) or was deadwood—either shirking the work, or doing shoddy work. They also provided a basis for assessing the progress of the class in developing uniform reporting standards that they would need in preparing the final written report.

"Progress Reviews" brought out into the light those major obstacles, including internal disagreements, which were hurting the project, and they demonstrated publicly who was cooperating with whom.

"Work Products" is a term intended to include all of those other written materials that represent evidence of what an individual or group was doing. These might be found in the project archives, or in courtesy copies provided the instructor, or in materials appended to the self-serving statements.

The "Self-serving Statements" were not required; however, most students did participate, at least to the extent of briefly summarizing the highlights of their contributions, including copies of materials as an appendix. A few students, particularly those who were most active and who were in leadership positions, would write extensive comments, sometimes on everyone in the class, identifying those few who were considered the hardest workers and the most dedicated and creative contributors. Many would identify a few classmates, including those who were very good and those who were very bad. One of the important functions these serve is to identify the contributions of those who are not in highly visible positions or activities.

The final written and oral presentations provided a sustained and detailed look at the project outcome. Usually clients and other stakeholders would provide significant evaluations. By going through many written drafts and several presentation dry runs, it becomes obvious as to which student is doing what, at least for some students. Those who are most involved here, as well as those most involved in other very visible project activities, are easy to identify; however, those behind the scenes, doing redactory work, or critiquing presentation dry runs come into view if one looks for them.

Much of the above information that is relevant is posted to the record the instructor has maintained on a daily basis of what each individual and each group has done. Using these notes as a start, the instructor assigns preliminary and individual and group grades (in pencil). Then an iterative process is followed by rereviewing all of the various sources, comparing each individual against the rest of the group and the class, and each group to the rest of the groups. This is an intense process, but it usually results in grades for the individual that have a reasonably explicit basis.

Summary

The major disadvantage of this process is that it consumes an enormous amount of time; it is, however, in that sense, analogous to the workload of a supervisor in the real world.

The major advantage is that it does not require the artificiality of decomposing the project tasks to facilitate the grading process or the artificiality of having a peer rating system.

My experience has been that the occasional disgruntled student when taken through the grading process in detail is usually relieved that the grade wasn't lower. My view is that this strikes a balance between the need to grade and the need to keep the project experience as close to reality as possible.

References

1. Brickell, James L., David B. Porter, Michael F. Reynolds, and Richard D. Cosgrove. Assigning Students to Groups for Engineering Design Projects: A Comparison of Five Methods, *Journal of Engineering Education*, Vol. 83, No. 3, July 1994, pp. 259–262.
2. Bronzino, Joseph D. Design and Teamwork: A Must for Freshmen, *IEEE Transactions on Education*, Vol. 37, No. 2, May 1994, pp. 184–188.
3. Byrd, Joseph S., and Jerry L. Hudgins. Teaming in the Design Laboratory, *Journal of Engineering Education*, Vol. 84, No. 4, October 1995, pp. 335–341.
4. Chapman, Graham M., and John Martin. Developing Business Awareness and Team Skills: The Use of a Computerized Business Game, *Journal of Engineering Education*, Vol. 85, No. 2, April 1996, pp. 103–106.
5. Koen, Billy V. Toward a Strategy for Teaching Engineering Design. *Journal of Engineering Education*, Vol. 83, No. 3, July 1994, pp. 193–201.
6. Moriarty, Gene. Engineering Design: Content and Context. *Journal of Engineering Education*, Vol. 83, No. 2, April 1994, pp. 135–140.
7. Pavelich, Michael J., and William S. Moore. Measuring the Effect of Experiential Education Using the Perry Model. *Journal of Engineering Education*, Vol. 85, No. 4, October 1996, pp. 287–292.
8. Schulz, Karl H., and Douglas K. Ludlow. Incorporating Group Writing Instruction in Engineering Courses. *Journal of Engineering Education*, Vol. 85, No. 3, July 1996, pp. 227–232.
9. Todd, Robert H., Spencer P. Magleby, Carl D. Sorensen, Bret R. Swan, and avid. K. Anthony. A Survey of Capstone Engineering Courses in North America. *Journal of Engineering Education,*, Vol. 84, No. 2, April 1995, pp. 165–174.
10. Dutson, A. J., R. H. Todd, S. P. Magleby, and C. D. Sorensen. A Review of Literature on Teaching Engineering Design Through Project-Oriented Capstone Courses, *Journal of Engineering Education*, Vol 86, No. 1, January 1997, pp. 17–28.
11. Croft, Frank M., Jr., Frederick D. Meyers, and Audeen W. Fentimen. An Algorithm for Evaluating Team Projects, *Engineering Design Graphics Journal*, Vol.59, No. 3, Autumn 1995, pp. 18–20.

CHAPTER 25

JOHN P. O'CONNELL

MARK A. SHIELDS

EUGENE R. SEELOFF

TIMOTHY C. SCOTT

BRIAN PFAFFENBERGER

Professional Development at the University of Virginia

Attributes, Experiences, ABET 2000 and an Implementation

Introduction

Since early 1995 a small committee of faculty and staff of the School of Engineering and Applied Science at the University of Virginia (UVa SEAS) has worked to define professional development values and objectives for students and to determine how they can be fostered in an undergraduate engineering curriculum. The committee developed a framework document outlining the key attributes and experiences of professional development (PD); met with alumni, business, faculty, and other representatives to solicit their input to the document; and, starting in fall 1995, pursued teaching collaborations to implement aspects of the professional development vision outlined in the document. Though UVa PD has been exclusively an engineering program, the involvement of two faculty of the SEAS Division of Technology, Culture and Communication ensured including a significant component of liberal education.

The expectation is that students possessing a significant measure and balance of these characteristics are most likely to become successful professionals.

Our view is that to nurture the attributes, students must have a rich variety of

TABLE 25.1. Relationship of ABET 2000 Criteria to UVa PD Attributes

ABET 2000 Criteria 3 Outcomes[4]

a) An ability to apply knowledge of mathematics, science, and engineering;
b) An ability to design and conduct experiments, and analyze and interpret data;
c) An ability to design a system, component, or process to meet desired needs;
d) An ability to function on multidisciplinary teams;
e) An ability to identify, formulate, and solve engineering problems;
f) An understanding of professional and ethical responsibility;
g) An ability to communicate effectively;
h) The broad education necessary to understand the impact of engineering solutions in a global societal context;
i) A recognition of the need for, and an ability to engage in, lifelong learning;
j) A knowledge of contemporary issues; and,
k) An ability to use the techniques, skills, and modern engineering tools necessary for engineering practice.

University of Virginia Professional Development Attributes

Graduates beginning their careers should have certain qualities	Related ABET criteria
Technological Capability	Know and be able to practice the fundamental technical facets of engineering (a, b, e, k).
Leadership/Cultural Competence	Become leaders in a diverse, complex world (h, j).
Industrial Readiness	Appreciate functions, dynamics and evolution of "industry"; understand the expectations about their roles, contributions, and attitudes (c).
Individual/Team Effectiveness	Understand themselves and others; thrive in diverse and ambiguous situations (d).
Ethics/Values/Service Commitment	Be dedicated to the highest professional and human values (f, h).
Communication Skills	Can inform others and make decisions in diverse contexts (e, g).
Career Vision	Begun moving in the direction of their life's work (i).

TABLE 25.2. UVa Professional Development Experiences

Students grow and develop confidence best in certain environments and situations.

Introspection/Self-Assessment	Who am I? How should I develop? Self-discovery and self-improvement.
Learning/Growing:	Building my competencies; recognizing my changes in technical and nontechnical knowledge and concepts; communication skills; challenges; varieties
Performing/Doing	Practice, practice, practice. Resource utilization, abstracting and analyzing; job searching; oral presentations; helping others; decision making and risk-taking, physical tasks (labs, equipment).
Leading/Following	What's my most effective role? Orientations and beginnings, closures and endings; goal-oriented teamwork; workings of professional, social, and service organizations
Employment/Service	Real work and how it's done. Outside the university in externships, internships, community service; inside in modules/courses, research and student organizations.
Interactions	Being and working with others. Students in academic settings, university personnel, professionals, general public.

experiences and environments. Table 25.2 gives a concise expression of these experiences; a complete discussion is in reference 3.

To be effective, experiences must be concretely connected to attributes. We illustrate this with a few examples phrased,

"As I graduate from SEAS, I have:

Enhanced My Technical Capabilities by:
- Analyzing the concept and design of a particular engineering object (a device, unit or system) to determine what decisions were made in its development and manufacture.
- Using a computer search to find references about the phenomena that would occur in a proposed object.

Raised My Leadership/Cultural Competence by:
- Supervising a team of students designing an engineering object, including market survey, specifications, financing, manufacturing, and marketing.
- Participating in a discussion about legislation and policy for affirmative action in an engineering industry.

Expanded My Industrial Readiness by:
- Visiting several engineering and manufacturing companies as an individual as well as in a group.

- Assessing the potential impacts of a novel object in cost, reliability, safety, and environment.

Increased My Individual/Team Effectiveness by:
- Conducting and communicating the results of a project requiring independent, individual research.
- Taking a personal characteristics inventory and learning how I can best function in team settings.
- Assuming responsibility for the results of a group project of several weeks' duration that demanded integration, application, and extension of the knowledge and processes learned in several classes.

Sensitized My Appreciation of Ethics/Values/Service by:
- Taking an active part in a student professional society.
- Visiting a secondary school and talking to the students about technology.
- Exploring several ethics-related case histories via role-playing followed by feedback from practicing professionals.

Sharpened My Communication Skills by:
- Comprehensively researching a technical subject and compiling a professional bibliography.
- Participating in a conference, including presentation of a technical paper with written proceedings.
- Explaining a complex technological topic in terms that the general public can understand.

Broadened My Career Vision by:
- Talking to several practicing engineers about what they do, how they chose their career paths, and what advice they would give aspiring young engineers and then presenting the summary in a class.
- Preparing a list of engineering functions for a particular industry and the requirements for its entry-level jobs, as well as its potential career paths.
- Having worked (summer, internship, co-op, full-time) for at least one company that hires engineers in my major.

The expectation is that living these experiences and assimilating their lessons should give graduates confidence, wisdom, and adaptability for lifelong service and accomplishment. Students exposed to this array of situations will better understand what professional life is and how they might live it to their greatest satisfaction and contribution. A program with this experience will touch the lives of all students; it can be organized and implemented.

The first part of this chapter describes the UVa Professional Development framework and how it relates to ABET 2000. The second part shows how we attempted to achieve the objectives in two first-semester courses of our school through an unusual collaborative teaching effort, including an assessment of how well the collaboration was able to fulfill its PD objectives. The chapter concludes by examining some broader lessons learned from UVa's model of professional development in undergraduate education.

Foundations and Benchmarks of Professional Development: The UVa Model

The UVa committee developed a framework document outlining the key attributes and experiences of PD; met with alumni, business, faculty, and other representatives to solicit their input to the document; and pursued teaching collaborations to implement aspects of the PD vision outlined in the document. The collaborations of 1995 are described in Pfaffenberger et al.,[1] while those of 1996 are discussed in Shields and O'Connell.[2]

The internal document, "Foundations and Benchmarks of Professional Development,"[3] identified seven *attributes* and six *experiences* that represent the key dimensions of PD that our undergraduate engineering curriculum should cultivate. Although these attributes and experiences were formulated before any of the participants were aware of the ABET 2000 criteria, we believe that our expressions are quite compatible with ABET 2000. Table 25.1 lists both the eleven outcomes of Criteria 3 of ABET 2000[4] and summarizes the UVa PD attributes. The ABET criteria connected to the attributes are given in parentheses. Comparison of the wordings shows that our PD framework is more comprehensive and possibly more ambitious; this is clearer in the full document.[3]

We anticipated the demands of ABET 2000 for both formulating objectives and for the educational process to achieve them. Our *attributes* were developed by a varied constituency of alumni, business, faculty, and staff. Then we described *experiences* (see Table 25.2) as general expressions of a process intended to achieve the objectives. Thus, we are confident that a UVa PD program can meet all of the expectations, and more of ABET 2000.

Collaborative Teaching and Professional Development at UVa

During the fall semester 1996, two of the authors (MAS, a sociologist of technology, and JPO'C, a chemical engineer) collaboratively taught two courses in an attempt to put the professional development model to work in first-semester undergraduate studies at UVa. One was a required core technical communications course (TCC 101) typically taught in sections of twenty-five students by faculty of the Multidisciplinary Division of Technology, Culture, and Communications, using a common syllabus but with specific assignments tailored to each instructor's disciplinary strengths. (This restricted the pool of students since approximately 25 percent of entering students are given advanced placement credit for this course.) The course introduces students to the uses of oral and written language communication, stressing their relevance to professionalism in engineering and applied science. Students learn how to search and retrieve technical information, write abstracts, essays, and reports, and give oral presentations for a variety of audiences. One key project assignment introduces students to fields of engineering and helps them choose a major. Other assignments address technical and humanistic aspects of engineering, technology, society, and ethics.

The other course, taken concurrently, was a required cross-disciplinary engineering design class (ENGR 164) normally taught in multiple sections of 35 students by engineering faculty from several different disciplines working from a common set of goals but not the same syllabus. These goals involve lectures, workshops, and five assigned projects to cover open-ended design case studies via individual and team designs; methodologies for computation, problem solving, and conceptual design; consideration of engineering economics, environmental aspects, quality, and safety; professional responsibilities and ethics; and career opportunities for engineers.

Student enrollments in TCC 101 and ENGR 164 sections are not normally coordinated and were not coordinated for the other sections taught during 1995–1996. (These courses were coordinated in a similar way in fall 1995 with four instructors.) Both years, assistance was received from the engineering dean's office to assign incoming students at random to the paired sections.

In the 1996 collaboration there were two groups of 28 students (totaling 12 percent of the entering class), each group taking a section of TCC 101 and a section of ENGR 164 from the authors. The section classes were of the same length on mostly the same days of the week, and the individual syllabi for the sections of each course were identical. As in the regular sections of each course, assignments ranged from short ones in resource utilization and generic communication skills to multiple-week projects analyzing information and designing alternative products and processes. Compared to the regular sections of each course, the paired sections were distinguished by generally more intensive student teamwork; more extensive assignments and in-class activities of broader socio-technological concern; joint formulation of project goals, statements, and activities of the courses; coordinated due dates; generally more extensive use of university library and electronic information resources; joint grading of team oral presentations; and more extensively coupled and systematic evaluations of the courses and teachers.

In terms of the PD attributes, our course objectives were selected in the order of importance of Table 25.3. Students were also taking two or more technically oriented courses such as chemistry, mathematics, and computers at the same time as our courses. This influenced our selection of attributes.

Assignments in both courses emphasized fundamentals of concise technical

TABLE 25.3. UVa PD Attributes in First-Year Course Objectives, Fall 1996

Engineering Design (ENGR 164) Objectives

1. *Expand Industrial Readiness:* Engineering analysis and conceptual designs, including issues such as safety, quality, reliability, and optimization.
2. *Sharpen Technical Communication Skills:* Written and oral reports describing recommended designs to meet performance requirements of open-ended problems.
3. *Increase Individual/Team Effectiveness:* Use recognized methodologies involving individual and collaborative work to formulate problems and report on projects.
4. *Broaden Career Vision:* Self-assessment and career investigation in choosing a major.
5. *Enhance Technological Capability:* Formulate and solve quantitative problems, including computations requiring conversions between systems of units. Apply basic computational techniques used in economic decision making, including interest, time-value of money, capital costs, annual costs, comparison techniques.
6. *Raise Leadership/Cultural Competence:* Understand professional responsibility through evaluating cases illustrating successful and unsuccessful designs.
7. *Nurture Ethics/Values/Service Commitment:* Analyze product liability issues, risk assessment, and cross-cultural practices.

Technical Communication (TCC 101) Course Objectives

1. *Sharpen Technical Communication Skills:* Learn principles and practices of concise technical communication for multiple audiences, effective use of high-quality information sources, graphical display of information for individual/team written reports, oral presentations, homework exercises, peer-critique, and revision.
2. *Broaden Career Vision:* Research engineering fields, interview faculty and practicing engineers, learn about career opportunities in each field, relate one's personal strengths to career options, and use knowledge to choose major intelligently.
3. *Increase Individual/Team Effectiveness:* Apply principles of cooperative work to major team projects, including team research, writing, and presentations. Learn how to adapt to different team members and problems, and know the key ingredients for successful teamwork.
4. *Raise Leadership/Cultural Competence:* Understand implications of crosscultural differences for exercising engineering expertise appropriately; analyze, in depth, one major engineering-related social transformation either in the U. S. or abroad.
5. *Nurture Ethics/Values/Service Commitment:* Analyze ethical aspects of product design, marketing, and impact, including how ethical professional judgment can be applied to real-world product development.
6. *Expand Industrial Readiness:* Develop appreciation for the importance of effective technical communication for professional success.
7. *Enhance Technological Capability:* Understand the "human-machine interface": how the technical features of devices and systems have nontechnical consequences.

writing, well-organized written reports and oral presentations with rich graphical materials, and extensive use of the library and the World Wide Web. Most projects, ranging in duration from one to five weeks, were carried out either in pairs or in teams of three or more.

In ENGR 164 the projects were finding and analyzing a current product liability case; physically unwrapping and disassembling an inexpensive consumer product (two workshops), and discussing its scientific principles, conceptual design, materials, economics, manufacture, maintenance, failure, disposal; brainstorming options, selecting viable concepts and doing detailed designs of ways to "make the UVa libraries more user-friendly"; researching and expressing in detail the range of impacts of commercial air transportation on the environment; designing a single piece of equipment or facility and a procedure for an instructor-selected aspect of flight, ground, and support operations that would minimize adverse effects while maintaining safety and economic viability in a global setting. The first assignment was for individuals. The next was in pairs, the third was in trios, and the last two multiweek projects were done by quartets. Each project had one or two written reports, with the final project also having twenty-five-minute group oral presentations and questioning. A field trip was taken to the regional airport with an orientation and tour with the airport operations manager. There were also two "hands-on" workshops analyzing the workings of a refrigerator and air conditioner. These workshops utilized the laboratories and participation of the laboratory instructor (TCS) for the Department of Mechanical, Aerospace and Nuclear Engineering.

In TCC 101, projects included several individual writing assignments to improve style, grammar, punctuation, conciseness, word choice, topic sentences, critical analysis, and argumentation. Some of these were peer-reviewed and subsequently revised. One key early paper was to weigh the ethical aspects of the Dow Corning silicone breast implants controversy in light of scientific-technical knowledge. Students also received instruction in organizing and delivering one individual and two-team oral presentations. Most classtime was used for cooperative learning workshops in which students discussed an issue, solved a problem, and/or gathered information in groups of three or more and then shared their group's results with the rest of the class. These exercises also improved students' abilities to give extemporaneous talks. Two large, multiweek projects were the other major focus of the course. The first had students work in eight teams of three to four members to gather information on each of the eight UVa undergraduate major fields.[5] In addition to using printed information sources, this project required each team to interview an engineering faculty member in the relevant field, as well as practicing engineers and senior engineering students. Teams gave fifteen-minute oral presentations on their research and wrote a detailed twenty-page team report on their project. Another multiweek team project in the communication course had student teams conduct extensive research on current global topics related to technology and human development, culminating in each team giving an oral/poster presentation and a twenty-five-page report. This project, lasting five weeks toward the end of the semester, was research-intensive and required students to apply virtually all of the skills they had learned earlier (see Shields).[6]

In both courses, students cooperated in all phases of the team projects from problem formulation and data gathering to report writing and oral presentations. Time commitments for several activities were fully shared between the two courses. These included two professional-level, team-based simulations of design and manufacturing, both involving model construction, led by a staff member of the Virginia Engineering Foundation trained in such activities. Also, a staff member of the Engineering School Office of Career Services used class meetings to administer and review personality and aptitude indicators, including the Myers-Briggs Type Indicator[7] and the Strong Interest Indicator;[8] to conduct a workshop on informational interviewing and describe career-oriented materials available in that office.

Evaluation

How well did the collaboration fulfill its professional development objectives? The summary answer to this question uses two sources of evaluation: the instructors' informal observations and formal student feedback from anonymous questionnaires completed near mid-term and during the final class meetings of ENGR 164 and TCC 101. More complete analysis is to be published, but we give some preliminary results here.

From the Instructors: ENGR 164 and TCC 101

Both instructors concluded that the collaboration was beneficial to meeting professional development objectives. The range and mixture of assignments and activities between the two courses meant that essentially all of the attributes of the PD model were fulfilled appropriately for beginning students. Table 25.4 shows that the joint teaching allowed weak connections to specific PD attributes in one course to be compensated by moderate or strong connections in the other.

From the Students—ENGR 164

Students in the design course were asked to assess how well the catalog description and course PD objectives were fulfilled. The aggregate mean ratings were "very well" for design and case studies and "well" for the other catalog items. The objectives met the best were technical communication and individual/team effectiveness, with ratings of "very well," while industrial readiness and leadership/culture/ethics were rated "well." Technological capabilities and career vision had ratings of "OK." Comparisons with the objectives in Table 25.4 indicate that the impressions made upon the students were generally consistent with the intentions of the

TABLE 25.4. Instructors' Assessments of PD Attributes in Their Courses

Attribute	ENGR 164	TCC 101
Industrial Readiness	Strong	Weak
Communication Skills	Moderate	Strong
Individual/Team Effectiveness	Strong	Moderate
Career Vision	Weak	Strong
Technological Capability	Moderate	Weak
Leadership/Cultural Competence, Ethics/Values/Service Commitment	Moderate	Moderate

instructors. There were significant differences between the responses of the two sections about some of the catalog items (design, case studies, and career opportunities) and all of the attributes. One group was always more favorable than the other, with the item mean scores differing by about the item standard deviation (which was about the same for both sections).

In addition to having students in the paired sections of the design course complete a final course evaluation, fifty students in other (unpaired) design sections completed the same questionnaire. In most areas, there appear to be beneficial differences in the paired sections, with the largest ones in the areas of case studies, technical communication, and leadership/culture/ethics/service. However, these differences may merely reflect variations in the approaches, goals, and effectiveness of ENGR 164 instructors rather than any direct benefits of collaborative PD teaching.

Finally, students also rated the importance of various activities to what they learned in ENGR 164. The highest-rated components at "Very Important" were "group projects," "electronic information," "oral presentations," and "workshops." The lowest-rated were readings ("Not Important") and lectures ("Somewhat Important"), with class activities and individual projects "Important." Students in the design course rated team-based projects more highly than individual projects. This reflects the relative emphasis in the course, which was chosen to provide opportunities and experiences that made direct connections to multiple attributes of professional development.

From the Students—TCC 101

Essentially all of students in the technical communication course completed a final course evaluation questionnaire. While the questionnaire did not ask students for direct ratings of specific PD attributes, it did ask them for detailed feedback on several aspects of the course related to PD. In particular, the student data indicate that the TCC 101 course provided rich experiences in support of three key attributes: Communication Skills (CS), Individual/Team Effectiveness (I/TE), and Career Vision (CV). The students indicated that the course helped them improve several specific skills: Give effective oral presentations (CS, I/TE); Use information sources effectively (CS, I/TE, CV); Work effectively as a team member (I/TE); Describe engineering fields (CV). Finally, students indicated the relative importance of several specific class components to what they had learned in the technical communication course. As in the paired ENGR 164 sections, students in the paired TCC 101 sections rated team projects as most important. The highest rated activities were: Poster Exhibition and Competition; Team Writing Projects; Oral Presentations; Use of Electronic Information Sources. In fact, the lowest-rated item was individual writing assignments, though there were several such exercises in the course.

When asked "Would you recommend a collaborative 101/164 section pairing to next-year's incoming students?," 72 percent of the students answered "yes." As one student put it: "TCC helped us to write more effectively and present effectively in ENGR 164. 164 helped us to understand the technical aspects of engineering." Another student wrote that "teamwork skills development occurred through assignments in both classes, and writing style improvements from TCC helped with ENGR 164 papers." Similarly, another commented approvingly that "oral presentation skills and group skills were used in both" [courses]. Most students emphasized the benefits of having the same classmates in two courses and of having coordinated course schedules that avoided conflicting deadlines on major projects.

By far, the major student complaint focused on the workload for the paired courses; this also occurred with the paired sections of the previous year. It was a widely held perception even among students who were positive about the advantages of pairing, yet especially pronounced among the 28 percent of students who would not recommend the pairing to future students. This response was much more prevalent in one section than the other. However, many of these students recognized that the learning benefits were substantial despite the heavy workload: "Although you do learn a lot from these two sections," wrote one student, "the workload in each makes it very difficult to get your work done in your other three classes." Likewise, as another student noted: "Actually, TCC and ENGR hurt other courses because we spent a lot of time for both classes. However, I think we learned many things from both classes." Still another student wrote that "the amount of work between the two courses was so overwhelming. It did not help me to do better in TCC, Design, or any other course. The overall amount of writing, though I hated it, helped improve my technical writing skills immensely." Student perceptions of the relatively heavier workload in the paired sections may well have been correct, based on some indicative but not definitive data collected at mid-semester. Those data suggested that students in the paired sections of TCC

101 and ENGR 164 were indeed inclined to rate their workloads as more time-consuming than students enrolled in unpaired sections of those courses. In any case, students obviously believed the workloads were greater; their beliefs, in turn, probably influenced their evaluations of the two courses.

Conclusions and Implications

Our experience suggests that first-semester engineering can be positively influenced with collaborative implementation of a carefully articulated PD vision and framework. The UVa expressions of attributes and experiences were very helpful in focusing decisions about content, activities, and assessment, even in two such disparate courses. The fact that these are consonant with ABET 2000 criteria was reassuring about the appropriateness of the course objectives.

The objective in pairing sections was to provide students with a "compleat" professional development experience; their formal evaluations of the courses confirm that this was largely accomplished. Even most of the students who would not recommend paired sections to future incoming students (because of the heavy workload) agreed with the value obtained. While many engineering educators believe (perhaps somewhat correctly) that first-year students lack the intellectual maturity and personal autonomy needed for optimal success in a rigorous curriculum, our experience is encouraging: many of our students seemed to rise to the challenge. In particular, the pairing of technical design and technical communication seems to provide a genuine synergy to reinforce and complement a shared set of PD goals. Further, most students said that they also benefited from the social integration of the two courses: stronger interpersonal ties and other aspects that supported cross-course teamings, such as the predictability and convenience of coordinating team meetings when meeting classmates twice a day.

What about the instructors (MAS and JPO'C)? Several conclusions stand out. First, the collaboration worked because of a shared vision of PD and its importance as well as a willingness and enjoyment of sustaining it by frequent interactions for planning and execution throughout the semester. Our sense is that both of these are not only desirable but essential. Further, by participating in the development of the shared PD vision as it emerged over months of intensive discussion, we were more able to implement a systematic model in our sections than might someone who came in "cold."

Second, we should keep both our PD efforts and ABET 2000 in proper historical perspective. "For more than a hundred years," Florman[10] observed, "educators have been trying to find an appropriate place for liberal learning in the engineering curriculum." That an engineer's professional development should include meaningful liberal education has long been part of engineering education reformism. From

Charles Mann in 1918 suggesting that liberal learning and engineering education be linked through professional development to the present, several reflections and reports discuss making the humanities and social sciences more integral to engineering education.[11] Among the most recent, the "Green Report" of 1994 calls for "a variety of models in engineering education," that "must take into account the social, economic, and political contexts of engineering practice; help students develop teamwork and communication skills; and motivate them to acquire new knowledge and capabilities on their own."[12] Our PD framework and the collaborative implementation of it in the first-year curriculum are thus part of a long tradition of engineering education reform.

Third, the teaching collaboration in fall 1996 provided something else novel and significant, namely, that *a very high degree of integration of technical and liberal engineering education is possible*.[13] Take, for example, our first attribute of PD, Technological Capability (TC): "Know and be able to practice the fundamental technical facets of engineering." The collaboration made clear to JPO'C and MAS that TC could be usefully broadened to include the capacity of engineers to integrate technical expertise, sociocultural analysis, and professional ethics in analyzing and solving real-world engineering problems.[14] Thus, the experience generated new faculty insights and cross-fertilization between the technical and nontechnical aspects of professional development that can be brought into students' professional development. It may be that visions of liberal engineering education work out better in practice than in theory.

Fourth, despite our satisfaction and enthusiasm, we do not underestimate the institutional and professional obstacles that impede broader cross-disciplinary linkages. Difficult as it is to find instructors in technical engineering fields, the social sciences, and the humanities with enough intersecting professional interests to find common teaching ground, a further (and perhaps bigger) hurdle is a lack of experience and support for collaborative teaching as a whole. In contrast to the PD framework that stimulated and guided our collaboration, there seems to be no formal model for any kind of collaborative teaching—let alone for a cross-disciplinary collaboration like ours. In fact, at this point we probably know much more about collaborative (or cooperative) learning than about collaborative (cooperative) teaching. Educators need to pay more attention to this if joint efforts like ours are to become more frequent and productive. Our experience was thus unusually fortunate—and perhaps fortuitous.

Finally, we all have reflected on what's next for the students. Our principal concern is the very few ways in which the rest of our undergraduate curriculum builds on this kind of first semester. Focus groups of students who met some months after the first year's collaborations mentioned this explicitly.[14] Perhaps consistent with this, a web survey of our 1996 students one year later indicated that several of them had an increased appreciation of the value of their experience with us compared to their end-of-the-term course evaluation. We noted at the time that getting widespread commitment to the PD vision will require considerable time and effort by faculty and a variety of administrative support mechanisms and that implementation at other levels will probably require exceptionally creative and adaptive teaching.

Coda: A Lesson Learned

In the intervening period, ENGR 164 has been reduced from four to two credits with much less instructor flexibility and none of the recommendations of the PD report have been adopted. Apparently the challenges we have described turned out to be too great for the necessary commitment to occur at this time. Whether SEAS's first accreditation under ABET 2000 will facilitate PD implemented remains to be seen.

Acknowledgments

The authors are grateful to many University of Virginia colleagues, especially Edmund P. Russell, Richard D. Jacques, Kristin A. Gildersleeve, William J. Thurneck, and C. J. Livesay for their involvement the 1996 paired sections of TCC 101 and ENGR 164.

References

1. Pfaffenberger, Bryan John P. O'Connell, Susan Carlson, Timothy C. Scott, and Mark A. Shields. Teaching professional development in the first-year writing course. ASEE Annual Conference, Session 2653, Washington, D.C., June, 1996.
2. Shields, Mark A., and John P. O'Connell. Professional development and collaborative teaching in an undergraduate engineering curriculum: A case study from the University of Virginia, 1997.
3. Copies of the full report can be obtained by contacting John P. O'Connell.
4. Peterson, George D. ABET Engineering Criteria 2000. Accreditation Board for Engineering and Technology, Inc., Baltimore, MD, 1995.
5. Pfaffenberger, Bryan and Mark A. Shields. Integrating communication skills, teamwork, and engineering literacy: The ECOP project. ASEE Annual Conference, Session 2561, Milwaukee, June, 1997.
6. Shields, Mark A. Enhancing cross-cultural understanding among engineering students: The Technology and Human Development Project. ASEE Annual Conference, Session 2660, Milwaukee, June, 1997.
7. Briggs, Katherine C. and Isabel Briggs Myers. Myers-Briggs Type Indicator™ Form, 1977. G Booklet. MBTI™ Form G Profile Report. Consulting Psychologists Press, Palo Alto, CA (1984).

8. Board of Trustees of the Leland Stanford, Junior, University, Strong Interest Inventory of the Strong Vocational Interest Blanks.™ Consulting Psychologists Press, Palo Alto, CA (1994).
9. Woods, Donald R. and P. E. Wood. The future of engineering education: A Canadian perspective. Presented to the New Approaches to Undergraduate Education VIII Conference, Kingston, ON, July 23–27, 1996.
10. Florman, Samuel C. Learning Liberally. *PRISM*, October, pp. 18–23, 1993.
11. Kranzberg, Melvin. Educating the Whole Engineer. *PRISM*, October, pp. 26–31, 1993.
12. The Green Report: Engineering Education for a Changing World. 1994. Report of the National Advisory Council, American Society for Engineering Education. http://www.asee.org/pubs2/html/green report.htm.
13. Shields, Mark A. 1998. Collaborative teaching: Reflections on a cross-disciplinary experience in engineering education. ASEE Annual Conference, Session 2461, Seattle, June.
14. Shields, Mark A. and John P. O'Connell. 1998. Technological capability: A multidisciplinary focus for undergraduate engineering education. ASEE Annual Conference, Session 1261, Seattle, June.

CHAPTER 26

O. ALLAN GIANNINY, JR.

A Century of ASEE and Liberal Education

(or How Did We Get Here from There, and Where Does It All Lead?)

Introduction

Why do we wish a retrospective look at humanities and social science in engineering education? It is a common ritual in celebrations such as the 100th to assume that we learn something about ourselves when we look over our past. Today, the Liberal Education Division has two sessions, one looking at our past, the other at the future. Both are fascinating conventional pastimes, and we may learn something about ourselves in the process. As always, this division functions within the context of the society, and we hope to have some relation to the broader interests of the field that we all serve. I have little notion that we can do more than caricature the past—or the future, but we may entertain ourselves with comments on our illusions.

To add a personal note, I attended my first meeting of ASEE in 1959, the good ole days, even the time of Camelot. At the time, I thought we were just ready to climb the great mountain, to realize the promise of a new era. Little did I suspect that we were already at the peak, perhaps a little past the peak, or maybe on a shoulder of the mountain. T. S. Eliot wrote,

> We had the experience but missed the meaning,
> And approach to the meaning restores the experience
> In a different form, beyond any meaning
> We can assign to happiness.[1]

and it would be a long time before there were better days.

In another session at this conference, Professors Terry S. Reynolds and Bruce E. Seely have noted that from the beginning, the objectives of ASEE have been to improve instruction in the classroom and to seek recognition from other professional societies and from government agencies as the spokesman for engineering education.[2] In this Division, the goals were slightly different. Our predecessors and we have sought to improve instruction at the classroom level but recognition could come only from engineering colleagues. Long ago, it became clear that the parent disciplines from which we come consider our efforts as misguided or irrelevant. For example, despite a century of effort in relation to engineering, there is no recognized genre of technical prose recognized by the Modern Languages Society of the National Council of Teachers of English.

To set the tone for this discussion, let us not forget the recent origin of the name, "Liberal Education Division." This name was chosen in the early 1980s to replace a somewhat more modest one, "The Liberal Studies Division." The name change is understood more readily when we recall that leaders of the division worried about irreverent use of acronyms. In an era marked by concern over political correctness, the acronym LSD seemed far more threatening than LED, which also had something of a jazzy, upscale, technical aura. The Liberal Studies Division (LSD) was formed in 1965, bringing together two former units, the English Division and the Humanities and Social Sciences Division, newly renamed from the Humanistic-Social Division, also dubbed irreverently, IUMSOC.[3] Before we get too sensitive about names, let us also remember that when ASEE was the Society for Promotion of Engineering Education, the irreverent called it the Society for Prevention of Engineering Education.

The formal organization of these divisions had evolved over years of exchange between both engineers and scholars from the disciplines represented. Every name change elevated the rhetoric, each time we claimed more profound status for the work. We moved from a description of the underlying disciplines to a defensive announcement of our goals. I do not think this is incidental. Nor is it trivial. Instead, it expresses a mood and an expected outcome. It also states a relationship with engineering areas that we may not wish to defend. If engineering disciplines were technologically determined, the Liberal Education Division was driven by social norms. This suggests that the role of our disciplines was not credited; we attempted to make up for that weakness by politicizing our name—and our position.

Let me add a further note. In education, unlike politics, it is almost impossible to argue against being *liberal*.

The Parent Divisions

In 1942, the Society elevated the English Committee to division status, and in 1945 the Humanistic-Social Committee became a Division. Though some members wished to combine the two groups, the English teachers objected, arguing that the two units could get two seats on the governing board of the Society and, hence, wield greater political influence in the Society. They prevailed, and the divisions remained separate for two decades.

At the end of World War II, many things about the Society changed, including the name. The Society for Promotion of Engineering Education became the American Society for Engineering Education of 1946. Some critics have placed the beginning of modern engineering education as occurring immediately following that war. Professor Alvin M. Foundation described the changes that affected the English Division.

The roaring whirlwind was the nationwide self-appraisal of the engineering profession, and the resultant emphasis on the "humanities," much of which lay outside English courses, especially in writing and speaking, and turned toward the more general courses in literature, sociology, economics, history, psychology, and the like. Many of the previous discussions (in the English Division) were suddenly dropped from the normal agenda, and the whole English Division approach had to be much changed or else to disappear entirely.[4]

Marriage of the two divisions had long been debated, but there were wide differences of opinion as to objectives. Members of the English Division saw their task as closely related to the professional work of the engineer; hence, they stressed a rigorous instruction in writing and speaking as these fields related to the profession. This utilitarian concern had been accepted as a necessary *skill* from the inception of the Society for the Promotion of Engineering Education in 1893.

However, members of the English Division brought from their parent discipline the understanding that the development of fluency and structure in a language requires practice in widespread reading; they also had to teach seminal literature of western civilization. Literature, whether fiction, philosophical, or poetic, they argued, gave models (some even used the term "case studies" of leadership, conflicting loyalties, and human responsibility of their actions).[5] These were the conditions that engineers would meet in their careers, and literature contained the metaphors and idioms of western thought. Members of the division sometimes used this subtle argument to justify their own teachings.[6] Language study embodies a worldview—elements of cosmology, epistemology, aesthetic, and ethical-base for understanding and using language. Embedded in the classical humanities, however, is the antitechnological proposition that human nature is largely unchanging; progress is an illusion. The engineering faculties wanted language to be understood as a tool. They were no more interested in the depths of the discipline than they were in the depths of physics and mathematics. They were interested in efficiency of instruction, and from their point of view the objective was to teach the rudiments of grammar, spelling, and punctuation.

HUMSOC

HUMSOC professors were expected to teach the historical, social, institutional, and economic aspects of engineering needed by engineer-managers—what came to be known in the 1920s and 1930s as "the second stem" of engineering education. Engineers expected these studies to undergird and support the technical points that needed to be made. HUMSOC faculty, on the other hand, saw their educational objectives as contributing to the "liberal" education of students. They were interested in the critical aspects of the history of technology, the history and sociology of science.

People like Lewis Mumford, Derek de Solla Price, and a generation of scholars were attempting to define the role of the engineer in society. Many of these scholars attempted to answer the question, "How did we get to where we are?" In some ways this was as difficult a question for engineers to answer as were those addressed by members of the English Division. There was a different philosophical posture; however, the social scientist was expected to share the engineers' notion of *progress* derived from the Enlightenment and Protestant Christian roots of engineering. They attempted to find a rationale for ideas of the improvement of the human condition.

Scientists who developed engineering in the nineteenth century seldom doubted that they were improving the spiritual potential of humanity by improving first the physical conditions under which men lived. This was the great romantic dream. In one sense, professors identified with the HUMSOC Division could show the development in benign ways, generally supporting the venture and the state of being to which engineers were leading the nation. The saving feature was that failures could be explained in terms of errors along the way to a higher order. Even as Aldous Huxley's *Brave New World* (1936) would not destroy the image—it was a fantastic fiction. Only after World War II was this position seriously challenged—in the terror of future warfare with nuclear and other hideous weapons of mass destruction.

This brief and perhaps oversimplified account sets the stage for combining the two units into a single Division. This *division* within the Division has been a spawning ground for academic offspring including such movements as the Society for the History of Technology; the National Association for Science, Technology and Society; the Society for Science, Technology and Human Values; the short-lived applied humanities movement; the Congress for Understanding Technology in Human Affairs; The Society for Literature and Technology; The Society for Liberal Arts and Technology; The New Liberal Arts; the Popular Culture Society; and others that are, if not offspring, at least the godchildren of the ASEE activity.

In some ways, the derived groups may be the most important contribution of this Division to higher education. Even developments that were unattractive to engineering faculties and did not support the prestige of humanities and social science faculties in engineering schools, grew, some initiating their own disciplines. We note these developments even as we address the more direct efforts within ASEE.

We have already foreshadowed the point that engineering educators have changed the emphasis of what is now the LED. In the earliest days, the Society emphasized utilitarian and pragmatic issues. Because engineers were called upon to place their ideas before industrial managers and government officials, they needed to understand the conventions of written and spoken language. They had to know the conventions of language precisely enough to exchange exact information with other engineers. In addition, they had to show managers and financial supporters the benefits of their work.

The nineteenth-century students of engineering in America were relatively homogeneous in background. They were children of mechanics and farmers; they needed to learn to work on tasks that were described only through the new sciences. The fields could not be learned by apprenticeship because the learning method took too long, and it lacked uniformity. The new engineers were needed in companies that required standardized approaches that could be understood only in quantitative terms; railroad engines had to be built, serviced, and repaired to material standard by people in many different locations. Tracks, bridges, and road beds had to be built to standards; electric power systems and telegraphy had to be compatible. Only by mastering the discipline of the systems could one be innovative or creative—the announced goal of engineering education.

In 1893, the Society for Promotion of Engineering Education was founded as an outcome of the World's Engineering Congress, held in conjunction with the Chicago exposition honoring the 400th anniversary of Columbus's landing in the new world. At that meeting, two of sixteen papers dealt specifically with topics related to the division today. One was on technical writing and the other on search and reading of technical literature. Both were given by engineers, and there appears to have been no representative of the humanities or social sciences in attendance. This is hardly surprising because engineers were making every effort to create a new form of higher education, centered on technology rather than on classical ideas or religious doctrine.

Engineers were interested in writing for a very practical reason. An engineer could no longer be a master craftsman, standing at a workbench and telling or showing the leading worker precisely what he or she wanted done. Engineers had to reduce their ideas to paper using English, mathematics, and graphics to convey their expertise to numbers of workers scattered over wide areas. Engineers needed to understand the conventions and the structure of these language systems so they could use them precisely and accurately.

In 1918, when Charles Riborg Mann published the first national study of engineering education, he described the early interests in engineering education that had led to the 1893 congress:

> It appears that from the beginning the engineering schools have had a clear conception of their functions. They themselves understood that their aim was increased industrial production, and that their special contribution to this end was systematic instruction in applied science. In addition, they believed that if this instruction were given in the proper spirit, engineering would become a learned profession and scientific research a recognized necessity.[7]

A generation later, William E. Wickenden wrote that engineering education in America was an outgrowth of a popular nineteenth-century movement to promote "the application of science to the common purposes of life."[8] Wickenden noted that engineering education grew in an autonomous fashion in America, *independent of statesmen and active practitioners, but controlled by scientists and educators*. The consequence, he claimed, was to make American engineering education widely accessible but a with narrow concept of the professional role, and lacking the support of programs of technician and apprentice training. This pattern led to a relatively poor definition of the engineering school. Without control or plan, Wickenden noted that engineering education had developed, between 1870 and 1900, a remarkable similarity, as imitations of early institutions like Cornell and The Massachusetts Institute of Technology.

SPEE was founded at the World's Congress of Engineering, called into session as part of the World's Columbian Exposition at Chicago in 1893. The discussions extended over a week, divided into groups for the different types of engineering. Among these groups, "Division E" addressed the theme of the education of engineers, and at the end of the week, this group decided to organize as the Society for the Promotion of Engineering Education (SPEE). This organization had its origin in a period of rapid growth of engineering education. Only thirty years earlier, at the end of the Civil War (War Between the States) had engineering education received its impetus from the first Morrill Act, which established the land grant colleges. Students of engineering education in that era report that between the years of 1867 and 1872 the number of engineering schools in the country grew from less than twenty to more than seventy. By 1893, the number had grown to well over a hundred. In this short period of time, engineering schools had experimented with different patterns of education and different courses. The professors who attended the 1893 meeting were the senior professors in the first full generation of engineering educators in America. Many had studied areas other than engineering as intensely as they had studied engineering. Most engineering programs were grounded in applied mathematics or in shop, and the schools adopted the posture of turning out students who would lead industrial development.

Professor Fountain noted that in 1893, the engineering schools were offering "scattered and fragmented courses of study." Their emphasis on mathematics and physics left humanistic subjects like literature and writing almost completely out of the curricula. Despite this pressure, however, J. B. Johnson, a professor of civil engineering at Washington University, presented a paper on "Methods of Training Engineering Students in Technical Literary Work." Johnson, who had served as the first editor of *The Engineering Index,* used "Literary" or "literature" to refer to any written material. His thesis was that engineering students should learn to use knowledge "where they find it," which meant from the literature. A second paper on engineering writing, on the presentation of engineering theses, was presented by Professor Gaetano Lanza, from the Massachusetts Institute of Technology. The two papers drew extended discussion.[9]

Thus, at the first meeting of SPEE, the paper included discussions of writing in the context of engineers' development rather than in the conventional "liberal arts" disciplines of the day. This first generation of engineering educators had come through schools like MIT and Cornell, in which many texts were imported from Europe, and though the professors in those institutions had been trained in a classical mode unknown to their students, they seldom did not consider it essential to their students.

Attention to writing continued, with papers on these topics about every second or third year. In 1896, the SPEE annual meeting in Buffalo, New York, included papers on ethics, modern language, and biology. Nevertheless, the president's address indicated that the curriculum was so crowded that "cultural" courses might have to be abandoned altogether. In 1901, a number of speakers called for better writing in engineering journals.

At the 1903 convention at Niagara Falls, T. J. Johnston, patent attorney from New York City, presented the first paper on "Engineering English." Mr. J. A. L. Waddell, a consulting engineer from St. Louis, presented a paper on the history of engineering as material for instruction. He and his colleagues at that meeting prepared a list of 340 book titles related to the history of engineering. In 1904, the annual meeting was held at St. Louis, in conjunction with the world's fair commemorating the centennial of Lewis and Clark's exploration of the American West, and at that time, several new members joined the Society, including editors and deans of liberal arts departments.

Debate over the curricular time assigned to English and humanistic studies continued. In 1905, the President of SPEE stated in his presidential address that he hoped that the time would soon come when all course requirements of English and foreign language could be abandoned.[10] There was little or no discussion of other disciplines in the humanities; they were considered to have little merit to the engineer.

First Professors of English Join SPEE

Just three years later, in 1908, the Society attracted interest among teachers of English. Assistant Professor J. Martin Telleen, from Case Institute, became the first professor of English to be admitted to the Society. His paper, "The Courses in English in Our Technical Schools," noted a lack of uniformity in English requirements for engineers and in the qualification of teachers. He argued that engineering students needed also to conduct literature research and to study artistic literature. Other speakers in the meeting discussed the need for training in public speaking and in writing for theses, which were required in most schools. In that same year, the University of Cincinnati instituted a cooperative program that led to active efforts for better training in English for engineering students. Editor T. A. Rickard published *A Guide to Technical Writing*, the first book on that topic.

In the years immediately preceding World War I, the Society showed increasing interest in writing instruction. In 1911, Professor Earle of Tufts College completed the first textbook for classroom use, *Theory and Practice of Technical Writing*. His paper, "English in the Engineering School at Tufts College" was presented by Dean Anthony of Tufts, who reported that Earle "was too busy completing his text to attend the convention." Attention to Professor Earle's topic was so great that notes and minutes of the sessions occupied sixty pages in the Proceedings. A year later, Earle attended the conference and reported on a study of practice in technical departments, noting that some colleges and universities required no English at all. In other schools there was an increasing tendency to teach English to engineers in sections separate from liberal arts majors.

In 1914, the SPEE established a system of committees to organize its work. Committees were established on the professional specialties, but also English. In that same year Mrs. Katharine Brown became the first woman teacher of technical writing to join the Society. Mrs. Brown was a professor of mathematics at Drexel Institute, and Professor Fountain thought she was more prominent as a writer of bridge columns in the newspaper than as a professor of technical writing.

By 1916 when the Society met at the University of Virginia, twelve English professors were members of SPEE, and some nonmembers, mostly from the host institution, attended. Several new members would become active in the Society during the period between the world wars, including Professor H. A. Watt of the University of Wisconsin, J. Raleigh Nelson of the University of Michigan, Miss Sada Harbarger of the University of Illinois, and C. Alphonso Smith of the University of Virginia.

In 1919, Miss Harbarger moved from the University of Illinois to Ohio State, where she was given charge of all English instruction to engineering students. Other new members were Karl Owen Thompson of Case Institute and Ray Palmer Baker of Rensselaer Institute. A year later, Professor H. L. Creek became a member of the English faculty at Purdue University and a member of the Society. His presence in SPEE marked the first purely land-grant college to have active members in the English group. All of the other English teachers to this time had taught in either private institutions or in universities with separate classes for engineers.

During this period, the English professors appear to have focused their attention on the promotion of English instruction in general. There seem to have been few standards of acceptable writing related to engineering except for the conventions: grammar, spelling, and punctuation. The professors lamented "the poor writing of engineering students" and spent a great deal of effort in the teaching of conventional mechanics of language. In their one concession to issues peculiar to engineering writing, they emphasized development of standard nomenclature for technical papers.

By the end of World War I, as engineering graduates were advancing to management levels, faculties were hearing from corporate managers that they should add courses in management, accounting, engineering economics, and corporate law, as well as English. This was the beginning of the engineering educator's concern about *social studies*.

The *Report of the Investigation of Engineering Education, 1922–1929*), directed by William E. Wickenden, summarized that need. The development of greater social insight and a larger sense of social responsibility should be a definite objective of the engineering profession if it is to gain recognition for more than its technical proficiency. At the same time it has need to improve its proficiency in dealing with problems of economy as related to the technical problems of engineering. The ability of the engineer to extend his influence in industrial organizations and in public life, to claim his due share of leadership and to discharge more adequately his function in society at large, now appears to hinge primarily on his attainment of greater competency and greater recognition on the economic and social side of his work. The public has ample confidence in his ability to deal with the material problems.[11]

Wickenden's report recognized two tracks for engineering study, one that would prepare graduates for technical careers and another for those students headed for managerial roles. This was the beginning of the notion of "two stems" of engineering education. Within a decade, SPEE reports would alter the concept beyond recognition.

First Meeting of English Teachers at SPEE

Fountain called the convention of 1922 "a major landmark of the English teaching within the Society." For the first time there was a separate meeting of English teachers. More than thirty professors attended, representing twenty universities and colleges; inquiries were received from sixteen other schools that were not represented in person. Many deans and department heads attended, and the English Committee was able to establish a regular feature column, "The English Department" in the bulletin of the Society.

The first English texts for engineering students were just gaining attention. These included H. A. Watt, *The Composition of Technical Papers* (1918), followed by three others in Karl Owen Thompson, *Technical Exposition* (1922); Miss Harbarger and her colleagues, *English for Engineers* (1922); and Ray Palmer Baker, *The Preparation of Reports* (1922).

In 1922, SPEE initiated the first nationwide study of engineering education, under direction of Wickenden, a mechanical engineer, Vice President of AT&T, and former dean at M.I.T. The first part of Wickenden's report, issued in 1926, advocated a system of "humanistic-social" studies throughout the four-year curriculum. His concept of humanities was limited to the materials and topics that supported engineers as managers in industry and government: accounting, business law, engineering economics, and engineering management. In a side comment, the report also noted that students at both Case Western Reserve and Purdue frequently took courses in literature without credit.

In 1926, the Society gave more recognition to the work in English as Miss Harbarger became the first teacher of English to be named to the General Council of SPEE. She served in that capacity for three years. English teachers were also giving more attention, as more than a hundred teachers attended an English conference at the University of Iowa, just before the regular convention of the Society. Enthusiasm was high among English teachers, but they received little tangible support from the Society. For example, in 1927, Miss Harbarger reported a national survey of freshman composition across the nation, funded at her own expense except for about $30 contributed by SPEE. Professor Fountain contrasted that level of support with a $30,000 allocation from the Carnegie Foundation for a study of the advisability of establishing technical institutes similar to those that Wickenden had seen in Europe.

The emphasis of the English teachers was changing, however. The papers began to reach beyond the narrowly utilitarian "technical writing." In 1928, the theme of papers of the English Committee was "English as a Developmental Experience."

In 1929, the theme was the problem of speech training for engineers. The meeting was at Ohio State, and, in a new venture, Miss Harbarger invited graduate students to participate. One paper was "English for Consulting Engineers," directed to the graduate students in attendance. Only the second textbook on report writing appeared that year: *Report Writing*, by C. G. Gaum and H. F. Graves, of Pennsylvania State College. There were also two general texts: *English and Science*, P. B. McDonald and *Expository Descriptions*, by V. Solberg.

The 1930 convention was the first to be international in nature, as it was held in Montreal, with participation from McGill University and Ecole Polytechnique. There was considerable discussion over the bilingual effect in the French school, where all texts were in English, but all lectures were delivered in French.

All this was prelude to a major summer school for English teachers at Ohio State University, in the summer of 1932, widely known as "The Great Conclave."[12] The conclave was made possible in part by the special interest of Wickenden in the teaching of English. Outstanding teachers participated as speakers and discussants in a spirited exchange. A highlight of the session was the development of an extended reading list, recommended for engineering students, Good Reading, edited by Atwood Townsend.[13] This major source book had a wide appeal, well beyond engineering students. It was adopted and revised regularly by the National Council of Teachers of English as a recommended source for all college students. Its publication in paperback continued until well past World War II. This was the first significant contribution from the English teachers of engineers to the broader academic community.

Another step in the recognition of the role of English occurred in 1934 when the first English sessions were held at the general convention. The Great Conclave accelerated the changing role of English professors, as they began to stress attention to wide reading experience as important to the writing and speaking performance of engineering students. These changes brought a convergence of interests among the teachers of English and advocates of a broader program of humanistic-social studies. There were some who wished to combine the two groups. However, the English teachers were resisting the forces beginning to suggest a merger of the two groups, relying on their mandate as teachers of necessary professional skills. They distrusted efforts to teach humanistic-social topics as both lacking in discipline and as efforts by the arts and science departments to draw engineering back toward the classical education.

Wickenden's report summarized curricular trends since 1870 in several areas of humanities and social sciences, included in engineering curricula: economics, history, foreign language, and English. In general, the time allocated to foreign language had been reduced from more than an academic year in 1870 to about one-eighth of a year in 1923. Economics had increased slightly at the expense of history, and English had held steady at approximately one-fourth of a year.

Accreditation of Curricula

Following the Wickenden study, the engineering professional societies initiated their first efforts to standardize the entry point into the profession by controlling engineering curricula. Dugald C. Jackson contrasted the studies by Mann and Wickenden, stating that Mann's study had emphasized the personal qualities developed by the students, whereas the Wickenden report stressed teaching personnel and curriculum.[14] The Engineers' Council for Professional Development (ECPD) grew almost directly from the so-called Founder Societies' response to the study led by Wickenden. The new Council had three tasks.

- To encourage pre-college programs to prepare students for the study of engineering.
- To accredit engineering curricula, thereby establishing a recognizable entry point into the profession.
- To support continuing professional development of newly graduated engineers during their first years in the workforce.

Whereas previous engineering curricula had developed initially from a loose consensus, the new accreditation program would assure a minimal level of competence by standardizing and upgrading the undergraduate curricula. The new criteria for accreditation included a statement about competence in writing and speaking, formally recognizing English instruction within the general frame of engineering education.

Accreditation affected English teachers also. Anticipating new accreditation criteria, subsequent English conferences were driven in two somewhat different directions. In the Society's meeting at Atlanta, there was a sentiment toward more work in literature and less in such fundamentals as spelling and composition. On the other hand there was pressure to develop stronger programs in English Departments to train students to teach writing.

At the Atlanta meeting, Dr. Creek of Purdue became the new chairman of the English Committee, replacing Miss Harbarger. He had come to believe that the task of teaching English to engineers required a major new effort in teacher training. He got permission from the SPEE council to initiate an extensive study of such teaching, to be supported by funds already promised from the Carnegie Foundation. The second approach was to provide special training and encouragement for young men just entering the profession. He had in mind the cooperative participation of a technical school with a liberal arts university, proposing that the Massachusetts Institute and Harvard University initiate such a joint program.

Creek persuaded the Society to establish a major committee, including professors from the liberal arts, to direct the national study. The prestigious committee was chaired by E. C. Elliott, President of Purdue University. Members included John G. Bowman, Chancellor of the University of Pittsburgh; H. S. Rogers, President of Brooklyn Polytechnic Institute; Harold Burris-Meyer, Dean of Humanities at Stevens Institute; A. M. Greene, Dean of Engineering at Princeton University; O. J. Ferguson, Dean of Engineering at the University of Nebraska; and John Mills of the Bell Telephone Laboratories.

To address the direct question of teaching English, the Committee established a regular column, "English Notes," in the Society bulletin, edited by Joseph L. Vaughan of the University of Virginia. The column was intended to build a strong case for instruction in English based on testimonials of presidents of major corporations and universities. For five years, until World War II ended the series, it presented regular contributions from industrial leaders, deans, and an occasional English professor. Creek led a separate investigation of English courses in engineering schools in the United States, with Vaughan as Vice Chairman and Greene as a member. In 1939 the results of these studies appeared in monthly installments in "English Notes."

Committee to Division

English

In 1941, the English Committee petitioned the Society to become a Division. There was a strong undercurrent of sentiment for the English Committee to join with the teachers of other humanistic courses to establish a broader Division, but the English professors preferred a separate division, and their position prevailed. A year later, in 1942, the change in status took place and Vaughan was elected the first Chairman of the new English Division. When he ceased to edit "English Notes" the column was discontinued. Creek's grandiose plan for a special training program to train English professors to teach engineers was a casualty of World War II and was never revived.

Hum-Soc

In the meantime, a group of faculty teaching humanistic courses were also seeking division status. They received funds from the Rockefeller Foundation to hold a special conference at Princeton University. The new Humanistic-Social

Division was formed in 1945, an outcome of a major study undertaken by Professor H. P. Hammond and published in 1940.[15] Hammond's study had argued for two "stems" of engineering education, the *scientific-technical* and the *humanistic-social*. By the end of World War II it was clear that the paths of the English Division and the Humanistic-Social Division were converging despite the objection of many members of the English Division. Both the intellectual interests and the social structure of engineering were driving the two areas together. Only the special interests of the leaders of the English Division were holding the two from merger. Individuals like Dr. James R. Pitman of Newark College of Engineering were highly active in both divisions and were working to bring the two together. J. L. Vaughan recalls that he and others had opposed a merger of the two Divisions, arguing that in the political environment of the Society, the two closely allied divisions could be more influential than if they had only a single voice.[16]

Post–World War II

Some critics have called the end of World War II the beginning of modern engineering education. At the end of World War II, the name of the society was changed from the Society for the Promotion of Engineering Education to the American Society for Engineering Education (ASEE).

Engineering education was undergoing a major self-appraisal, as was the nation. Wartime success of technical developments led by physicists—radar and nuclear weapons, for example—made clear that engineers needed more science in their education, simply to deal with a new level of technical sophistication.

For the past forty years, there have been regular studies of the role of humanities and social sciences in engineering education. There have been grants from foundations and from governmental agencies to initiate new curricula. During that same period there have been relatively ineffectual efforts to develop this part of the curriculum in as orderly a form as found in the technical areas. Yet the curriculum has proved to be rugged and stubborn in its resistance. One feature has been the widespread requirement that all engineering students include approximately 13 to 20 percent of the undergraduate curriculum in this area. This minimal requirement has been written into the criteria for accreditation of curricula, but in actual curricula it receives minimal attention. Beyond the quantitative requirement there has been little agreement as to what should be taught in that area and even less quality control. Add to these issues the reluctance of engineering faculty to take aggressive control of this part of the curricula and the retreat of humanities and social science faculties from active participation, and the current void is understandable.

There was also a new emphasis on the "humanities," stimulated largely by the shock of the war, including the use of nuclear weapons at its end. English courses stressing writing and speaking appeared to be narrowly confined when confronted with more general courses in literature, economics, sociology, history, psychology, and the like. The result was that topics related to teaching techniques and methods disappeared from the usual agenda. As we have seen, Fountain noted that the English Division faced a rapid change. Through many people participated in both, membership in the English Division grew more rapidly than that in the Humanistic-Social Division.

Fountain called the era "The Age of Idealism," in which industrialists and university administrators brought about change in the shape of universities. Barriers between programs were demolished and teachers began to approach their students through "the Great Books" inspired by the University of Chicago and other liberal arts institutions. Fountain wrote, "Almost overnight, engineering students found themselves studying psychology, philosophy, or even religion instead of grammar and spelling." The attempts to broaden the intellectual base of education clearly made Creek's prewar plan for the special instruction for English professors in engineering schools obsolete. Fountain wrote, "the time had passed for such specialized training; the philosophy of the period was taking further acceleration toward what was coming to be called 'general studies.'"

In 1950, Creek wrote a history of the English Division, "Working Backward and Forward on English in ASEE." He gave a hopeful view of the future, warning, however, against allowing courses in civilization from becoming indoctrination, and pleading that imaginative literature be retained in English courses, and that composition courses have a large measure of humanistic quality.

L. E. Grinter edited the major report in the series, *Report on the Committee on Evaluation of Engineering Education,* published June 15, 1955. The report recommended:

> Strengthening work in basic sciences,
> Identifying and including six engineering sciences,
> Integrat[ing] study of engineering analysis, design, and engineering systems,
> Including elective subjects
> - to develop special talents of individual students,
> - to serve the varied needs of society,
> - to provide flexibility of opportunity for gifted students.
>
> Continuing a concentrated effort to strengthen and integrate work in the humanistic and social sciences into engineering programs,
> Developing a high level of performance in the oral, written, and graphical communication of ideas,
> Encouraging experiments in all areas of engineering education,
> Strengthening graduate programs,

Maintain[ing] faculties at a high level of professional and intellectual capacity, and

Prepare for a major increase in engineering enrollments

About instructional goals, the report said:

> ... the engineer should be a well-educated man. He must be not only a competent professional engineer, but also an informed and participating citizen, and a person whose life expresses high cultural values and moral standards. Thus, the competent engineer needs understanding and appreciation in the humanities and social sciences as much as in his own field of engineering. He needs to be able to deal with the economic, human and social factors of his professional problems. His facility with, and understanding of, ideas in the fields of humanities and social sciences not only provided an essential contribution to his professional engineering work, but also contribute to his success as a citizen and to the enrichment and meaning of his life as an individual. . . . This requires that the faculties of the humanities and the social sciences regard the teaching of engineering students as challenging and rewarding, and that engineering faculty members adopt an appreciative and understanding attitude toward their colleagues in the liberal arts.[17]

The language of this report appeared in the accreditation criteria were subsequently used by ECPD (and later by its successor the Accrediting Board for Engineering and Technology, ABET), spelling out the acceptable and unacceptable courses that embody the content they supported in the general statement.

The fields of humanities and social studies from which some courses must be selected include history, economics, and government, wherein knowledge is essential to competence as a citizen; and literature, sociology, philosophy, psychology, and fine arts, which afford means for broadening the engineer's intellectual outlook. The Committee has found no reason to disagree with the recommendations of previous ASEE Committees that about one-fifth of the curriculum should be devoted to humanistic and social studies.

Such nonengineering courses as accounting, management, industrial finance, marketing, and personnel administration may well be valuable components of a particular curriculum, but being essentially technical in content, they do not adequately fulfill the main purpose of the program in humanities and social studies.

> The course should be designed to liberate him from provincialism, whether geographical, historical, or occupational and to give him a sense of the satisfaction that he can gain later in life by adventuring more deeply into the areas of ciritical and creative thought represented in the humanities and social studies. His capacity to make sound qualitative judgements should be developed so that he may distinguish that which is good from that which is mediocre.[18]

General Education in Engineering, The Gullette Report

A year later, ASEE published its most ambitious planned curricular plan for the humanities and social studies: *General Education in Engineering, A Report of the Humanistic-Social Research Project*. The project, directed by Professor George A. Gullette, head of the Department of Social Studies at North Carolina State College, anticipated close cooperation between faculties of engineering and liberal arts to develop a *"designated sequence of courses extending throughout the four undergraduate years"* [emphasis added].

The report was prepared by a committee largely composed of teachers in the humanities and social sciences but who were also active in teaching engineering students. They were clearly intending to remove the aura of "service courses" from their area. In doing so, they also removed the direct interest of engineering faculty. The curriculum would be split between *professional education*, which would include all of the mathematics, science, and engineering courses, and *general education*, which was the responsibility of the humanities and social sciences. By implication, the report left language composition and speech in an academic "limbo" somewhere between the two areas. After review and implied acceptance by representatives from industry, engineering faculties, and colleagues in the humanities and social sciences the report was released. Professors could participate with enthusiasm in creating general statements of policy yet do nothing on their own campuses to carry out those statements. They identified a number of courses that illustrated possible points of departure in development of new programs.

A few schools were innovative, some because of principle, others sought some advantage to balance weaknesses in their own institutions. The largest influence, however, probably came from the conversion of land grant colleges to full-fledged universities. The dominant model was either the comprehensive state university or the elite private university. They quickly turned their attention to development of conventional programs in the humanities and social sciences. The proposed integrated programs materialized only in part and in a handful of schools.

At about the same time, ASEE issued a series of reports related to other topics, including graduate studies in engineering and the engineering sciences. Taken together, the reports describe a major reinvention of engineering education in America. However, the proposals about humanities and social sciences were carried out in only a few campuses. Competition for time in the curriculum, and priorities for increased science and mathematics, claimed the attention of engineering educators.

One sign of the difficulty faced by the humanistic-social researchers is the defensive statement of philosophy of the course sequence: "The humanities and social sciences must be understood as professional disciplines in their own right." Engineers avoided use of the term *science* to denote what is now called social or behavioral science. Cautiously, their writings referred to *social studies*. *Humanistic* sounded like a doctrine, derived from *humanism*.

Failure of the schools to follow the recommendations of the Gullette report also arose from the fact that there were no disciplines—no communities of recognized scholars to support the most innovative proposals. Despite the plea that disciplines in the humanities and social sciences had to be recognized as professional in their own right, there were no disciplines to undergird some of the proposals in the report. The special programs cited were largely the product of individual faculty members who had developed their own cross-disciplinary interests. As a result, several groups of professors began to develop groups to investigate some of the underdeveloped areas.

Many of the humanists were soon either retired or were engaged in developing degree programs for arts and sciences students in their own new universities. Other scholars began to organize research groups that would soon grow into entire scholarly societies and to new disciplines, relating technology to the humanities and social sciences. While these disciplines became sources of exciting new scholarship, they had little effect on engineering education, and most were separated from ASEE.

A notable response occurred in 1958 when several members of ASEE established a group to study the history of technology. Melvin Kranzberg, John Rae, Thomas Parke Hughes, and Lynn White, Jr. started what became the Society of the History of Technology (SHOT). Thirty-five years later, the discipline of history of technology is joined by the sociology of technology, the philosophy of science, the history of science, popular culture, programs in science, technology and society, science, technology and human values, literature and technology, the new liberal arts, and other newer fields of disciplines for understanding and interpreting the work of engineering. None of these disciplines has remained intimately involved with engineering education, but all grew from the postwar environment that made technology more accessible to the general population.

Liberal Learning for the Engineer [1968]

For a brief period in the late 1960s, in the days when campuses were the centers of the so-called counterculture, some engineering students placed a strong emphasis on social awareness. Some faculties developed engineering programs that emphasized the political, sociological, and psychological aspects of engineering. Entire schools of engineering emphasized the social responsibility of engineering, drawing in persons from the social sciences. There were also attempts to introduce engineering ethics at a serious level that could attract attention of philosophers. Technological policy studies entered briefly on the scene, as did efforts to make the public aware of technological ideas and institutions. Yet despite these temporary glimmers, little has changed in the instruction of engineering students in the areas of humanities and social sciences.

In 1964-1966 ASEE conducted another study, intended to alter the nature of engineering curricula. The report, Goals of Engineering Education, published in 1966, proposed once again that engineering education should be a five-year program, with a preprofessional degree at the end of four years. The extra year of preparation would be used to increase the number of courses in science and mathematics, humanities and social sciences, and more sophisticated technical courses. As we know, the report had little or no effect on curricula.

The reasons for the general failure of the changes to transform engineering education are complex, and they lie beyond the scope of this discussion. However, there are a few hints. First, there was the external assault on engineering faculties from the outside. Some humanists took on a missionary zeal. People like Lewis Mumford had looked at modern technology and seen a monster. Technology, they believed, had gained a momentum that exercised unlimited contol over the minds of westerners.

Technological determination, they believed, ruled the world. They accepted the proposition that "whatever can be done, will be done." They embraced the distorted form of Franciscan thought expressed in the religious plaque, "Lord, give me the patience to endure the things I cannot change, the courage to change the things I can, and the wisdom to know the difference between the two." Note that the work *can* supplants the word *should* that expressed the Franciscan view. This was also the time when the western world was driven by a cold war mentality. Values like patriotism, capitalism, and entrepreneurism were unchallenged, and the political correctness could not be challenged. Many humanists also suspected that engineering professors wished, in the depths of their souls, that the humanist inquiries would simply go away, for it was in those areas that the "correct" values were questioned. Even in an academic forum such challenges could be disconcerting to the young men, and later women, who entered engineering study. Moreover, the police mentality needed to maintain security of research efforts looked unkindly on challenges to national loyalty.

The Olmsted report reviewed approaches to humanities and social sciences for engineers in terms of four sets of objectives:

1. *Utilitarian objectives*—"the importance of writing and speaking, along with the skills of managing people."
2. *Cultural objectives*—"an emphasis on subject matter—concepts, principles, and methodologies, either of separated disciplines or in an interdisciplinary program. The purpose [being] to induct the student into some larger segment of the culture."

3. *Developmental objectives*—focus attention less on subject matter than on the development of the person. [These objectives would present] matters on the humanities and social sciences not as ends in themselves but as means for developing in the student certain personal and intellectual qualities presumably not fostered to the same extent by the more technical parts of the curriculum.
4. *Contextual objectives*—"focus on the student in his role as an engineer . . . and the engineer as an agent of social change, who must live his professional life within a human context. An important purpose of the humanities and social sciences is to make him fully aware of this context, so that he may operate more effectively within it."[19]

While these objectives were not new, they made explicit the approaches to humanities and social sciences in engineering curricula. They codified the approaches acknowledging that many schools used some combination of them. The study team rejected the utilitarian and cultural approaches as falling short of the concept of "liberal education." Like the writers of the report a decade earlier, they advocated comprehensive programs that could draw on engineering content as an area for study. However, they found no comprehensive programs of the sort envisioned by the earlier team. Like Gullette, they found a few schools that offered one or two courses that attempted to relate humanities and social sciences to engineering, but there was hardly a "movement" in that direction.

They expected that from the study of disciplines of humanities and social sciences, students would gain an understanding of the cultural context for their engineering studies or could address a separate problem, the personal development of one who happened to be an engineer. By this time, engineers had persuaded foundations to underwrite curriculum development. Substantial funding came from the Alfred P. Sloan Foundation, National Science Foundation, and other sources for experimental curricular developments, but to little avail. Some schools were transformed by the infusion of funds. A good case is that of Worcester Polytechnic Institute, which was rescued from near financial disaster. By developing a new orientation to engineering, with integrated humanities and social sciences, Dean William Grogan and his colleagues were able to develop a program that remains a model for the report. Others were unable to continue their efforts after grants expired or current faculty were no longer available.

Punderson House Conference (1970)

In 1970, the National Science Foundation (NSF) supported a conference at the Punderson Manor House, in Newberry, Ohio, entitled Action Programs to

Implement the Olmsted Report, "Liberal Learning for the Engineer."[20] Echoing the conferences of the English group of the 1930s, the conclave brought together thirty-five invited leaders, representing seventeen schools, and three foreign observers, to develop concrete proposals for experiments at a variety of institutions. Interest in the conference was high, as shown by applications from thirty-six institutions. The conference was directed by Professor Henry Knepler and the secretary was Professor Dorothy Lambert, both from the Illinois Institute of Technology. Held only days before ASEE met at Ohio State University, this conference raised hopes of its leaders that, at long last, there was a concerted effort to alter the path of engineering education.

A new term arose in the report, "applied humanities and social sciences." The analogy to applied science was striking as a slogan but it did not survive. The term had been discussed in earlier meetings, but the most detailed statement appeared in a 1968 report by W. H. Davenport and J. P. Frankel, from the Department of Engineering, UCLA. The report presented a thoughtful, but largely pessimistic summary of the responses of engineering schools to both the Gullett and Olmsted reports. They offered a UCLA plan to engage students in liberal arts courses that would:

- give the student opportunities to develop the kinds of habit and attitude required for lifetime learning.
- enable the students to recognize, analyze critically, and to form wise opinions about important social problems. (The study of liberal arts should enable them to gain a broader understanding of human social organizations and their interactions with technology, and from that a more relaxed and more knowledgeable operation as a professional engineer.)

This program was proposed, but never reported as having been put into place.[21] Dorothy Mack Lambert, Chair of the Liberal Studies Division in 1971–1972, reported to the Division in 1972 a summary of membership in the Division. She was looking to see who might be the leaders of efforts to carry out the programs discussed at the Punderson House Conference. Her conclusion was that there was a very small number of workers. Using the newly computerized lists of membership, she counted 462 members, of whom 137 chose LSD as their first choice of a division connection. She called them the "core members." Of this group, 36 were presidents, deans, directors, or other administrators; 20 were department chairmen; 15 were professors emeritus; 11 were engineers; and 3 or 4 were graduate students. The remaining 50 to 55 were full-time teaching staff in the humanities of social sciences.

Of the 365 engineering and technical schools represented in ASEE, she noted, only 65 were represented, and only 14 schools had as many as 3 members. She listed them: Clarkson, IIT, Lowell, Michigan, Newark, Purdue, San Jose State,

Virigina, [University of] Washington, Worcester, Georgia Institute of Technology (Southern), Vermont Tech, Wentworth, and GMI.[22] It would be interesting to discover the background of the Liberal Education Division today.

Evaluation of Liberal Learning for the Engineer, Five Years Later (1974)

A follow-up study of effects of the Olmsted report in 1973 and 1974 showed that once again there was debate about powerful ideas, leading to "development of new courses in otherwise conventional schools of engineering."[23] Twelve projects, supported by grants from the Alfred P. Sloan Foundation, were attempting major alterations of curricula. Overall, however, most schools continued to accept the "two stem" concept, offering electives in the humanities and social sciences with little or no integration. The report concluded that the Olmsted Report provided guidance to those schools that wished to change their programs but had little effect as a motivation toward change. However, there was no sign of a national movement; educational institutions across the nation were reacting against the chaotic period of cultural unrest in the late 1960s. A distinct turn toward conventional disciplines and traditional curricula was already well under way.

Reevaluation of Engineering and Technology Studies (REETS) (1980)

Studies of the issues raised in the earlier reports were coming one after the other, almost without a break. Beginning in 1977, the ASEE Board of Directors instructed their long-range planning committee to develop a plan for implementation of recommendations of the REETS recommendations.[24] This was a Society document, reflecting the attempt by ASEE (described by Reynolds and Seely) to hold the ground as the recognized spokesman for engineering education. Though there was little consultation with institutions, once again, ASEE identified needs in education. Several topics from the Olmsted study were mentioned:

- curricular issues related to humanities and social sciences,
- technology assessment, and
- communication with the public.

All bore on the notion of integrated effort of humanities and social sciences with engineering. However, there were now few humanists or social scientists who were interested in engineering education; departments in colleges of arts and sciences turned their attention to other matters that introduced more autonomy and less risk to innovation in an area they did not control. Moreover, engineering faculties had little interest in making space for humanists in their research or on their faculties. Issues such as social awareness and responsibility were either being addressed by engineers or were being introduced in other forms. Even as the general society was pressing for recognition of such concerns as corporate liability for energy depletion, pollution, worker safety, and ethical responsibilities of engineers, engineering schools were treating these topics as technical topics if they were addressing them at all.

Unfinished Design (Association of American Colleges) (1988)

By the mid 1980s, virtually all efforts to establish special curricula to increase social sensitivity of engineering students had vanished. The standard engineering curriculum contained a minimum of one-half academic session in the humanities and social sciences. In most schools the engineering faculty had little or no control over the choices made by the students. Instead of seeking lofty ideals as expressed in the reports since 1955, the question faced by most students was "what is the minimal requirement." The courses were the standard offerings of the liberal arts departments. Very few were special courses that related engineering to the areas of interest. If there was to be no specially developed curriculum, could students be guided to take standard courses that would form coherent patterns?

In 1980, when ECPD had been reorganized, the accreditation assignment went to a new agency. the Accreditation Board for Engineering and Technology (ABET). The criteria for humanities and social sciences in the curricula called for a minimum of one-half year of the normal curriculum. In most schools this would be some fifteen to eighteen semester hours. Courses were to be chosen from a pattern that supported interests of the student, but should make a recognizable pattern. Students were still prohibited from including technical courses, ROTC (in most cases), skills courses, and business courses. The collection of courses could include a "distribution" of interests reflecting different disciplines, but at least one sequence of courses should reflect "depth" in a field. Such a loose set of criteria left a wide range of choice to the student and made the management of the program almost beyond reasonable control. Once more, ASEE attempted to take the lead in the educational area, this time with help from the Association of American Colleges. The approach was to be a soft hand of persuasion rather than the hard fist of control.

A consortium of ASEE, the Accrediting Board for Engineering and Technology (ABET, successor to ECPD), and the Association of American Colleges (AAC), undertook an action program from 1986 to 1988, directed again at the humanities and social science portion of the engineering curriculum.[25] This study was the first in a series by AAC to consider the role of humanities, liberal arts, and social sciences in the professions.

They undertook an elaborate statistical study of data from sample schools to determine what curricular choices the faculty thought students were making and what they were actually choosing. The deeper questions are those of students' motivation in their choices, influences on the choices, and the advising that entered into the effort. As might be expected, the choices made by most students were not those mentioned in the high-minded reports. Many were selecting courses that offered the least risk, the least effort, and the least educational benefit.

The project director, Dr. Joseph L. Johnston, from AAC set forth the results of this study in a report, Unfinished Design. This report was directed to deans and faculties who make decisions about the choices students must make. In addition, the team prepared a self-help handbook for use by students who received little or no advising help. Copies of the booklets were distributed to more than a hundred schools who sent representatives to a special conformance in Baltimore in 1968.

Conclusion

ASEE has come virtually full cycle in its first century, at least with respect to the humanities and social sciences. The high point of humanities and social science education for engineers came in a single decade in the post–World War II era, bounded by the efforts leading to the reports, General Education in Engineering (1956) and Liberal Learning for the Engineer (1968).

A century of activity leaves the curricular role of humanities and social sciences on the fringes of engineering education. The curriculum may be only the tip of the iceberg, but in this case, the tip scarcely breaks the surface. The engineering student faces a highly scientific and quantitative program of study. Many are headed for graduate study, and a large percentage of those head for a graduate business school or a law school. Others opt for advanced engineering science, and even for research careers. In some way, this pattern reflects the proposals of the Goals of Engineering report, a five- (maybe six or seven) year program to the first professional degree. Yet that extended study is in a professional context, and little of it imparts as much of a critical point of view as it does a technical view.

There are also a new set of social, political, cultural, and even economic issues to be faced by the student who comes to engineering study in the postcold war era. Global corporations, new waves of immigrants into western nations, religious differences on a global scale, changing political alignments, cultural differences arising from new expectations—the list could go on. These are conditions that will determine the work role, the workplace, the home, or the nomadic path of the contemporary student in engineering.

Nearly one-quarter of today's engineering students are women, whereas most all were men twenty-five years ago. Population estimates suggest that Caucasian male students may make up as little as 15 percent of the engineering school population by the turn of the century. International and minority students will fill the schools. Engineering students who came from relatively homogeneous backgrounds are long since graduated.

Engineering educators will face new challenges in establishing the values base needed for these persons during their careers. Add to this cultural mix of students the likely pattern of global employers, and it is evident that issues will be confronted unlike those that were present when middle-class white male students, often first generation of the college-bound in their families, prepared to enter American industry. Most will attend universities; land-grant agricultural and technical colleges were converted to universities at least a quarter century ago. Only a few schools have remained strictly engineering schools, and even they have established special arrangements with nearby colleges to extend their special curricula.

In summary, during its first century, ASEE has been generally frustrated in its attempt to use curricula to demonstrate the social and cultural aspects of engineering, or the social and cultural responsibility of professionals in the field. In some ways the attempts to apply an engineering design solution to the curriculum has failed in this area.

Reliance on electives has given little ground for comfort. Gone are the days when professors and students voluntarily formed reading and discussion clubs to consider ideas and issues outside the curriculum. Efforts to influence students directly, as through the AAC pamphlets, have only recently been attempted, and there is no indication as to how that may affect future students. Many observers now suspect that the crosscultural and social aspects of engineering work will provide the basis for the new education. Perhaps the new mix of students will have its own effect, and learning will occur outside the classroom.

If we have expected "progress" as a result of our efforts, perhaps we should turn again to T. S. Eliot's observation.

> The moment of the rose and the moment of the yew-tree
> Are of equal duration. A people without history
> Is not redeemed from time, for history is a pattern
> Of timeless moments
> . . .
> And the end of all our exploring
> Will be to arrive where we started
> And know the place for the first time.

References

1. Eliot, T. S., *The Dry Salvages*.
2. Terry S. Reynolds and Bruce E. Seely, Striving for Balance: A Hundred Years of the American Society for Engineering Education.
3. Fountain, Alvin M. The English Division of the American Society for Engineering Education: A History, North Carolina State University, unpublished. [distributed to members of the English Division of ASEE] June 26, 1961. p. 21. I draw heavily on Professor Fountain's account for the early history of the English Division.
4. Joseph L. Vaughan and colleagues turned to Odysscus not only for literary merit but also for studies of leadership, to Macbeth for ambition, to Lear for geriatrics.
5. Some even used C. S. Forester's hierarchy of meaning to teach language—literal, allegorical, analogical, and anagogical. In sum, at its best, this was a philosophical approach to the study of language.
6. Charles Riborg Mann, A Study of Engineering Education, New York: Carnegie Foundation for the Advancement of Teaching, Bulletin 11, [1918], p. 11.
7. Report of the Investigations of Engineering Education, 1923-1929. Vol. 1, p. 807.
8. In an admittedly provincial note, we would forget that Lanza was the son of a Professor of Modern Languages, and had distinguished himself in the classics and mathematics as a student in the University of Virginia from 1867 to 1871.
9. The literature in many fields was printed in foreign German or French, so the language requirement was related to the technical content, not to "liberating" concepts.
10. Report of the Investigation of Engineering Education, 1923-1929, Vol. 1, pp. 90-91.
11. The Wickenden report was used to encourage various foundations to fund summer sessions to raise the level of teaching of topics throughout the engineering curriculum. The 1932 summer session was devoted to the teaching of English.
12. Good Reading, NY: The New American Library (Mentor Book, M 19). Reprinted several times by The Committee on College Reading, sponsored by the National Council of Teachers of English. The fifth printing was dated 1951.
13. Dugald C. Jackson, *Present Status and Trends of Engineering Education in the United States*, NY: Engineers Council for Professional Development, 1939, p. 34.
14. H. P. Hammond, "Aims and Scope of Engineering Curricula." SPEE, 1949.
15. Personal conversation with Joseph L. Vaughan, March 23, 1993.
16. T. S. Eliot, "Little Gidding."
17. Evaluation., p. 12.
18. Evaluation, pp. 16-17. The courses to be arranged "to facilitate emphasis on the interrelationships that exist among the fields of knowledge." The proposed sequence of courses "should occupy about 20% of the student's time." The report anticipated a "vertical sequence and the interdisciplinary approach to subject matter." The Committee explicitly omitted business training and ROTC from this area.
19. *Engineering Education* Vol. 59: No. 4, December 1968, p. 318.
20. The Punderson House Conference: Action Programs to Iimplement thre Olmstead Report. "Liberal Learning for the Engineer," June 17-20. NSF Grant #GY7206, by Dorothy Lambert.
21. The Applied Humanities, EPD 3-68, May 1968.
22. Dorothy M. Lambert, LSD—WHAT ARE WE? Presented at ASEE Annual Meeting, Lubbock, Texas, June 19-23, 1982.

23. Liberal Learning for the Engineer: An Evaluation Five Years Later, 1974. O. Allan Gianniny, Jr., Ed.
24. Final Report From REETS Action Committees, [American Society for Engineering Education. [1979]. Unpublished Board of Directors document. See especially Recommendation #3) "Programs and Curricula" Recommendation #3 "Programs and Curricula"; Recommendation #7, "Communication with the Public," and Recommendation #8, "Engineer in Society.
25. Unfinished Design (Association of American Colleges) (1988).

Contributors

Charles C. Adams is Professor of Engineering and Chair of the Engineering Department at Dordt College in Sioux City, Iowa. In addition to graduate degrees in mechanical engineering and education, he holds a Ph.D. in Philosophy of Engineering Education. He has worked in the aircraft industry, taught mathematics and science in a secondary school, and in 1979 was responsible for starting the engineering program at Dordt College.

Ann Brown (retired) taught English and directed the Writing Assistance Program at North Carolina State University. She currently raises cattle and horses on her 300-acre farm, and fixes her tractor when it breaks down.

John Brown is Associate Professor, Technology, Culture and Communications, School of Engineering, at the University of Virginia.

Donald L. Decker is retired, formerly Professor, Mechanical Engineering, Rose-Hulman Institute of Technology.

Scot Douglass is an Assistant Professor of Comparative Literature and the Herbst Program of Humanities for Engineers at the University of Colorado, Boulder. He received a Ph.D. in Comparative Literature, has a Th.M. from Dallas Seminary and a B.S. in Cellular Biology from the University of Arizona.

Samuel C. Florman is a civil engineer, and a principal in Kreisler-Borg Constuction, New York, New York. He has written *The Existential Pleasures of Engineering* (1976, 2nd ed 1994), *Blaming Technology*(1981), *The Civilized Engineer*(1987), and

Engineering and the Liberal Arts (1968), in which the Prologue of the present volume first appeared.

O. Allan Gianniny, Jr. is Professor Emeritus, Technology, Culture, and Communications at the University of Virginia.

Michael E. Gorman is Professor and Chair, Technology, Culture and Communications at the University of Virginia.

Craig Gunn is the Director of the Communication Program in the Department of Mechanical Engineering at Michigan State University. He is currently the editor of the *CED Newsbriefs* and the MCCE *Co-op Courier* and recently co-authored a textbook entitled *Engineering Your Future*.

Joseph R. Herkert is Associate Professor of Multidisciplinary Studies at North Carolina State University, Raleigh, NC. He teaches in the Science, Technology and Society Program and is Director of the Benjamin Franklin Scholars Program, a dual-degree program in engineering and humanities/social sciences. Dr. Herkert is a Past-President of the IEEE Society on Social Implications of Technology.

John Krupczak is Associate Professor, Physics and Engineering, Hope College, Holland, MI.

Marshall M. Lih is director of the National Science Foundations's Division of Engineering Education and Centers.

Heinz C. Luegenbiehl is Head of the Department of Humanities and Social Sciences and Professor of Philosophy and Technology Studies at Rose-Hulman Institute of Technology. He has served as Chair of the Liberal Education Division of the American Society for Engineering Education and Vice-President of the Humanities and Technology Association, and held various visiting professorships, including a current one at Kanazawa Institute of Technology in Japan.

Steve Luyendyk has taught English at North Carolina State University, and presently works in industry as a technical editor.

Matthew M. Mehalik is a Ph.D. candidate in Technology, Culture and Communications at the University of Virginia.

Carolyn R. Miller is Professor of English at North Carolina State University, where she directs the technical and professional writing courses, as well as teaching rhetorical theory and criticism. She also directs the Center for Communication in Science, Technology, and Management. Her primary research interest is in the rhetoric of science and technology. She was 1996–98 President of the Rhetoric Society of America and in 1995 was elected Fellow of the Association of Teachers of Technical Writing.

Ronald L. Miller is Professor of Chemical Engineering and Petroleum Refining at the Colorado School of Mines (CSM). He received his Ph.D. in Chemical and Petroleum Refining Engineering from CSM in 1982. Dr. Miller has been active in curriculum revision and educational research at CSM for over a decade. He chairs CSM's assessment committee and has been active in the engineering education and assessment communities.

Kathryn A. Neeley is Associate Professor of Technology, Culture, and Communication at the School of Engineering and Applied Science a the University of Virginia. She is past chair of the Liberal Education Division of the American Society for Engineering Education and is past president of the Humanities and Technology Association.

John P. O'Connell is H. D. Forsyth Professor of Chemical Engineering and Coordinator of the Professional Development Program in the School of Engineering and Applied Science of the University of Virginia.

Barbara M. Olds is Principal Tutor of the McBride Honors Program in Public Affairs for Engineers and Professor of Liberal Arts and International Studies at the Colorado School of Mines where she has taught for the past seventeen years. She has participated in a number of curriculum innovation projects and has been active in the engineering education and assessment communities. Dr. Olds has received the Brown Innovative Teaching Grant and Amoco Outstanding Teaching Award at CSM and was the CSM Faculty Senate Distinguished Lecturer for 1993–94. She was a Fulbright lecturer/researcher in Sweden in 1999.

David F. Ollis, Distinguished Professor of Chemical Engineering at NC State University. A past chair of the ASEE LED division, his teaching includes research proposal writing for engineering grad students, and technology literacy for non-technical majors.

Edward Alton Parrish is currently President of Worcester Polytechnic Institute and Professor of Electrical and Computer Engineering. He is a Fellow of the Institute of Electrical and Electronic Engineers (IEEE) and a Fellow of the Accreditation Board for Engineering and Technology (ABET). Currently, he serves on the boards of

ABET and the IEEE Foundation. He is listed in *Who's Who in Engineering* as well as many other such registries and is a licensed Professional Engineer in the Commonwealth of Virginia and in the State of Tennessee.

Leslie C. Perelman is Director of Writing Across the Curriculum in the Program in Writing and Humanistic Studies at the Massachusetts Institute of Technology and an Associate Dean in the Office of the Dean of Undergraduate Education. He teaches classes in technical communication, was Project Director for a National Science Foundation grant to develop a model Communication-Intensive Undergraduate Program in Science and Engineering. Dr. Perelman has written on technical communication, the history of rhetoric, sociolinguistic theory, and medieval literature.

Bryan Pfaffenberger is an anthropologist and Associate Professor in the Division of Technology, Culture and Communication in the School of Engineering and Applied Science of the University of Virginia.

Lance Schachterle joined WPI as an assistant professor of English in 1970, and has taught a variety of courses and projects in American and British literature. He received his Bachelor of Arts degree from Haverford College in 1966 and his Ph.D. from the University of Pennsylvania in 1970. He has published a variety of studies on Dickens, Cooper, and Pynchon, as well as on liberal and engineering education, and served as the first president of the Society for Literature and Science. From 1984 to 1993, he chaired Interdisciplinary Studies and Global Programs at WPI, where he has been assistant provost since 1995.

Timothy C. Scott is Associate Professor and Director of Laboratories for the Mechanical and Aerospace Engineering Department in the School of Engineering and Applied Science of the University of Virginia.

Eugene R. Seeloff retired from the position of Assistant Dean for Career Planning and Placement in the School of Engineering and Applied Science of the University of Virginia.

Mark A. Shields is a Sociologist and Assistant Professor in the Division of Technology, Culture and Communication in the School of Engineering and Applied Science of the University of Virginia.

Julie M. Stocker is a former student of the Department of Systems Engineering, University of Virginia.

Charles W. N. Thompson is a professor of industrial and management sciences, with teaching and research foci in field research methodology, systems engineering and analysis, information systems, and project management. His experience in systems design and engineering management includes periods at Wright Field U.S. Air Force and a large Chicago electronics manufacturer.

Edward Wenk, Jr. is professor emeritus of engineering and public affairs at the University of Washington and is the author of *Making Waves: Engineering Politics, and the Social Management of Technology* (University of Illinois, 1996). Wenk was the first Congressional science advisor; served as science advisor to Presidents Kennedy, Johnson, and Nixon; and is currently a member of the National Academy of Engineers.

WPIStudies

WPI Studies is sponsored by Worcester Polytechnic Institute, the nation's third oldest independent technological university. WPI Studies aims to publish monographs, edited collections of essays, and research tools and texts of interest to scholarly audiences. WPI Studies accepts manuscripts in all languages, and is especially interested in reviewing potential publications on interdisciplinary topics relating science, technology, and culture. WPI Studies is edited by a board of WPI faculty from many disciplines. The board is chaired by Lance Schachterle, Assistant Provost and Professor of English, to whom potential authors should direct their inquiries (WPI, Worcester, MA 01609).

To order other books in this series, please contact our Customer Service Department at:

800-770-LANG (within the U.S.)
212-647-7706 (outside the U.S.)
212-647-7707 FAX

or browse online by series at:
www.peterlangusa.com